AF553780

ANIMAL ENDOCRINOLOGY

ENCYCLOPAEDIA OF ENDOCRINOLOGY - II

ANIMAL ENDOCRINOLOGY

By

Manju Yadav

Lecturer
Department of Zoology
M.M.H. College
Ghaziabad (U.P.)
(India)

DISCOVERY PUBLISHING HOUSE PVT. LTD
NEW DELHI-110 002

Published by:

DISCOVERY PUBLISHING HOUSE PVT. LTD.
4383/4B, Ansari Road, Darya Ganj
New Delhi-110 002 (India)
Phone : +91-11-23279245; 23253475; 43596065
E-mail : discoverybooksindia@gmail.com
discoverypublishinghouse@gmail.com
namitwasan9@gmail.com
web : www.discoverypublishinggroup.com

***First Published:* 2007**
***Reprinted:* 2022**

ISBN: 978-93-5056-570-4 (Set)

ISBN: 978-81-8356-272-0

Animal Endocrinology

Printed at:
Infinity Imaging Systems
Delhi

Preface

The present title **Animal Endocrinology** provides a definitive and comprehensive review of all aspects relating to hormones in different animals including wild and domestic species. It discusses the intimate physiology of the endocrine system itself and describes the role of hormones in the processes of nutrition, osmoregulation, colour change, calcium metabolism and reproduction. Subject matter has been presented in a readable way emphasizing the differences between species from a functional aspects. It not only contains a wealth of information concerning the endocrinology of the various species but also contains more than enough basic information concerning the biosynthesis, metabolism and mode of action of the hormones for the book to be read and understood without referring to a text book of basic endocrinology. This book has been written primarily for use as a text book by under-graduate, as well as graduate students.

Efforts have also been made to make the presentation lucid and accessible for beginning students, emphasizing major concepts while in corporating the latest experimental findings and theories throughout. The aim is to enthuse the readers with this active and exciting area of research and to lay a solid foundation on which further study of its various facts may be based.

Though, the author has taken special care to present a current account, yet he is fully aware about limitations and the readers may come across the mistakes of various types. For all types of mistakes author extends his due apology.

There can be no claim to originality except in the manner of treatment and much of the information has been obtained from the books and scientific journals available in the different libraries.

The author expresses his thanks to his friends and colleagues whose continue inspirations have initiated him to bring out this book.

The author expresses his gratitude to Mr. Wasan and staff of M/s Discovery Publishing House for their whole hearted co-operation in the publication of this book.

Author

CONTENTS

1

INTRODUCTION

Endocrinology is the study of certain glands known as endocrine glands and how these glands regulate the physiology and behaviour of individual animals and populations. The endocrine system is basic to the abilities of an organism to adapt to its environment in both an ecological and an evolutionary sense. It is a chemical link between the environment and the organism.

At one time physiologists viewed the nervous and endocrine systems as discrete regulatory units each with their own brand of chemical messenger. The neurotransmitter released from neurons was easily distinguished from a hormone secreted into the blood. This view began to change when it was discovered that neural activity could influence release of hormones from endocrine glands such as the pituitary and that hormones could in turn alter activity in the nervous system. Soon it was shown that some classic neurotransmitters could be released into the blood and function as hormones. Similarly, we learned that classic hormones were being used as neurotransmitters in the nervous system. Modified neurons were observed to produce special hormones that were released directly into the blood. Hypotheses appeared suggesting that many endocrine cells and neural tissues were derived embryonically from the same tissue. These observations gave birth to the field of *neuroendocrinology* which has become a dominant research area in the field of chemical regulation. This development has prompted some investigators to propose that endocrinology should be considered as a branch of neuroendocrinology.

The endocrine system may be conceptualized as a series of reflexes much like the more familiar neural reflexes. The most simple reflexive unit in the endocrine system is the *endocrine reflex*. It is similar to the familiar neural reflex involving a stimulus, an integrating unit and an effector. A change in the amount of some substance circulating in the blood causes an endocrine cell to release a hormone into the blood that ultimately brings about a return of the circulating substance to "normal" levels. The *neurohormonal reflex* involves modified neurons within the central nervous system that secrete special hormones into the blood in response to information collected by neural receptors. This reflex may deal not only with information obtained from internal receptors but also information obtained from the external environment. The *neuroendocrine reflex* is the most complex of these reflexes involving neurohormonal control over activities of specific endocrine glands. It integrates external and internal information into precise regulatory control over major physiological processes such as reproduction, growth and metabolism.

Glands and Glandular Secretion

There are many different kinds of glands in vertebrates that are highly specialized for secretion. Classically, they have been segregated into two kinds: endocrine and exocrine. The vertebrate endocrine system includes a variety of endocrine glands that secrete (synthesize and release) into the blood specific chemical messengers known as *hormones*. These chemical messengers travel via the blood to specific target tissues where they cause changes in the activities of the target tissue cells. Glands that secrete their products directly into the blood are termed *endocrine glands*. In recent years, this definition has been expanded to include the observation that some hormones are secreted into the cerebrospinal fluid. The pituitary gland, thyroid gland and the

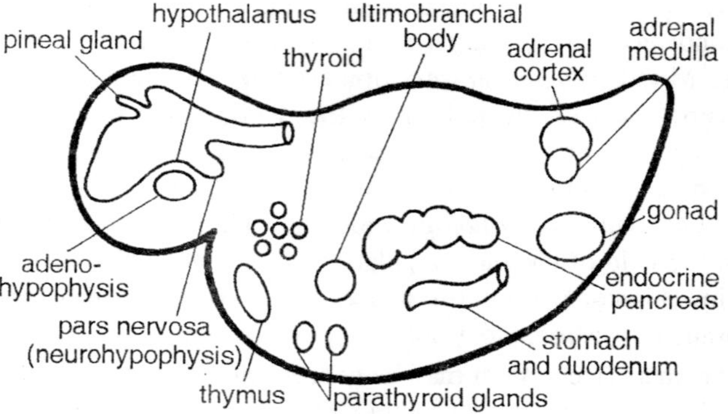

Fig. 1.1. Generalized endocrine structures in vertebrates.

adrenal cortex are examples. *Exocrine glands* secrete their products into ducts through which the secretions pass to an internal or external surface. Salivary glands, sebaceous glands, sweat glands and mammary glands are all examples of exocrine glands.

Frequently glands are characterized according to the manner by which their secretory products are liberated. *Merocrine* secretion involves the process of *exocytosis*, which is essentially a reverse of pinocytosis or phagocytosis. The membrane-bound secretory granule or droplet containing the intracellular secretion product fuses with the cell membrane (plasmalemma), and the contents are dumped outside the secretory cell. Such a process causes no damage to the secretory cell and is typical of most endocrine cells as well as of certain exocrine cells (for example, the enzyme-secreting cells of the pancreas). Some exocrine cells slough the apical portion of the cell (that portion nearest the duct) containing the secretory products. This type of secretion is termed *apocrine* and is characteristic of cells of the mammary gland. Still other glandular cells may die and undergo lysis upon releasing their stored products. Sebaceous gland cells exhibit this type of secretory pattern, which is termed *holocrine*. Finally the term *cytogenous* has been proposed for situations in which live, viable cells are "secreted" by organs such as the testes and ovaries, which release spermatozoa and ova respectively.

Neurosecretion: Neurohormones

The nervous system is usually viewed as a complex of neurons, most of which secrete specific chemicals, *neurotransmitters*, from their axonal endings. Such a neuron secretes neurotransmitter substance into an extracellular space (the synaptic cleft) through which it diffuses to bind with specific receptors in the postsynaptic cell (which may be another neuron, a muscle cell or a gland cell). Many different kinds of neurons have been identified in the central and peripheral nervous systems, each secreting its particular neurotransmitter. These neurons are frequently designated according to the chemical nature of their neurotransmitter such as cholinergic neurons (acetylcholine), adrenergic (norepinephrine), aminergic (amines such as dopamine), peptidergic (peptides), serotonergic (serotonin) and purinergic (purines). The pharmacological investigation of vertebrate nervous systems largely deals with the specific roles for these various neurònal types and the actions of drugs as they influence one or another type of neuron.

The neurotransmitter substance produced by a specific neuronal type can be partially identified by means of the electron microscope.

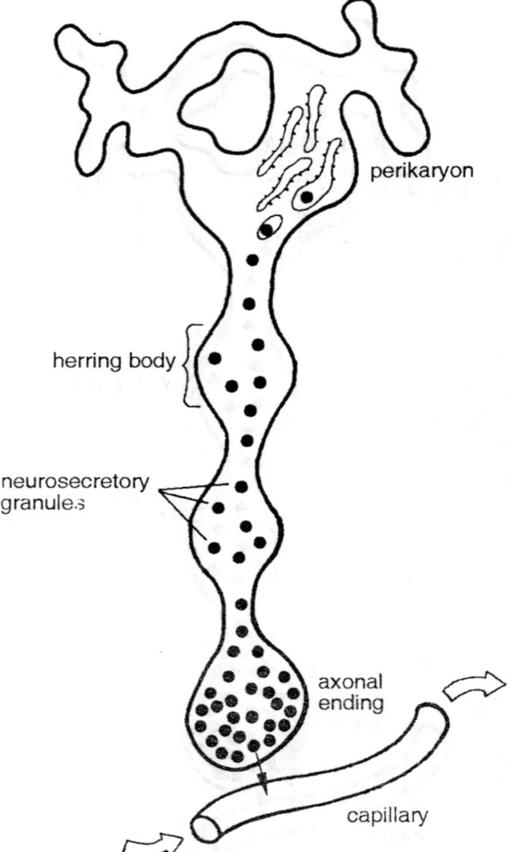

Fig. 1.2. Generalized neurosecretory cell.

Each neurotransmitter is stored prior to release within small membrane -bound droplets (synaptic vesicles) located in the expanded axonal tips. Cholinergic synaptic vesicles have a diameter of 30-45 nm and are not electron dense, whereas adrenergic synaptic vesicles are larger (approximately 70 nm diameter) and exhibit electron-dense cores.

A number of years ago it was discovered that another type of axonal secretory product associated with certain highly specialized neurons could be identified at the light-microscope level by selective staining techniques (such as paraldehyde fuchsin, pseudoisocyanin and the Gomori chrome-alum hematoxylin procedure). Later, with the electron microscope, these secretory products were identified as large electron-dense granules (100-300 nm diameter). Anatomical studies suggested that these neurons release their secretory products directly into the blood vascular system. These neurons are termed *neurosecretory neurons*, and the dyes that they selectively accumulate are termed *neurosecretory stains*. Their secretory products are peptidergic substances (small peptides) or amines termed *neurohormones*. The process of secreting neurohormones is termed *neurosecretion*. A large number of neurohormones have been chemically characterized and others have been proposed. Phylogenetically neurohormones constitute the first type of endocrine system to appear in animals and the only endocrine-secreting mechanism in many invertebrate groups.

The neurosecretory (NS) neuron is capable of conducting an action potential, although duration of action potentials measured from NS neurons is longer than that of "ordinary" neurons. The secretory granule of the NS neuron consists primarily of the neurohormones plus a "carrier" protein, called a *neurophysin*. Although several neurophysins have been identified, their role in neurohormonal secretion is not clear. They may be related to the process of synthesizing the neurohormones.

An aggregation of perikarya (neuronal cell bodies) within the central nervous system is known as a *nucleus*. Perikarya of NS neurons typically occur in groupings called *neurosecretory nuclei*. Such nuclei are demonstrated easily with the use of NS stains. Generally NS neurons secreting a particular neurohormone occur in the same NS nucleus.

The classical NS neuron described above must now be modified somewhat. In the vertebrate brain at least two types of NS elements have been described on the basis of cytological details. The classical NS neurons are characterized by large cytoplasmic granules (100-300 nm diameter) in the axon terminals. The secretion products of these neurons are peptide neurohormones. Other NS neurons typically exhibit smaller granules (about 70 nm diameter). At least some of these NS neurons secrete amines such as dopamine (hence termed *aminergic*), but others secrete specific peptides (termed *peptidergic* neurons).

Neurosecretory tracts consisting of bundles of NS cell axons also can be stained within the central nervous system by NS stains. The endings of NS axons may accumulate in a highly vascularized area known as a *neurohemal organ*. Neurohormones are stored in these neurohemal organs prior to release. These organs usually stain intensely with NS dyes because of dense accumulations of NS material.

Neurosecretory dyes appear to stain the carrier proteins (neurophysins) in the secretory granules but will also stain other cellular inclusions and organelles such as other protein granules, mitochondria and endoplasmic reticulum. Consequently one cannot be sure that these dyes are staining only NS material by using this approach. A further caution in the use of NS staining procedures is related to the occasional observation that "stainability" of NS material does not always parallel neurohormonal content as determined by biological tests. High stainability may be correlated with high levels of the neurohormone on one occasion but with low levels on another. Many neurosecretory neurons release more than just their reputed neurohormone. Secretory granules in these cells may contain other peptides, enzymes and amines which are released along with the hormones.

Neuroendocrine Responses

The cells of endocrine glands may be influenced by neurohormones circulating via the blood vascular system. Scharrer has emphasized that in addition to direct innervation of endocrine cells by neurons forming the typical secretomotor junctions associated with neurotransmitters, some NS neurons form "neurosecretomotor junctions" with endocrine cells. These junctions are usually of the peptidergic

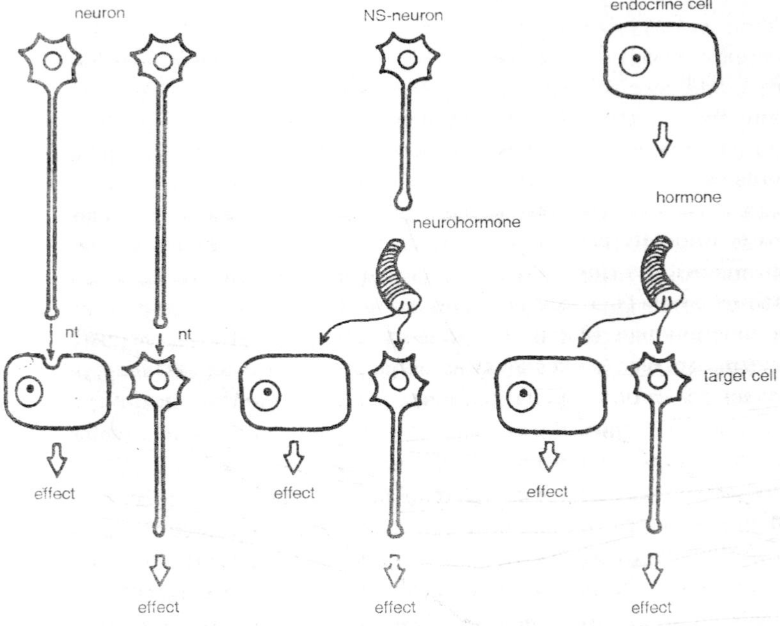

Fig. 1.3. General patterns of chemical regulation by neurons, NS neurons and endocrine cells. nt, neurotransmitter.

type, although some are apparently aminergic. Scharrer has described another variation on the general theme of NS influences on endocrine cells in which NS neurons release their secretory products into an extracellular matrix, thereby relying on diffusion to accomplish the actual contact with an endocrine cell. Any of these variations involving NS neurons and endocrine cells constitutes a neuroendocrine control mechanism.

Types of Chemical Regulators

The knowledge that individual neurons can function like endocrine cells, that many known neurotransmitters are released into the blood where they function as classical hormones and that widely distributed cells release chemical regulators into extracellular fluids has altered the field of endocrinology. Chemical regulators of the latter type that diffuse short distances are designated as *paracrines*. Internal chemical regulators occur as a continuum from classical neurotransmitters to paracrines to classical hormones. Attempts to provide rigid definitions for these various types of regulators may be counterproductive, for evolution has assured that we cannot provide discrete categories. These types are simply categories for convenience and are not meant to be mutually exclusive. Epinephrine, for example, is a neurotransmitter when released from a neuron into a synaptic extracellular space. It is a paracrine substance if released into a generalized extracellular space through which it diffuses to a nearby target cell. When released into the blood from the adrenal medulla, epinephrine is a hormone.

The situation is complicated further by the existence of other chemical messengers. *Parahormones* are one category of such compounds; they are specialized substances that may be produced by a variety of cellular types and often have rather generalized effects. Histamine and angiotensin II might be considered parahormones because of their widespread effects on smooth-muscle contractions. Carbon dioxide produced by all respiring cells has generalized effects on many tissues, including rather specific effects on the respiratory centers in the brain.

The so-called *semiochemicals* form another category of chemical messengers that cannot be classified strictly as hormones. They are often produced by specialized cells but are secreted outside the body and influence the physiology and behaviour of other individuals. Some of these semiochemicals may be modified steroid hormones, which helps justify their inclusion under the umbrella of endocrinology.

A variety of intracellular chemical messengers have also been identified. The prostaglandins and certain cyclic nucleotides are important intracellular messengers that are involved in the mechanisms of action for hormones that are bound to the plasmalemma and do not enter the cytoplasm of their target cells.

Chemical Nature of Hormones and Neurohormones

Vertebrate hormones and neurohormones may be proteins, polypeptides, derivatives of amino acids or lipids. Most of the lipid hormones are steroids derived from cholesterol or fatty acid derivatives such as the prostaglandins. Steroid hormones differ further from polypeptide hormones and those derived from amino acids in that they are not generally stored in glandular cells but are released immediately after their synthesis. Secretion of the nonsteroidal hormones and the neurohormones can be readily separated into synthetic and release phases that may be regulated separately.

The mode of action of hormones is also different, correlated in part to their chemical and physical properties. Water-soluble hormones tend not to enter their target cells but bind to the plasmalemma, whereas steroids and thyroid normones (derived from the amino acid tyrosine) are nonpolar molecules and readily move across cell membranes. Transport of steroids and thyroid hormones into cells may not be simple diffusion but may involve a membrane-mediated process. These latter hormones require binding proteins in the blood plasma to facilitate their transport and reduce their rate of removal from the blood, metabolism and excretion.

Cytological Features and Hormone Synthesis

Cells responsible for elaboration of peptide hormones possess a well-developed rough endoplasmic reticulum where hormone synthesis takes place. The hormonal products are stored in membrane-bound secretion granules in the cytoplasm. Peptide hormones are synthesized on ribosomes as part of a larger peptide called a *preprohormone*. A special terminal segment of the preprohormone is a short amino acid sequence called a *singal peptide*. The presence of this segment allows the newly synthesized peptide to enter into the cisterna of the endoplasmic reticulum. In the process, the signal peptide is enzymatically removed leaving a large peptide called a *prohormone*. This peptide travels through the endoplasmic reticulum to the Golgi apparatus where it is packaged into secretory granules. Addition of carbohydrates may occur in the Golgi when the final products are glycoproteins. Enzymatic cleavage of the prohormone occurs within the secretory granules. This cleavage produces the specific hormone and a separate peptide fragment. The neurophysins may be produced as the non-hormonal fragment of a prohormone.

Synthesis of proteins and peptides is controlled by the activity of nuclear genes or, in certain cases, mitochondrial genes. A given gene determines the sequence of messenger RNA that will be transcribed. Messenger RNA leaves the nucleus and travels to the ribosome where it directs the specific sequence by which amino acids are linked to construct a particular polypeptide. By identifying the sequence of amino acids in the hormone as well as in its preprohormone and prohormone fragments, it has been possible to construct artificial genes which can be inserted into cultured cells or cells of intact animals. Once incorporated into their genome, these cells can synthesize large quantities of specific hormones. Already, the synthesis of genes for producing growth hormone, insulin and many others has been achieved. Such genetic engineering offers the promise that some day similar gene insertions may alleviate clinical problems associated with underproduction of a peptide or protein hormone.

It is the storage granules in endocrine cells that are responsible for their unique affinities for various dyes and their identification at the light microscope level. Furthermore, depending upon their size, shape, or both, these storage granules may provide cytological markers for particular cellular types at the ultrastructural level.

Steroid-secreting cells are characterized by abundant smooth endoplasmic reticulum for steroidogenesis and a well-developed Golgi apparatus for packaging lipid droplets. Steroidogenic cells can be identified histochemically by localization of specific enzymes involved in steroid hormone synthesis (for example, Δ^5, 3β-hydroxysteroid dehydrogenase). Quiescent steroidogenic cells may contain large cholesterol-positive droplets in their cytoplasm, but as a rule most of the steroid hormones synthesized from cholesterol are released into the blood as they are produced and are not stored. Enzymatic activity or cholesterol content is an index of hormone synthesis.

Histological Organization of Endocrine Cells

Endocrine cells are grouped or clumped in characteristic arrangements, although in certain instances isolated cells may be diffusely distributed among other cellular types such as occurs in the gastrointestinal tract. A common arrangement for endocrine cells is in folded sheets or *cords* of cells as found in the pituitary or the adrenal cortex. In some cases separate clumps of endocrine cells or *islets* may be found embedded within another tissue. The endocrine pancreas consists of islets of cells dispersed among the exocrine-secreting tissue of the pancreas. Other endocrine cells are arranged in hollow balls of cells or *follicles*. In this arrangement there is only a single layer of endocrine cells in each follicle surrounding a central space filled with fluid or colloidal material secreted primarily by the endocrine cells themselves. Such a follicular organization is characteristic of thyroid glands.

Internal Rhythms in Endocrine Secretion

Hormones are released episodically or in bursts (phasic secretion) rather than at a constant rate (tonic secretion). Many of these bursts show a decisive diurnal, monthly or seasonal cyclicity or rhythm. Diurnal fluctuations are being observed commonly in all kinds of animals as a consequence of the development of precise techniques for monitoring blood levels of hormones. The daily to seasonal variations in reproductive hormones during the menstrual cycles of higher primates and the estrous cycles of other mammals are well known to students of zoology. Not only do hormone levels vary, but the sensitivities of target tissues to the same hormone dose may also exhibit diurnal or seasonal fluctuations.

The establishment of variations in secretory patterns of endocrine glands on a daily or seasonal basis or both, as well as rhythmic variations in sensitivities of target tissues, means that additional caution is required in interpreting experimental results, especially when no effects of a given treatment are observed or when two investigators obtain opposite results in the same system. Indeed the response observed to one hormone may depend upon the rhythm of another hormone.

The relationship of cyclical exogenous (external or environmental) rhythms to migration, reproduction and other phenomena is of great interest to all types of zoologists. Endocrine glands are proving to be important mediators of cyclic environmental phenomena that program physiological events in other endocrine glands. Many studies, for example, have shown that alterations in photoperiod can influence endogenous (internal) endocrine rhythms dramatically. The discovery of these endogenous hormonal rhythms and the involvement of exogenous or external factors to set internal timers has opened research into one of the most exciting areas of endocrinology and will alter the direction of future endocrine research.

Hormones and Age

Throughout the life of any vertebrate, developmental changes are occurring with respect to maturation of hormonal secretory patterns and to the responsiveness of particular target tissues. Many of the changes associated with development and aging may be caused by hormones themselves. Fetal tissues may respond differently than neonatal tissues to the same hormone. Young animals may respond differently than neonates. The secretory patterns for certain hormones may be altered by the attainment of sexual maturity, and they may be altered again after the animal ceases to be reproductively active (a rare event in nature but common in humans and some domesticated animals). The role of growth hormone is very different in a growing animal and an adult animal, for example. The action of growth hormone on certain growing tissues can be altered markedly by the secretion of gonadal steroids at puberty. Consequently it is important to consider developmental ages of animals when interpreting endocrine data.

Mammalian Hormones

The major hormones and glands of the mammalian endocrine system are listed below.

The central figure of the mammalian endocrine system is the *hypothalamo-hypophysial axis*. The *hypothalamus* is the ventral portion of the brain immediately beneath the thalamus and above the *pituitary gland* or *hypophysis*. Neurosecretory centers in the hypothalamus produce neurohormones that independently control release of several different hormones from the pituitary. These hypothalamic NS centers are in turn influenced by neurons from higher neural centers in the brain. The hypothalamic neurohormones are termed releasing hormone(s) (RH) or release-inhibiting hormone(s) (RIH). They are sometimes designated as the *hypothalamo-hypophysiotropic hormones*.

The pituitary hormones include *prolactin* (PRL); *growth hormone* (GH); *thyrotropin*, or thyroid-stimulating hormone (TSH); *luteinizing hormone* (LH); *follicle-stimulating hormone* (FSH); *corticotropin*, or adrenocorticotropic hormone (ACTH); *melanophore-stimulating hormone* (MSH); and *lipotropin*, or lipotropic hormone (LPH). These hormones are termed collectively the *tropic hormones* since several of them influence the activities of other endocrine glands. Thyrotropin controls the synthesis of thyroid hormones *triiodothyronine* (T_3) and *thyroxine* (T_4) by the thyroid gland. Corticotropin regulates synthesis and release of corticosteroids such as *cortisol* and *corticosterone* by adrenal cortical cells. Luteinizing hormone and FSH influence the gonads (testes, ovaries) by stimulating gametogenesis (gamete production) and synthesis and release of male sex hormones or *androgens* (for example, *testosterone*) and female sex hormones or *estrogens* (for example, *estradiol*-17β) and *progestogens* (*progesterone*). The other tropic hormones do not have endocrine targets. Prolactin influences lactation by the mammary gland. Growth hormone has metabolic effects on muscle, adipose cells and other tissues. Pigment cells in the skin are targets for MSH, and LPH produces metabolic effects on adipose cells.

Two neurohormones produced by hypothalamic centers are stored in a special region of the pituitary, the *pars nervosa*. When released, *oxytocin* stimulates contraction of uterine smooth muscle and initiates labor. Oxytocin also causes contraction of modified cells in the mammary gland that cause milk ejection following suckling by the young. The *vasopressins* are similar molecules to oxytocin that enhance water reabsorption in the kidney and have been termed antidiuretic hormones.

In addition to the hypothalamo-hypophysial hormones there are a number of important endocrine glands that are not influenced by neurohormones although they may be influenced by direct innervation with sympathetic or parasympathetic fibers. Parathyroid glands produce *parathyroid hormone* (PHT) which, together with *calcitonin* (CT) from the thyroid, regulates calcium balance. The *endocrine pancreas* produces at least two hormones that control intermediary metabolism, *insulin* and *glucagon*. A variety of cellular types distributed in particular regions of the digestive tract produce a series of *gastrointestinal hormones* including *gastrin*, *secretin* and *pancreozymin-cholecystokinin* (PZCCK). In recent years the secretions of the *thymus gland* have achieved recognition as hormones in controlling differentiation of immunologically responsive tissues.

Two organs derived from neural tissue are well-known hormone sources. The *adrenal medulla* releases both *epinephrine* (adrenaline) and *norepinephrine* (noradrenaline) into the circulation. Norepinephrine binds primarily to target tissues with alpha-type receptors on the cell surface (plasmalemma). Epinephrine may bind to either alpha-type or beta-type receptors and consequently may influence different targets than norepinephrine. The *pineal gland* produces *melatonin* as well as several peptides that appear to deserve hormonal status. These substances may influence thyroid and reproductive functions in mammals, and the pineal gland may be an endocrine transducer that mediates the effects of photoperiod on reproductive cycles.

Three hormones are made by the kidney. Renin is secreted by the kidney and influences secretion of *aldosterone*, a hormone involved with regulation of sodium and potassium balance, by the adrenal cortical cells. The second involves production of *erythropoietin*, a hormone controlling formation of red blood cells (erythropoiesis) in the bone marrow. The kidney converts a derivative of vitamin D to the third hormone, *1,25-dihydroxycholecalciferol*. This hormone is responsible for stimulating calcium uptake in the small intestine.

Inactivation and excretion of hormones occur in many target tissue cells. In addition, the liver, primarily, and the kidney, to a lesser extent, also metabolize and excrete hormones via the bile or urine respectively.

Origin of Endocrine Cells

Endocrinologists have recognized four separate embryonic origins for hormone-secreting cells exclusive of the NS cells that are derived from neural ectoderm. Cells of the pituitary gland that secrete peptide and protein tropic hormones, such as thyrotropin and corticotropin, classically have been believed to develop from oral (non-neural) ectoderm. The steroid-synthesizing cells of the gonads and adrenal glands are mesodermal in origin. Each adrenal medulla is a modified sympathetic ganglion and is derived from neural ectoderm. Finally the majority of endocrine glands appear to have their origins from the endodermal primitive gut, and they all synthesize protein hormones, polypeptide hormones, or modified amino acids.

A considerable revision in thinking followed the discovery of a peculiar property of many hormone-secreting cells previously thought to characterize only cells that secrete biogenic amines such as epinephrine. This property is known as *amine content* and *amine precursor uptake* and *decarboxylation* or simply as APUD. Cells possessing APUD characteristics were originally proposed by Pearse to originate from the *neural crest*, which is a series of small masses of neuroectodermal cells located adjacent to and distributed along the neural tube of very young vertebrate embryos. The neural crest gives rise to the sympathetic ganglia, the adrenal medulla, all of the melanin-producing pigment cells of vertebrates, the branchial skeleton of the head region and other structures. The discovery of APUD characteristics in many endocrine cells led to the elegant demonstration of a neural crest origin for the CT-producing cells that migrate to the ultimobranchial body, which eventually becomes incorporated into the thyroid gland of mammals. Parathyroid cells that secrete PTH arise from neural ectoderm in the frog *Rana temporaria*, and a similar origin may occur in birds and mammals. Later, this definition was broadened to encompass cells originating from other neural and even non-neural tissue. For example, endocrine cells of the gut, pancreatic islets, adenohypophysis, hypothalamus, placenta, thymus and others were included. Although some of the pituitary cells lack the APUD characteristics, Pearse argues that because of their purported origin from neural ectoderm they should be included.

Although there is considerable debate concerning the common origins of all of these cellular types, it is agreed that almost all secrete regulatory peptides. Some consider that the possession of APUD characteristics and certain enzymes is probably related to similar metabolic requirements and not to common embryonic origins.

Fujita proposed the *paraneuron concept* to recognize paracrine secreting cells that have properties in common with neural cells. The origins of paraneurons have not been established. They do have APUD characteristics, but it is not clear whether they are derived from ectoderm or endoderm. Their name was coined in the belief that, although they are not neurons, they do receive environmental input and release their secretory products into extracellular spaces. It has been hypothesized that paracrine hormones, neurohormones and hormones all had a common origin. This hypothesis is strengthened by the discovery that many vertebrate regulatory peptides are present in unicellular organisms and that many invertebrate regulatory peptides are present in the vertebrates. Moreover, peptides that serve as neurotransmitters in the central and peripheral nervous systems may also function as neurohormones, paracrines and/or hormones in the same animal.

Regulatory peptides may have been basic to the first living cells. During evolution, they may have been adapted in various ways; first as paracrine secretions, later as neurotransmitters, neurohormones and finally as endocrine hormones.

Mechanisms of Hormone Action

This brief account of possible mechanisms of hormone action is designed only to introduce the topic so that these points may be considered in reference to hormones discussed in the following chapters. More will be said about how specific hormones produce their effects as actions of each hormone on its target cells are considered.

In order that a hormone produce any effect on a target cell it must be bound specifically by the target cell either intracellularly or at the cell surface. Those cells that respond to a given hormone synthesize *receptors* that selectively bind only the appropriate hormone. Nontarget cells lack these receptors. Steroid and thyroid hormone receptors are located intracellularly, and these hormones must pass through the plasmalemma before being bound. Polypeptide, protein and amine hormones bind to receptors in the plasmalemma and are believed not to enter the cell. Recent studies, however, have unequivocally demonstrated that polypeptide hormones attached to their receptors enter target cells. It appears that the entrance of the hormone-receptor complex is related to enzymatic degradation of the complex. It is possible that "delayed" actions on enzyme synthesis that take place much later than the initial actions are related to internalization of the hormone-receptor complex.

Hormones may bring about their effects through a variety of mechanisms and may influence a variety of different cellular activities. Many hormones influence specific synthetic or catabolic enzymatic reactions. Certain hormones act to stimulate or inhibit release of other hormones. One hypothalamo-hypophysiotropic hormone is known to inhibit release of GH from the pituitary, whereas another is believed to stimulate its release. Epinephrine is a known inhibitor of insulin release from the endocrine pancreas, which is an action entirely separate from its effect of increasing blood glucose levels through alteration of liver metabolism. Permeability changes in membranes may occur as a consequence of binding of hormone molecules to receptor molecules embedded in the plasmalemma. The consequence of binding might alter the conformation of the membrane, allowing for ionic fluxes to occur across the membrane. Movements of ions may increase or decrease metabolic reactions in the cell, or they may alter the influx or efflux of some important nutrient. One hypothesis for explaining all of the many actions of insulin is based entirely upon changes induced in membrane structure and hence permeability following binding of insulin to the receptor molecule in the plasmalemma.

Hormones do not generally initiate reactions in target cells de novo but only influence the biochemical machinery already there. One of the first biochemical events measured in hormone target cells was an increase or decrease in activity of certain enzymes. Other chemical substances are known to affect enzyme activity, but the mechanism by which hormones operate is different. For example, two known stimulators of the level of tryptophan pyrrolase (a liver enzyme) are found to increase the in-vivo activity in different ways. Cortisone, a steroid hormone produced by the adrenal cortex, stimulates synthesis of this enzyme, but the amino acid tryptophan inhibits degradation of the enzyme, which also raises the effective level of enzyme in the liver cells.

The common biochemical target for many hormones is frequently some *rate-limiting step* in a biochemical pathway. Generally, complex enzyme systems possess at least one rate-limiting step; that is, one enzyme-activated transformation that can operate as a valve or control step. Once activated, this rate-limiting reaction occurs and allows the synthesis of some substrate that starts a sequence of transformations. Hormones alter the rate of reaction by influencing availability of the enzyme responsible for catalyzing that rate-limited step. This action of a hormone on the rate-limiting step may be due to an increase in the synthesis of an enzyme or an effect on the activity of enzymes already present. In the latter case the hormone might activate or inactivate an enzyme already present or simply alter the permeability of the plasmalemma such that additional substrate becomes available, which in turn stimulates enzyme activity. Research on the mechanisms of hormonal action on enzyme systems has centered around these three basic phenomena: transport of substrates, new enzyme synthesis and activation of inactive enzymes.

Membrane Transport and Enzyme Activity

In some cases the action of the hormone may be simply an effect on transport. Transport regulation would determine the availability of certain substrates, cofactors, etc., essential for activity of specific enzymes. Thus hormones might increase or decrease the activity of specific enzyme systems via the availability of some essential substrate or cofactor. Insulin, for example, accelerates lipogenesis in adipose cells in part by facilitating entrance of glucose into the adipose cells. Intracellular glucose is metabolized to glycerol and other substances necessary for the synthesis of triglycerides (fats). In muscle cells, insulin stimulates uptake of both glucose and amino acids, which respectively increase glycogen synthesis and protein synthesis as a result of increased substrate availability.

Gene Action and Enzyme Activity

Following the lead of microbial investigations on the genetic apparatus and control of enzyme synthesis, a "repressor" or "inducer" role for hormones has been sought in eukaryotic cells. Several workers studying the peculiar polytene salivary chromosomes of dipteran larvae observed that "puffs" appeared in particular regions of these chromosomes correlated with particular stages of development. Ecdysone, the insect molting hormone, stimulated formation of a certain puffing pattern (that is, formation of puffs at specific loci along the chromosome), suggesting an action of ecdysone on chromosomal events.

These puffs were shown to be active sites of DNA-dependent RNA synthesis. Analogous experiments were performed with chromosomes of newt oocytes. These cells exhibit attenuated chromosomes with peculiar loops consisting of single DNA strands extending outward from the body of the chromosome. These chromosomes are termed lampbrush chromosomes (after the brushes used for oil lamps that they supposedly resembled), and the loops have been shown to be sites of DNA-dependent RNA synthesis like the puffs of the dipteran polytene chromosomes. ^{3}H-estradiol-17β, a vertebrate steroid hormone similar in structure to ecdysone, binds specifically to some of the unwound loops of lampbrush chromosomes, and subsequently a number of steroids have been shown to bind to nuclei of their target cells.

The effect of cortisone on synthesis of the enzyme tryptophan pyrrolase is mediated via DNA-dependent RNA synthesis that is inhibited by actinomycin D. Such stimulation of RNA synthesis has been shown by a variety of experiments to be an integral part of hormone-induced responses in target cells. Estradiol stimulates RNA and protein synthesis of a specific egg protein, ovalbumin, in chicken oviducts. Chick-oviduct RNA extracted following estradiol treatment can direct synthesis of ovalbumin in a cell-free system.

Once it was shown that steroid hormones affect nuclear RNA synthesis, investigations were launched to identify earlier events in hormone action, in other words, what happens during the interval between arrival at the cell and the stimulation of nuclear events. One of the central questions examined was whether hormones have a direct or an indirect action on RNA synthesis. These studies have not only elucidated some of the early events in the stimulation of enzyme synthesis by hormones but have also identified how activity of existing enzymes can be influenced by hormones. Most of the studies with steroid and thyroid hormones have centered around the binding of hormones to intracellular receptors, while the majority of studies with polypeptide and amine hormones have emphasized binding to receptors in the plasmalemma and activation of a plasmalemma-bound enzyme, *adenyl* (adenylyl, adenylate) *cyclase*.

Mechanism of Steroid Action on Protein Synthesis

Binding of steroid hormones to target tissues is essentially the study of protein receptor molecules and behaviour of a hormone-receptor complex in activating enzyme synthesis. Such receptor molecules have been found to be highly specific for a given hormone and are not present in significant amounts in nontarget cells. Receptor molecules for estradiol-17β extractable from known target tissues will not bind other steroid hormones appreciably, including the biologically inactive 17α-isomer of estradiol.

Jensen proposed a model for the binding of estradiol-17β and the role of the hormone-receptor complex in stimulation of nuclear RNA synthesis in the uterus. This model has undergone several revisions. Simply stated, estradiol (E) enters through the cell membrane and is bound to a *receptor molecule* to form a *hormone-receptor complex*. The complex migrates to and enters the nucleus where it binds to a nuclear *acceptor site*. This hormone-receptor complex has a molecular weight of approximately 70,000 daltons, but it may form a dimer (MW = 130,000) before being translocated to the nucleus. After binding of the hormone-receptor complex to the nuclear acceptor site the complex is believed to activate a specific gene, resulting eventually in new synthesis of hormone-specific enzyme. This action on enzyme synthesis may be achieved by blocking the synthesis of a specific repressor molecule or by inactivating the repressor itself. New messenger RNA (mRNA) and translation of messenger are enhanced, and an increase

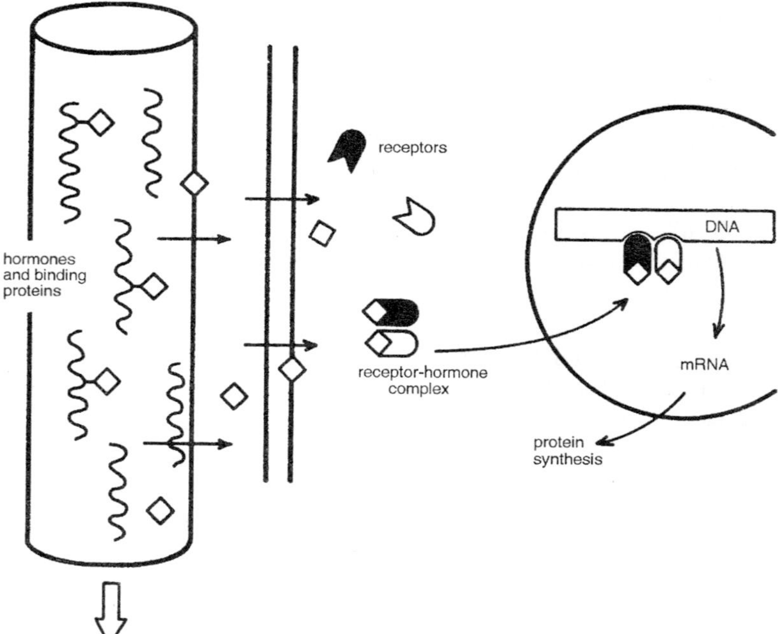

Fig. 1.4. Model of the proposed mechanism of action for steroid hormones.

in enzyme synthesis occurs. After activating the synthesis of mRNA the hormone-receptor complex is degraded metabolically, and the gene may again be repressed. In addition to stimulating synthesis of enzymes associated with the overt actions of estradiol on uterine tissues, it appears that estradiol also stimulates synthesis of new receptors so that more estradiol may be bound by the cell.

The actions of other steroid hormones (corticosteroids, progesterone and androgens) have not been studied as extensively, but their mechanisms of action appear to be very similar to estrogens. They may bind to cytoplasmic receptors prior to translocation and binding to nuclear chromatin. One difference has appeared in many androgen-sensitive tissues in that the major circulating androgen, testosterone, must be enzymatically converted to another androgen, 5α-dihydrotestosterone or DHT, before it can be bound to the receptor. Androgens may also be converted to estrogens, and this alteration may be important in the actions of androgens on the central nervous system.

The Jensen hypothesis should be treated as a working hypothesis for steroid hormone action since many aspects of the model as presented here require experimental verification, and considerable revision may still take place. For example, the observations that unoccupied estrogen receptors may occur in the nucleus of intact cells suggests that the demonstration of receptors in the cytoplasm of cell homogenates (an integral part of the Jensen hypothesis) could be artifactual. Nevertheless the Jensen hypothesis has been of immeasurable value in stimulating research and furthering understanding of steroid hormone actions.

Additional Mechanisms for Steroid Action

Estrogens not only stimulate growth of the uterine lining (endometrium) but also cause increased vascularity (hyperemia). A number of investigators believe that the hyperemic action of estrogens is mediated by an estrogen-dependent release of histamine, a potent local vasodilator. A local action for estrogens possibly involving histamine has been reported also for the ovary of a lizard.

Alterations of the properties of the plasmalemma by corticosteroids has been reported for cultured mammalian cells. Corticosteroids stimulate new enzyme synthesis in these cells and also produce changes in cell adhesiveness and in electrophoretic and antigenic properties. The importance of these observations for in-vivo action of corticosteroids is not clear.

Thyroid Hormones and Gene Function

The mechanism of action for thyroid hormones appears to be analogous to the Jensen hypothesis for steroids, although some major differences are noted. Some evidence indicates that thyroid hormones are not bound to cytoplasmic receptors but bind directly to nuclear receptors. Once bound to the nucleus, thyroid hormones stimulate DNA-dependent protein synthesis similar to that produced by steroid hormones in their target tissues. Others provide evidence for mitochondrial receptors for thyroid hormones. Following binding of thyroid hormones to the mitochondrial receptors the synthesis of mitochondrial protein is stimulated.

Enzyme Activation by Hormones: The "Second Messenger" Hypothesis

The discovery of the effect of epinephrine (the "emergency" hormone released from the adrenal medulla) on the formation of the cyclic nucleotide *cyclic adenosine 3',5'-monophosphate* (cAMP) and its subsequent role in cardiac muscle responses resulted in the awarding of a Nobel Prize to E.W. Sutherland in 1972. Studies by Sutherland and his co-workers showed that the first measurable effect of epinephrine was a fourfold increase in the level of cAMP within 3 seconds after the hormone was administered. The formation of cAMP from ATP is catalyzed by the enzyme *adenyl cyclase* associated with the plasmalemma and a guanine-nucleotide-binding protein. Cyclic AMP activates an enzyme, *protein kinase*, in the cytosol. This protein kinase phosphorylates other enzymes, causing them to assume or lose catalytic activity. The changes in catalytic activity of these enzymes determines the cellular events that characterize the actions of the given hormone. Epinephrine, glucagon and FSH are among those hormones whose action is mediated via a cAMP-dependent mechanism. Other peptide and polypeptide hormones that stimulate cAMP are LH, TSH, GH, ACTH, MSH, some hypothalamic releasing hormones, vasopressin and gastrin. The hormone might be considered the first messenger, and cAMP produced following the binding of the hormone to the plasmalemma would be the *second messenger*.

Another "second messenger"

Guanyl (*guanylyl*, *guanylate*) cyclase converts guanosine triphosphate to a cyclic nucleotide, *cyclic guanosine 3', 5'-monophosphate* (cGMP), and this enzyme is activated by acetylcholine (ACh) in heart muscle. Heart rate may be depressed by addition of ACh or cGMP, whereas cAMP or adrenergic substances accelerate heart rate. The mechanism

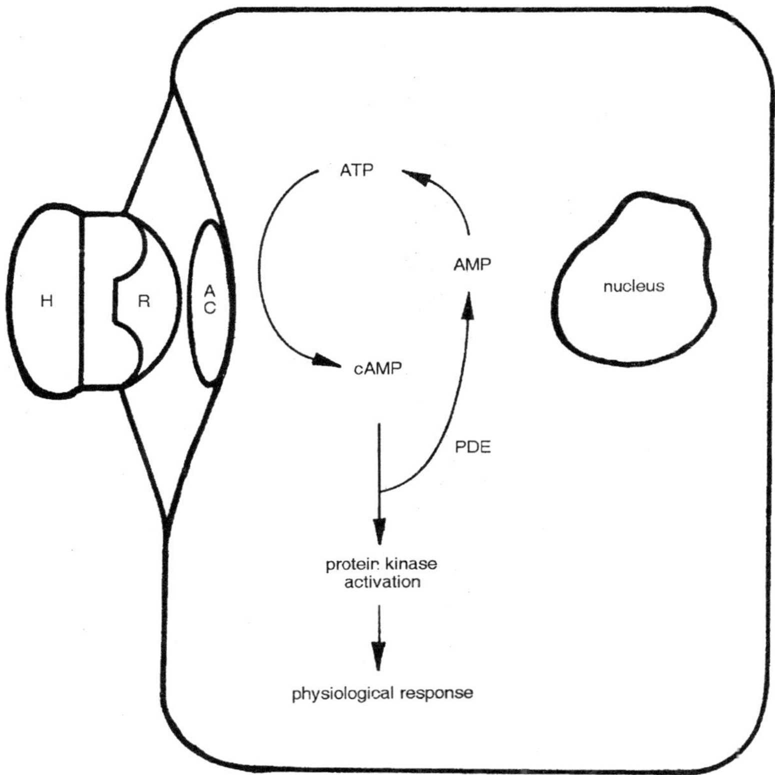

Fig. 1.5. Schematic representation of the role of cAMP in the mechanism of peptide and protein hormone action. H, hormone; R, receptor; AC, adenyl cyclase; PDE, phosphodiesterase.

of GH release from the pituitary involves activation of cAMP by a specific GH-releasing hormone (GHRH). A second hypothalamic neurohormone (somatostatin or GH release-inhibiting hormone [GHRIH]) has been found to inhibit GH release. This GHRIH elevates cGMP levels and depresses cAMP levels. Similar observations have been reported for liver, where both glucagon and epinephrine stimulate glycogen conversion to glucose via a cAMP-dependent mechanism. Insulin stimulates cGMP formation and depresses cAMP levels. One hypothesis suggests that the hormone receptor is separate from the adenyl cyclase which occurs on the cytoplasmic side of the plasmalemma. A component of the enzyme is a guanine nucleotide-binding protein (GNBP). When guanine triphosphate (GTP) is bound,

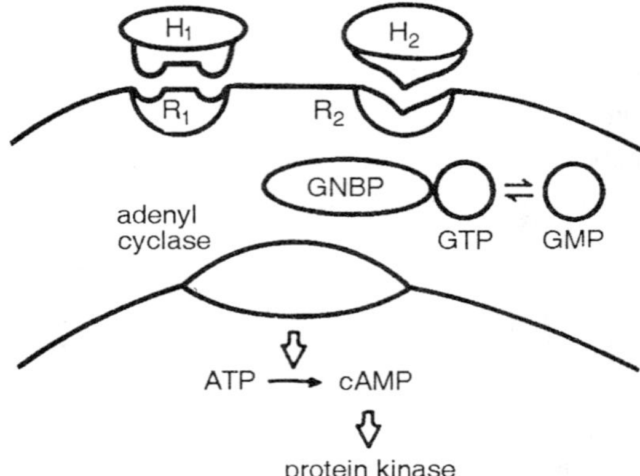

Fig. 1.6. Current hypotheses concerning the mechanism of action for protein and peptide hormone (H_1) suggest the interaction of the receptor (R_1) with a guanine nucleotide binding protein (GNBP) is necessary to activate the enzyme adenyl cyclase.

the GNBP interacts with the catalytic unit of adenyl cyclase and cAMP is formed from ATP. Hydrolysis of GTP to cGMP frees the GNBP and adenyl cyclase activity is reduced. A hormone could also act by altering the ability of the GNBP to bind GTP.

There is reason to suspect that the cytoskeleton of the target cell may be involved in the mode of action of peptide hormones. Microfilaments are associated with GNBP, adenyl cyclase, and the hormone-receptor complex. Any factors that might alter the cytoskeleton could influence hormone action.

Phosphodiesterase

Another enzyme complex, *phosphodiesterase* (PDE), is responsible for degradation of cAMP to 5'-AMP, which has no activity in this system but is used in the resynthesis of ATP. The activity of PDE is readily blocked by drugs known as methyl xanthines, such as caffeine and theophylline. Methyl xanthines potentiate the action of hormones by extending the biological life of cAMP, and they have proved useful for investigating the mechanism of action of many hormones in addition to epinephrine. Theophylline is used clinically to potentiate the actions of endogenous epinephrine. The discovery that there are at least three forms of the enzyme PDE in liver cells may provide clues to the interactions between these nucleotides. Only one of these PDEs is specific for cGMP, and the other two are specific for cAMP. Of the last two, one is activated by low levels of cGMP. Thus a hormone that elevates cGMP levels could reduce effective levels of cAMP and antagonize processes dependent on cAMP.

A "third messenger"?

A number of small lipids known as prostaglandins have been implicated as agents that influence cyclic nucleotides in hormone target tissues. Prostaglandins inhibit cAMP production in adipose (fat) cells, which can be stimulated by hormones like epinephrine. They also stimulate cAMP formation and progesterone synthesis in a manner similar to the effects of LH on the mouse ovary. One prostaglandin stimulates cAMP and causes venous vasodilation in the cow, whereas another one stimulates cGMP formation and brings about vasoconstriction. These compounds or related substances may be confirmed as mediators in the mechanism of action for peptide hormones, but more research is needed.

Role of Calcium in the Action of Hormones

Calcium ions have been implicated in presynaptic and postsynaptic events associated with neural transmission and action potential generation; in regulation of glycogenolysis (hydrolysis of stored glycogen) within muscle cells; in synthesis of glucose from noncarbohydrate sources (amino acids, fats) in kidney and possibly in liver cells; in the action of MSH on pigment cells of the skin, of vasopressin on the toad bladder, of PTH on bone cells, and of hypothalamo-hypophysiotropic hormones, to name a few.

The actions of calcium ions may be related to microtubular contractions involved with such processes as migration and exocytosis of secretion granules (for example, in stimulating hormone release) or intracellular movements of specialized organelles (for example, dispersal of pigmented organelles throughout the cytoplasm of melanophores). Activation of numerous enzymes caused by certain hormones and mediated via cAMP-protein kinase mechanisms may also be influenced by the availability of intracellular calcium ions. Binding of polypeptide hormones to the plasmalemma of a target cell apparently results in an increase of intracellular calcium that is essential for the general action of the hormone.

TECHNIQUES IN ENDOCRINOLOGY

Numerous techniques developed in other areas of physiological and biochemical research have been applied to the study of endocrine systems. The basic approach to determine if a suspected tissue may produce a hormone is extirpation and replacement. This classic

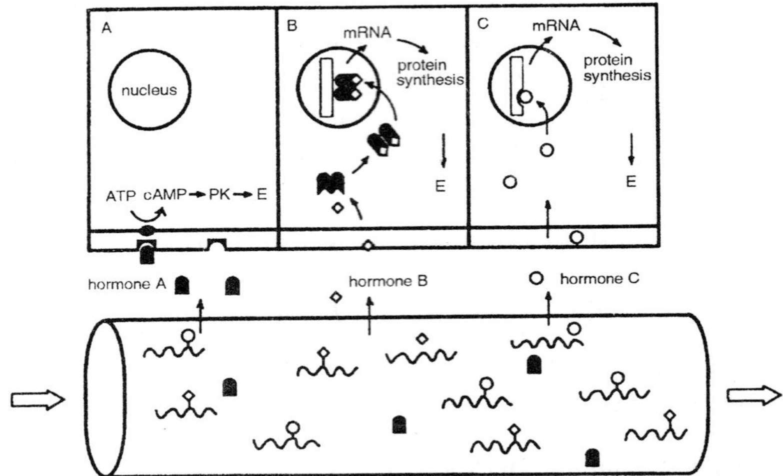

Fig. 1.7. Mechanisms of actions for peptide (A), steroid (B) and thyroid hormone (C) on target cells.

approach is described here along with two "borrowed" approaches: bioassay and radioimmunoassay. Bioassay is discussed in some detail here and elsewhere in the text because of its historical importance and because of its continued importance for validating other techniques such as radioimmunoassay. Bioassay techniques are the only approaches available for quantifying hormone levels in some systems. Radioimmunoassay is one of the most sensitive probes available to endocrine researchers for monitoring changes in hormone levels. Consequently it is essential to understand the theory of its use and its limitations.

Classic Technique

Historically the technique for identifying an endocrine gland has involved (1) *extirpation* (removal) of the putative gland and (2) *observation* of the effects of removal; (3) *replacement* of the gland by implant or by injection of a homogenate or extract prepared from the putative gland or by injection of some purified chemical substance and (4) *observation* to see whether the consequences of extirpation are reversed to the "normal" or pre-extirpation condition. This procedure (extirpation-replacement) is often referred to as the classic technique of endocrinology although it is still in common use.

Bioassay

The next step following the classic procedure to establish the existence of a hormonal substance has been identification of the active chemical principle responsible for the effects of the gland. This procedure requires development of some suitable bioassay system so that the investigator has a simple, quantitative response system in which to test extracts and chemical fractions for biological activity. The bioassay need not be conducted in the same organisms in which the gland is found. For example, in the mature pigeon a structure termed the *crop sac* produces a cytogenous secretion called *crop milk* that is regurgitated and fed to the young birds while in the nest. This secretion is abolished in hypophysectomized animals (extirpation of the hypophysis or pituitary gland) but can be restored by subcutaneous injection of pituitary extract directly over the crop sac (replacement). Crop milk is secreted only in that portion of the crop sac directly beneath the injection site. This crop-milk material can be scraped from a given area around the injection site, dried, weighed and compared quantitatively to a similar site in the same crop sac beneath an injection of saline or of an extract prepared from some other tissue. (The pigeon crop sac is conveniently proportioned into two halves such that the test may be made on one side of the bird and the control injection given on the other hemi-crop sac of the same individual.) The amount of crop-sac secretion (dry weight of the unit area for comparison) increases proportionately to the amount of pituitary extract administered over a defined range. The crop-sac bioassay allowed endocrinologists to identify the pituitary crop-stimulating factor as the hormone prolactin. The pigeon crop sac also responds to prolactins or pituitary extracts from amphibians, reptiles and other birds and mammals but not to fish preparations.

Because biological systems are subject to fluctuations induced by environmental conditions such as temperature, it is essential that conditions for the bioassay be rigidly defined. For example, the crop sac bioassay may be done in 6-week-old inbred White King pigeons to reduce variability in the response due to genetic differences between individuals and other strains of pigeons and to insure that prepubertal pigeons are used.

An important feature of a good bioassay system is the dose-response relationship, that is, the relationship between amount injected and degree of response. Some bioassays are designed so that one identifies the minimal quantity of extract that will cause a response in the system. It is desirable, however, to have a bioassay in which there is a minimum dose that will produce a response with quantitatively dose-related increments in response up to some maximum. Frequently supramaximal doses will produce no additional response, and in some cases may reduce the response (toxicity effect).

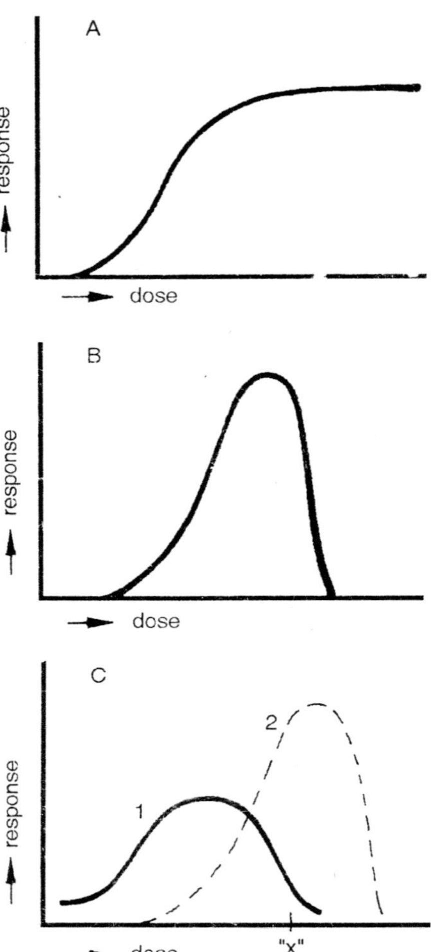

Fig. 1.8. Some examples of dose-response relationships. A—Dose-response relationship in which additional increases beyond a maximum effective dose produce no increase in response. B—Increasing dose above maximum effective dose causes inhibition of response. C—Different tissues (1 and 2) in a given animal may exhibit different dose-response relationships. A dose of "X" is inhibitory to tissue 1 but stimulatory to tissue 2.

One frequent error of researchers is failure to ascertain a dose-response relationship before deciding what dose of a hormone, inhibitor, etc., is to be used. Rather, they choose a dose arbitrarily or make a guess from published literature. Effective doses of hormones are generally in the microgram (μg, 10^{-6} g), nanogram (ng, 10^{-9} g) or picogram (pg, 10^{-12} g) range, and circulating hormone levels are expressed as a weight to volume of plasma or serum (0.1 μg/ml, 10 ng/dl). Hormone doses that approximate levels that can be experienced by the animal are often termed *physiological*, whereas excessive or supramaximal doses are referred to as *pharmacological*. Although pharmacological doses may produce marked effects, it is difficult to ascertain the biological significance of such data.

Knowledge concerning minimally effective doses of hormones is especially important when one is considering the effects of one hormone on another. A given hormone may produce a *priming effect* (increase in sensitivity of target tissue to another hormone), inhibit release or action of another hormone, increase release of another hormone or act cooperatively to produce an enhanced effect. This last action may involve either an *additive effect* or a *synergistic effect*. An additive effect occurs when the action of a particular dose of two hormones produces a response that is equal to the sum of the two hormones working alone (1 + 1 = 2). The term synergism is loosely applied to situations in which the combined effect is greater than additive, implying a different kind of cooperative effort (1 + 1 = 3).

Pharmacological doses of seemingly purified hormones may produce effects characteristic of the contaminants. For example, stimulation of the thyroid gland of the European crested newt following administration of massive doses of mammalian PRL is most likely due to contamination with TSH. Similarly, induction of metamorphosis in tiger salamander larvae by injection of FSH is due to the presence of TSH as a contaminant.

Radioimmunoassay

Radioimmunoassay (RIA) is being employed widely in comparative studies to measure blood levels of specific hormones. This technique allows direct and more accurate assessment of the endocrine state of the organism, whereas many other techniques such as histological examination or application of exogenous hormones tend to be indirect, inconclusive or both. RIA is essentially the same in principle and method as competitive protein-binding assays and the newer highly specific radioreceptor assays.

An antibody produced against any antigen (in this case a hormone) has a binding site on the antibody molecule that is specific for the antigen, much like the binding site of an enzyme for its substrate. A given amount of antibody (a given number of antibody molecules) quantitatively possesses a given number of binding sites for antigen. For purposes of this discussion, one antigen-binding site per molecule of antibody will be assumed. The antibody, however, cannot distinguish between an antigen that contains radioactive atoms (that is, labeled atoms) from one that does not. The assay technique is based upon the assumption that, at equilibrium, increasing quantities of cold hormone will displace proportionately increasing amounts of hot hormone (radioactively labeled) from the antibody, because there will be competition for a limited number of binding sites. A standard curve can be prepared by placing increasing amounts of cold hormone in separate reaction tubes with a constant number of antibody molecules and constant quantity of labeled hormone. The quantity of labeled hormone bound to the antibody can be accurately determined by precipitating the antibody and measuring the amount of radioactivity bound to it. Unbound labeled hormone is discarded with the supernatant. The radioactivity measured in the precipitated antibody is inversely

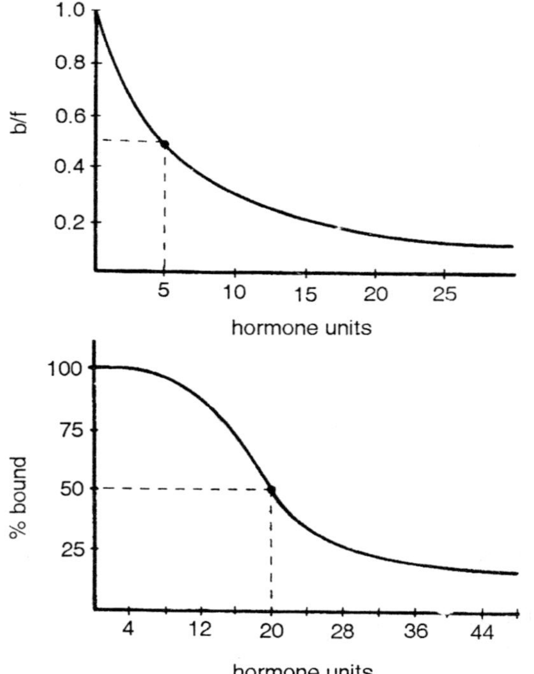

Fig. 1.9. Standardized curves for radioimmunoassay determination of hormone concentrations.

proportional to the amount of cold hormone added. These data may be plotted in various ways (ratio of bound to free label, percentage of total label added as measured in the antibody precipitate, the reciprocal of the radioactive counts per minute bound to the antibody, etc.) to yield a dose-response curve. The hormone concentration of an unknown plasma or serum sample can be assayed in the same manner and the quantity of hormone extrapolated from the standard curve using the amount of label bound to the antibody.

RIAs require a pure source of hormone in order to prepare an antibody against it as well as a source of radioactively labeled pure hormone. Furthermore, the system must be tested rigorously for specificity to determine whether molecules closely related chemically but lacking the same biological activity of the hormone assayed might also displace hct hormone from the antibody. The problem of a purified source becomes particularly acute when one is investigating endocrine systems of nonmammalian vertebrates in which the chemical structures for most protein and polypeptide hormones are unknown and the hormones are not available in pure form. Frequently such hormones may have activity similar to mammalian hormones, but there are often important biological differences reflecting variations in the structure of the hormones themselves. The use of antibodies prepared against mammalian peptide and protein hormones to measure nonhomologous molecules in nonmammals results in uncertain interpretations. Immunological data may not parallel data obtained by biological assay, and the tendency should be to accept the bioassay data.

The determination of receptor levels is considered by many to be important in understanding the status of a given system. If receptor levels in the target tissue are low, there are fewer sites to bind hormone molecules and less effect is produced. In many systems, there is evidence that binding of the hormone to the target cells results in a reduction in the number of receptors. This "*down-regulation*" reduces the observable hormone effects on the target when circulating levels are constant or even increasing. Later, production of new receptor is initiated, and the target tissue becomes more responsive. Special biochemical procedures have been developed to measure the levels of specific receptors in target tissues using techniques similar to those described for RIA.

The difficulties encountered with the application of mammalian RIA techniques to non-mammals may be circumvented by development of radioreceptor assays that employ hormone-specific receptor molecules isolated from the animal's own target tissue in place of the antibody. The principle of the assay is still the same, but the problem of not knowing the structure of the animal's native hormone becomes irrelevant. Although this approach also has its technical difficulties, it is of value to investigations of a comparative nature.

Endocrine systems must constantly be considered as dynamic states. Hormones are continuously being synthesized and released and are being lost from the circulation through metabolic inactivation, excretion or both. The effectiveness of a hormone and its circulating levels will be determined not only by rates of synthesis and release but also by rates of inactivation and excretion. Consequently it is helpful to know *turnover rates* or the *biological half-life* (time for half of a given population of hormone molecules to be inactivated, excreted or both) in assessing endocrine states. Turnover rates for peptide hormones are very short and can be measured in minutes. Steroids have much longer half-lives in blood because they are bound to specific binding proteins and are not removed as rapidly. Their half-lives are usually measured in hours. Thyroid hormones are more tightly bound to binding proteins in the blood than are steroids. For example, the half-life of thyroxine in humans is about 7 days.

The presence of binding proteins adds another complication in measuring hormone levels. An equilibrium is established between bound hormone and unbound or free hormone. It is the free hormone that is considered by most researchers to be available for entering tissues and therefore the important parameter to measure. Not only is the rate of secretion of hormones important, but the circulating levels of binding protein may influence the availability of free hormone. Thus, the levels of binding protein may be important, too. Another hypothesis has been propose that suggests the binding protein actually facilitates entry of the hormone into target cells, and free hormone is destined for excretion. This hypothesis is based in part on the presence of binding protein within steroid target cells.

The limitation of using blood level data as a diagnostic tool may be illustrated further by considering attempts to localize the source of an abnormal endocrine state. The demonstration of unusually low or high circulating hormone levels does not distinguish between changes in secretion rates and excretion rates for the hormone, and additional tests are required. Similarly, are hormone levels higher in a fish adapted to 10°C than in one adapted to 20°C because the former fish is secreting more or because the rates of excretion are markedly reduced at lower temperatures?

In spite of the cautions listed above, RIA has provided the endocrinologist with a highly sensitive probe with which to investigate dynamic endocrine events. When used with appropriate methods and experimental designs, RIA techniques are providing considerable insight into the functioning of the vertebrate endocrine system, bringing about major revisions in some established concepts while strengthening or rejecting others.

High Performance Liquid Chromatography

One tool has facilitated greatly the purification of peptides high performance liquid chromatography or HPLC. This method is capable of cleanly separating peptides differing by a single amino acid. Differences in small molecules like steroid hormones allow their separation by HPLC as well. By coupling HPLC with RIA, it is possible to separate potentially contaminating molecules from the fraction to be assayed. If suitable sensitivity can be achieved, HPLC validated with RIA and bioassay may someday be used for determination of circulating levels of hormones.

Man-made Hormones

Positive identification of a hormone with a physiological response involves the artificial synthesis of that hormone. Recent advances in biochemical techniques have made it possible to synthesize a new peptide shortly after verification of its structure. The final verification involves administration of the synthetic hormone and comparison of its actions to data obtained using the procedures discussed earlier. Once the true nature of the hormone is known, biochemists produce structurally altered forms called *analogues*. By comparing the actions and metabolism of these analogues they can better understand how the structure of the hormone relates to its effects. Some synthetic analogues are many times more potent than the naturally occurring hormone and prove very useful in both experimental and clinical work. Inhibitory analogues block actions of the natural hormone by binding to the same receptor but not effecting the anticipated actions.

2

ENDOCRINOLOGY OF PROTOCHORDATES

The Phylum Chordata comprises the Subphylum Vertebrata and three other Subphyla which are collectively known as the Protochordata. The latter are of exceptional interest, because they combine certain features of vertebrate organization with other characteristics which provide something of a link with the Echinodermata, the group of invertebrates to which the vertebrates are now thought to be most closely related. Thus, we may reasonably expect that endocrinological investigations of the group will extend our understanding of the origin, organization and functioning of vertebrate hormonal systems.

The best-known protochordates are the Cephalochordata, represented by *Amphioxus*, and the Urochordata (Tunicata), including the sessile ascidians (sea-squirts) and various pelagic forms such as the salps. All of these are filter feeders, and they resemble each other in the possession of a large filtering pharynx, along the floor of which runs a groove or channel, the endostyle. This organ is provided with pairs of longitudinal glandular tracts, and it is generally held that it contributes to the mucus-like secretions with which the animals trap small particles of food material. An organ closely resembling it, and clearly homologous with it, is found in the ammocoete larva of the lamprey, the most primitive surviving vertebrate. It has long been known that at the metamorphosis of this larva a part of the endostyle is transformed into the thyroid gland, and it has been established by Gorbman and Creaser (1942), by Leloup and Berg (1954), and by Leloup (1955) that even in the larva the endostyle is trapping and binding iodine and is

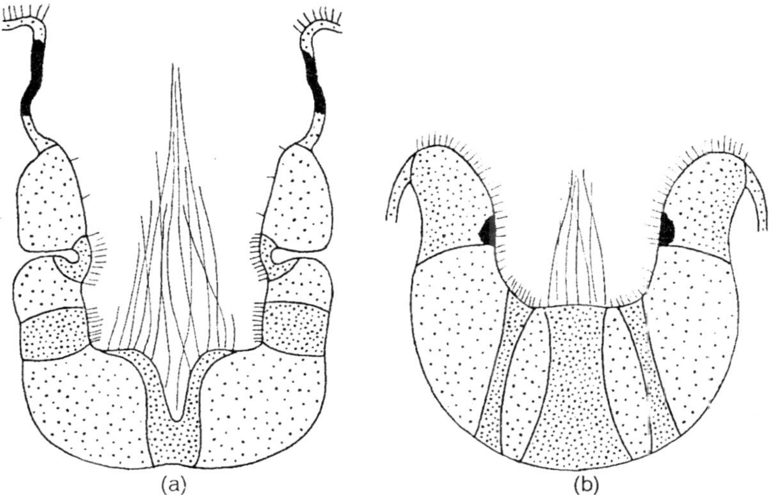

Fig. 2.1. Diagram of the endostyles: (a) Ciona; (b) Amphioxus.

thereby forming monoiodotyrosine, diiodotyrosine, thyroxine, and probably triiodothyronine. We are thus led to enquire whether such thyroidal biosynthesis may not already be established in the endostyle of the protochordates; if it is, there is then opened up the possibility of learning something of the fundamental functions of the thyroidal hormones before they have been overlain by the extensive specializations of the vertebrates.

As regards *Amphioxus*, Thomas (1956) has recently shown that the endostyle of this animal does, in fact, contain organically bound iodine, demonstrable in autoradiographs, prepared from animals which have been immersed in sea water containing I^{131}. I have been investigating this situation further with a view to comparing his results with the observations which we at Nottingham were simultaneously and independently making on the Tunicata. He was inclined to ascribe some at least of the iodine binding in the endostyle to the more dorsal of the two pairs of glandular tracts. My autoradiographs, however, make it clear that the centre of iodination lies at, or near to, the surface of a group of cells lying immediately above these tracts; the latter do not themselves appear to be involved in iodination, and in this respect the situation resembles that already established for the endostyle of the ammocoete larva.

I have further been able to show that in *Amphioxus* the cells concerned produce a clearly defined secretion; this may reasonably be regarded as providing the molecular basis for the binding process and it is thus of obvious interest to compare it with thyroglobulin. Like the latter, it is strongly PAS-positive, but certain other tests reveal considerable differences. The significance of these is not easy to assess, owing to the lack of precise information as to the chemical composition of so-called muco-substances. It can be said, however, that epithelial mucins which are characteristic of alimentary tracts, and which we should reasonably expect to find in the pharynx of *Amphioxus*, have as a major component the carbohydrate polymers known as acid mucopolysaccharides or, as Kent and Whitehouse (1955) would prefer, as amino-polysaccharides, these being carbohydrates which contain the 2-aminosugars glucosamine and galactosamine. Secretions of this type react positively with mucicarmine and with alcian blue, and show the characteristic red colour of gamma metachromasia with toluidine blue.

The thyroglobulin of thyroid colloid, being a glycoprotein, is negative to these three tests, but the secretion of the iodine-binding cells of the *Amphioxus* endostyle is positive. This may well mean that in the latter animal we are witnessing an early stage in the evolution of thyroidal biosynthesis, one in which the process is associated with the presence of an alimentary muco-substance. I am not, of course, suggesting that the iodine is bound to a carbohydrate polymer. The position seems to be that aminopolysaccharides are often, and perhaps always, associated loosely with some protein; it is presumably to the latter that binding takes place, and the possibility that it is itself a thyroglobulin-like substance is not excluded by these tests. The point of interest is that iodine binding seems here to be evolving out of an essentially alimentary secretory process.

A further important conclusion which can be drawn from *Amphioxus* results from the clearly observable fact that cells other than those associated with iodine binding react positively to the tests for epithelial mucins, notably a group of cells lying at the base of the groove. No bound iodine has been found to be associated with these, however, and we can thus conclude that the binding is not a random result of any generalized property of mucins, but is a specialized property of one particular group of cells. From this it would seem to follow that the binding is a biochemically purposive act, giving rise to a secretion which is biochemically significant for the animal. This secretion, there is reason to believe, is mixed with the food material in the pharyngeal lumen, and can be absorbed through the wall of the alimentary canal.

In my published analysis of the binding of iodine in the endostyle of the ascidian *Ciona* it was suggested that the binding was actually taking place in an iodination centre in a clearly defined area of nonciliated epithelium which lies immediately above the three pairs of glandular tracts. The presence of iodine on the ciliated lip of the endostyle, and elsewhere on the pharyngeal epithelium, was explained as being the result of its adhesion to the ciliated cells which move the food-trapping mucus secretion over the pharyngeal wall, the bound iodine being incorporated into this secretion on its discharge from the endostyle. On this interpretation, the organization of the endostyle of the ascidians is fundamentally similar to that of *Amphioxus*, and subsequent work, as yet unpublished, has entirely confirmed this view both for *Ciona* and for other ascidian genera.

Good examples of the latter are provided by *Dendrodoa* and *Botryllus*, which, as members of the order Pleurogona, have a more specialized organization than *Ciona* and are regarded by the systematist as widely separated from it. In *Dendrodoa* a region corresponding to zone 7 is clearly defined, although it is less extended than in *Ciona*; its cells contain PAS-positive material, and in autoradiographs bound iodine is associated with them as well as the ciliated lip of the endostyle. *Botryllus* is particularly striking, for in this well-known colonial form the individual zooids are minute in size, ranging up to only 4 mm in length, as compared with the 25 mm of *Dendrodoa* or

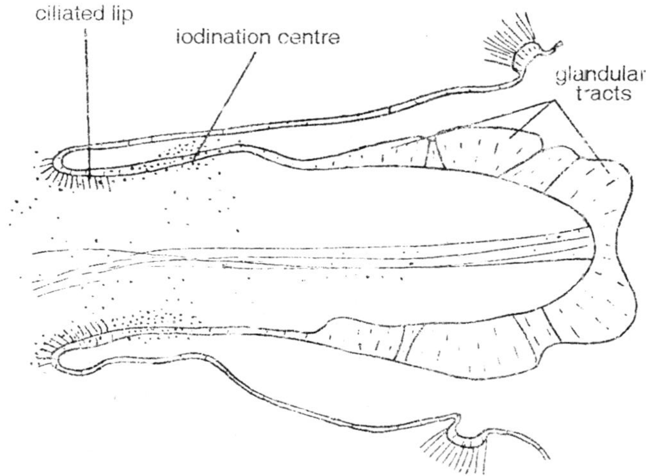

Fig. 2.2. Diagram of an autoradiograph of a transverse section of the endostyle of Botryllus.

the 12 cm or more of *Ciona*. Despite this, the same regions are distinguishable in the endostyle, and clear autograph images indicate an iodination centre in the upper part of zone 7, whereas elsewhere, on the ciliated lip and over the pharyngeal wall, the signs of bound iodine are irregular and may be quite inconspicuous.

At present the nature of the secretion of the iodine-binding cells is less easily definable in the ascidians than it is in *Amphioxus*, and individual genera may prove to vary in this respect. In *Ciona*, as I have pointed out elsewhere, it bears some resemblance to the secretion of the corresponding cells of the endostyle of the ammocoete larva. In *Dendrodoa* it is PAS-positive, but probably differs from the corresponding secretion in *Amphioxus* in being negative to alcian blue and to mucicarmine. In *Botryllus* PAS-positive droplets are sometimes distinguishable, but often there is no sign of secretion at all, probably, in part, because of the very small size of the cells and the difficulty of securing satisfactory fixation of them. This matter is still under investigation, but in the meantime it may be well to emphasize that the adult ascidians are probably not on the direct line of ascent to vertebrates, and are presumably much less closely related to the latter than is *Amphioxus*. Current theory derives the vertebrates, and also *Amphioxus* itself, by neoteny from the ascidian larval stage, although the further possibility that both *Amphioxus* and the *Tunicata* may have arisen independently from the third protochordate Subphylum, the Hemichordata, cannot be excluded. On either view it would seem that adult tunicates must have pursued their own independent specializations, which might in some cases he expected to parallel those of vertebrates and in others to diverge from them.

An important result of these studies on tunicates has been the demonstration of the presence of bound iodine in other parts of their bodies. There have been earlier reports of its occurrence in the stolonic sep'a of *Perophora annectens*, and in the canals connec'ing the zooids of *Botryllus*, but I have been particularly impressed by its accumulation in large quantities in the tunic or test of all forms I examined. This tunic is a tough protective coat which is secreted over the body and was shown by Cameron (1914) to be very rich in iodine. Autoradiographs show that some at least of this iodine is organically bound and that it is actually very much more conspicuous here than it is in the endostyle. The main concentration is found to be in a narrow zone at the extreme surface, an important fact that at once brings into focus a striking contrast with the situation obtaining in the endostyle, for there is no evidence that the binding of iodine in the tunic is related to any particular groups of cells. It is true that wandering cells derived from the mesenchyme are conspicuous throughout the tunic, but there is no constant accumulation of them in the surface layer, and there is quite certainly no epithelial layer developed there.

This matter also is still under investigation, but in the meantime certain properties of the tunic seem to merit attention. Its matrix is well known to contain a high proportion of cellulose, a condition unique in the animal kingdom. An early analysis of the composition of the tunic of *Ciona* showed it to be composed of 60.34 per cent cellulose, 27 per cent nitrogenous material, and 12.66 per cent inorganic material. According to recent work of Peres (1948a, b), the main mass of the tunic consists of cellulose associated with some glycoprotein, but the surface is formed of a distinct cuticle-like layer of pure protein. Here again there may well be variation from genus to genus; it appears, for example, that in *Clavelina* the cuticle is well-developed, particularly in the stalk region, whereas in *Ciona* it is much more inconspicuous. These facts certainly suggest, however, that we are dealing here with the phenomenon which has been shown to characterize a wide range of invertebrates, the association of iodine binding with the laying down of exoskeletal scleroproteins.

At this stage of the analysis it is clearly necessary to inquire as to the nature of the iodinated products which arise in the protochordates, for this has an obvious bearing on the question as to how far we are here dealing with an hormonal situation. As far as *Amphioxus* is concerned, the only positive evidence comes from the work of Sembrat (1953), who implanted 40 dried endostyles into one axolotl and 65 into another, and thereby produced metamorphic reduction of the fins and gills. Since control implants of muscle had at most only slight effects, this was regarded as giving some indication of the presence of thyroid hormone in the endostyle. My own material of *Amphioxus* has not been sufficiently plentiful to enable me as yet to study the problem in this animal, but I have applied radiochromatographic methods to *Ciona*, using Gleaşon's (1953) *tert*-amyl alcohol-2*N* ammonia solvent.

This work is not yet complete, but preliminary results have given satisfactory demonstrations of the presence of diiodotyrosine and thyroxine in extracts of the tunic, for using added carriers for identification, definite radioactive spots can be shown to coincide with spots of those particular substances, as revealed by the ceric sulphate-arsenious acid reaction. Triiodothyronine is probably also present, but it has not yet been possible to separate out diiodothyronine either from the extracts or from carrier mixtures. The activity of aqueous extracts of the tunic is weak, but treatment of the tissue with *N* NaOH at 50°C for 5 hours gives extracts of very much higher activity, a result which provides some support for the view, suggested above, that the bound iodine is associated with structural proteins. The radioactivity of aqueous extracts of the endostyle is very low, as is to be expected from the weak autoradiographs given by this organ, but the same radiochromatographic procedure has given good evidence for the presence in these extracts of diiodotyrosine, and it is highly probable that at least thyroxine is also present.

Perhaps at this point one might conveniently summarize the conclusions which seem to emerge from this review, always bearing in mind that these particular investigations are still in progress, and that the arguments should at this stage be regarded as primarily a reasoned directive for further research. As far as the endostyle is concerned, iodine binding is already established in it at the protochordate level of evolution, and it is established in a localized region of the organ in such a manner as to suggest that it is biochemically purposive and not a product of random iodine uptake. The cells concerned produce a secretion which, at least in *Amphioxus*, is iodinated at or near to the cell surface, as it apparently is in the cells of the thyroid gland of vertebrates. There is presumably little storage, and the iodinated secretion is discharged and becomes mixed with food cords, probably to be absorbed through the epithelium of the alimentary canal. It is highly probable that this iodine binding is indicative of an essentially thyroidal biosynthesis, and one may surmise that the replacement of the mucin-like secretion of *Amphioxus* by thyroglobulin may have provided a more efficient molecular basis for this purpose and for the storage of its products. It is interesting in this connection to note that Hooghwinkel et al. (1954) have interpreted thyroglobulin as a firm compound of mucopolysaccharide and protein, although as Gross (1957) has pointed out, their results seem to conform also to the more usual view that it is a glycoprotein. One may further surmise that this cytochemical evolution may well have been determined by the shortage of iodine which the chordates would have encountered when they migrated from marine to freshwater habitats, and this may also account for what seems to be a greater area of iodinating epithelium in the endostyle of the ammocoete larva as compared with that of the protochordates. Here, then, would seem to be an ecological adaptation, analogous to those which have been considered elsewhere in this symposium, but operating on a geological time scale.

As regards the tunic, the association of bound iodine with cuticular structures is, as I have already mentioned, by no means novel, for it has been recorded in annelids, arthropods, and molluscs. This association may well be a biological accident, with the degree of fixation and the yield of thyroxine being determined by the disposition of the tyrosine residues in scleroprotein molecules, and it has been plausibly argued that the thyroid gland might have evolved as a consequence of the ancestors of vertebrates becoming biochemically dependent on iodinated amino acids which were initially made available to them by such accidental means. It would be premature to develop this argument far in our present state of knowledge, but it is quite clear that conditions in the tunicates lend some support to it. Their tunic is often richly provided both with wandering mesenchyme cells and with blood vessels, so that it is by no means unlikely that some of its iodinated products might be released and transported through the body, and thus become available for utilization. Further, and perhaps more important, the tunic is not a passive tissue, but is constantly being secreted by the epidermis and lost from the surface, so that there would seem to be rather a good possibility of these ciliary feeding animals ingesting, and subsequently digesting, some of their own iodinated products, the more so in that they are often gregarious or colonial in habit. It is surely far from fantastic to suggest that in some such way tunicates might have become biochemically dependent upon these iodinated products at an early stage of their evolution; the iodinating properties of the endostyle might have arisen thereafter as an adaptation for the secretion of these substances in a form more readily available and in a position from which they could more easily be assimilated. At the moment, however, this can be no more than a suggestion, for it carries the implication that *Amphioxus* and the vertebrates evolved from tunicates, and we cannot be certain that this was, in fact, so. From this point of view it is therefore particularly important that we should learn more as to the significance of the association of bound iodine with the dermal glands of the Enteropneusta, for this rather puzzling group of Hemichordata appears to antedate the Tunicata and might have been ancestral to them.

I have deliberately devoted most of this review to a consideration of thyroidal biosynthesis, because it is the field in which I have been most directly concerned, and because it effectively illustrates the type of problem which the protochordates present to the endocrinologist. Other aspects, and particularly those related to reproduction, have been reviewed by Dodd (1955), who has drawn attention to the uncertainty of the evidence which has from time to time been held to establish the homology of the neural gland of the Tunicata with the pituitary of vertebrates. However much we may sympathize with the view that it is impossible to doubt this homology on morphological grounds, the fact is that we are still in no position to substantiate it on biochemical or physiological grounds, although we are equally unable to deny it. Dodd (1958) has now satisfied himself that no gonadotrophin, assayable by mouse or male toad methods, exists in the neural complex of breeding *Ciona*, nor is there any convincing evidence for the presence of vasopressin or of melanophore-expanding hormone. He certainly finds some support for the contention of earlier workers that an oxytocic substance is present, but the significance of this is quite obscure, for he has been able to show that its properties are very different from those of mammalian oxytocin, whereas a very similar substance can be extracted from various parts of the bodies of starfish and lugworms. Here, then, is a field in which recent work, in complete contrast to that on the endostyle, has failed to substantiate earlier views.

At the present time the clearest positive evidence as to the function of the neural gland of tunicates is that it has powers of phagocytosis and that it is capable of taking up finely divided particles which have entered the inhalent siphon in the incurrent stream of water. It seems well to emphasize once again in this connection that if vertebrates are derived from ascidians it is from their larvae and not from the adults, and that although the larvae possess an endostyle they certainly do not possess a neural gland, although the neotenous appendicularians are said to have a well-developed ciliated tubercle, this being the structure on which the duct of the gland opens in the adults of the other groups. It may, then, be more prudent to think of the neural gland not so much as a forerunner of the pituitary, but as an independent specialization of the Tunicata, although, of course, it might still be the expression of some genetic potentiality common to all the lower chordates.

Brambell and Cole (1939) have suggested that some such common potentiality for the production by invagination of an anterior pit-like organ may explain the development of the ciliated organ which they discovered at the base of the proboscis of the Enteropneusta. It might also account for the existence in *Amphioxus* of Hatschek's pit, and I should like to emphasize the potential importance of the latter, which we are now restudying at Nottingham, for there are excellent morphological and embryological grounds for homologizing it with the adenohypophysis, as Goodrich (1917) first demonstrated. Its function has often been said to be the secretion of mucus to aid in the trapping of food, but its cell structure is so elaborate that it is difficult to feel satisfied with such a simple interpretation. It is said to have no nerve supply, but some of its cells have long-drawn-out distal ends which project into the lumen of the pit, and one wonders whether they may not be sensitive to substances in the water current passing over them. Without wishing to indulge in too easy-going speculation, one cannot but wonder whether the evolutionary history of this organ may not have some analogy with the story of the endostyle outlined above. Just as a mucus-secreting pharyngeal organ, with powers of iodination, seems to have evolved into a glycoprotein-secreting endocrine gland, so perhaps might a mucus-secreting stomodaeal pit, sensitive to passing substances, have evolved into another glycoprotein-secreting gland, sensitive to materials reaching it in the blood. Here at least is an obvious field for further investigation.

Where, it may be asked, is one then to look for the homologue of the neurohypophysis? There is, in fact, in the floor of the cerebral vesicle of *Amphioxus* a group of peculiar cells referred to by earlier workers as an infundibular organ. Olsson and Wingstrand (1954) have shown that these are nerve cells containing Gomori-positive granules, although since they appear to secrete a Reissner's fibre they have something in common with the subcommissural organ of vertebrates, which is composed of ependymal cells secreting a Gomori-positive material. Clearly we need to know more as to the function of these cells, for Franz (1923) believed that they were light-sensitive. According to his interpretation, they respond to shadows cast upon them by the so-called "eye-spot" which is situated in front of them, in the anterior wall of the cerebral vesicle, and he believed them to be an essential element in the orientation of the animal to light and shade.

This particular problem has been reviewed in later publications by Olsson (1958), and further study will be needed before we can hope to resolve it or, indeed, any of the other problems which I have included in this review. I can only hope that my account of them will have served to demonstrate what I believe to be the great interest of the protochordates, and the possibilities which they offer us of enlarging our understanding of vertebral hormonal mechanisms, always provided that we approach these animals with due regard for their habitat and their specialized mode of life.

3

Hypothalamo-Hypophysial Complex

The majority of endocrine activity in vertebrates is generated through the hypothalamo-hypophysial axis. Under neurohormonal directives from the hypothalamus, the hypophysis secretes hormones that control secretion by the thyroid gland, adrenal cortex and the gonads. In addition, the hypophysis secretes several other hormones.

The hypothalamo-hypophysial axis consists of the neurosecretory (NS) portions of the hypothalamic region of the brain and the hypophysis or pituitary gland. During development of the brain the diencephalon differentiates into three regions: Most of the diencephalon becomes the thalamus, a major relay station between higher and lower portions of the brain. The floor or ventral portion of the diencephalon becomes the hypothalamus, containing various NS nuclei, which are sources for neurohormones. The third portion, the epithalamus, is derived from the roof of the diencephalon and gives rise to the endocrine epiphysial complex that includes the pineal gland.

The pituitary gland or hypophysis develops through fusion of a ventral growth from the diencephalon, the *infundibulum*, with an ectodermal sac known as *Rathke's pouch*, and is located directly beneath the third ventricle of the brain. The third ventricle is a cavity continuous with the other ventricles of the brain and the central canal of the spinal cord. It is filled with a fluid known as cerebrospinal fluid.

The term hypophysis is derived from *hypo* under (the brain) + *physis* growth. Its alternate name, pituitary gland, is derived from *pituita* slime or phlegm. The pituitary was believed to be the source

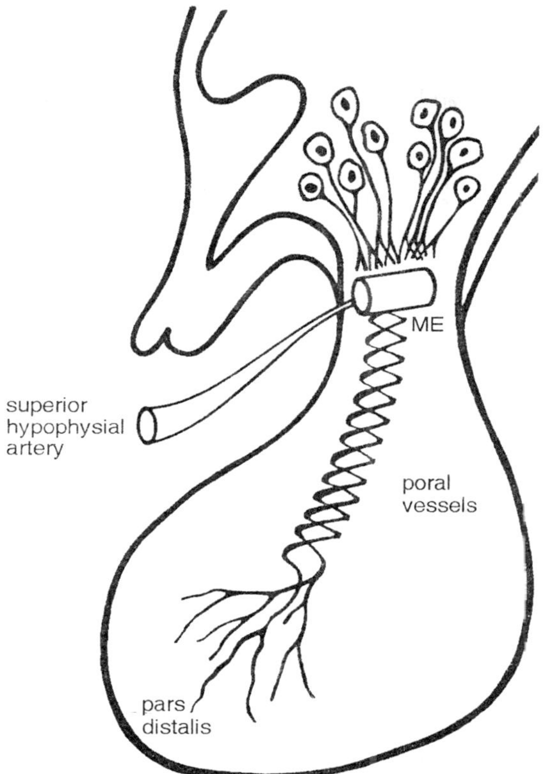

Fig. 3.1. The mammalian hypothalamo-adenohypophysial system. ME, median eminence.

of phlegm, one of the four humors of the body proposed by Galen centuries ago. The other humors were blood, black bile and yellow bile.

The mammalian hypothalamo-hypophysial axis will be discussed first and followed by descriptions of nonmammalian vertebrates. Although in an evolutionary sense such a discussion should begin with agnathan fishes, the mammalian system is better understood and provides the nomenclature with respect to structures, hormones and functions that are applied to the nonmammalian systems. The entire field of endocrinology has branched in similar fashion from mammalian investigations to nonmammalian.

Mammalian Hypophysis

The hypophysis of adult mammals is located ventral to the brain just posterior to the optic chiasma, and it remains attached to the

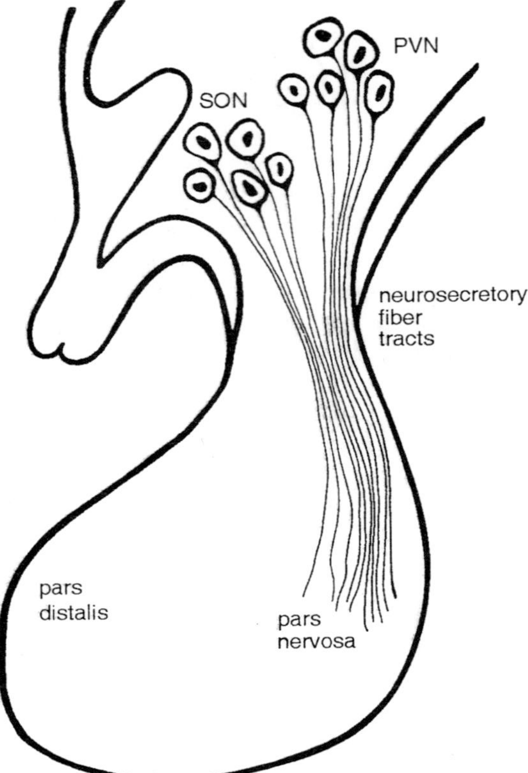

Fig. 3.2. The mammalian hypothalamo-pars nervosa system. Axonal fibers from the neurosecretory neurons of the supraoptic nucleus (SON) and paraventricular nucleus (PVN) travel through the infundibular stalk to the pars nervosa where their secretory products, the neurohypophysial octapeptide hormones, are stored.

hypothalamus by a stalk-like connection. Embryologically the hypophysis has been described as having two separate origins: The *adenohypophysis* develops from Rathke's pouch, which has been thought to develop from oral (non-neural) ectoderm. The adenohypophysis is an epithelial structure (*adeno* gland) and can be subdivided into three regions: the *pars distalis*, *pars tuberalis* and the *pars intermedia*. These regions are distinguished by their cytological features as well as their anatomical relationships to the remainder of the pituitary gland, the *neurohypophysis*. The infundibulum growing out ventrally from the floor of the diencephalon gives rise to the neurohypophysis, whose name reflects its neural origin. Two distinct subregions can be identified in the neurohypophysis: the *median eminence* and the *pars nervosa*. The median eminence consists mostly of aminergic axonal endings. The majority of axonal endings in the pars nervosa are peptidergic. An extensive vascular portal system, the *hypothalamo-hypophysial portal system*, develops between the median eminence of the neurohypophysis and the adenohypophysis. This portal system carries blood from the median eminence directly to the epithelial cells of the pars distalis.

The classic concept of the neurovascular link between the hypothalamus and the pituitary gland (the hypothalamohypophysial portal system) is based upon the pioneering anatomical studies of Wislocki. Presumably blood containing the hypophysiotropic regulating substances flows from the median eminence to the adenohypophysis, and the venous drainage from the latter carries pituitary hormones into the general circulation. Data gathered over the last several years, however, suggest that there may be significant blood flow from the adenohypophysis to the hypothalamus as well, and such a pathway may prove to be important in actions of pituitary hormones on the central nervous system.

Frequently in mammalian studies, the terms anterior lobe and posterior lobe are used to represent two divisions of the pituitary gland. These terms will not be used in this text since they do not refer to the same components of the gland as do adenohypophysis and neurohypophysis. Furthermore, the "posterior lobe" includes cells derived from both Rathke's pouch and the infundibulum, whereas the preferred terminology divides the pituitary according to its origin from Rathke's pouch and the infundibular process.

Recent reexamination of the hypothalamo-hypophysial axis supports a common origin from neuroectoderm for Rathke's pouch and the portion contributed by the diencephalon. According to these studies, Rathke's pouch, which gives rise to the adenohypophysis and its associated hormones, is of neural origin and has the same embryonic origin as the hypothalamus and the neurohypophysis. A neural origin for the adenohypophysis would be consistent with the hypothesis that all peptide hormone-producing cells are of neural origin and belong to the amine content and amine precursor uptake and decarboxylation (APUD) series. Because of the proposed origin from neuroectoderm, Pearse and Takor Takor include all tropic hormone-producing cells as part of the general APUD series although no amines can be demonstrated in certain cellular types. The neuroectodermal origin for adenohypophysial cells is further strengthened by the demonstration of pituitary hormones in the brain.

Adenohypophysis

Pars tuberalis

The pars tuberalis consists of a thin layer of cells projecting rostrally (anteriorly and dorsally) from the adenohypophysis, and it is in contact with the median eminence of the neurohypophysis. The blood vessels of the hypothalamo-hypophysial portal system pass near or through the pars tuberalis en route to the pars distalis.

The pars tuberalis has long been termed a portion of the adenohypophysis of unknown function but characteristic of all tetrapod vertebrates. Recent evidence seems to imply that it is only an extension of the pars distalis related primarily to reproduction.

Structurally the cells of the pars tuberalis are connected to the cerebrospinal fluid of the third ventricle in the brain through cellular processes originating in modified ependymal cells known as *tanycytes*. It has been suggested that tanycytes may selectively remove materials (neurohormones?) from cerebrospinal fluid and transfer them to cells of the pars tuberalis, causing the latter to release their stored products (?). Although this is a highly speculative idea, such an interesting anatomical relationship demands some imaginative research to provide

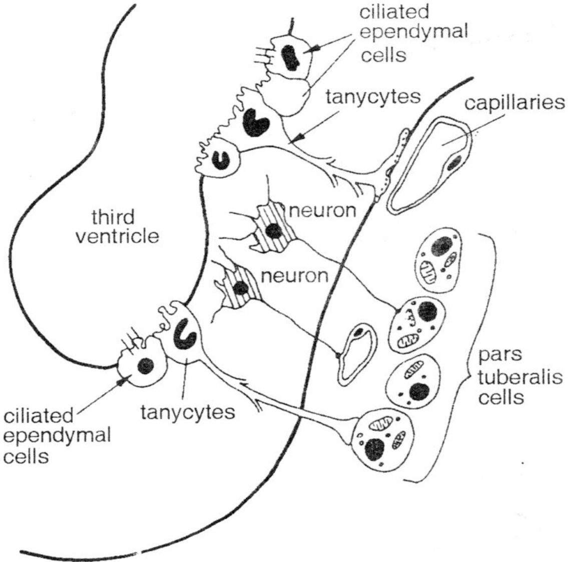

Fig. 3.3. Relationship of modified ependymal cells lining third ventricle to the pars tuberalis.

a better understanding of both tanycytes and the cells of the pars tuberalis.

Pars intermedia

That portion of the adenohypophysis that makes contact with the pars nervosa of the neurohypophysis is defined as the pars intermedia. Indeed, formation of the pars intermedia occurs only if physical contact takes place with the infundibulum during development. In some species the pars intermedia becomes separated from the remainder of the adenohypophysis by a cavity or cleft. Only one cellular type appears in the mammalian pars intermedia as identified by selective staining procedures, and it is supposedly responsible for secretion of the peptide hormone *melanophore-stimulating hormone* or *melanotropin* (MSH). Cells associated with the skin and known as melanocytes in mammals synthesize a brown pigment, melanin, under the influence of MSH. This peptide hormone has also been identified in the hypothalamus, although its importance there is not clear.

Pars distalis

The major portion of the adenohypophysis is designated as the pars distalis. A variety of cellular types can be identified in the pars distalis by selective staining procedures. These various cellular types are responsible for secretion of seven pituitary hormones: *corticotropin*, or adrenocorticotropic hormone (ACTH); *thyrotropin*, or thyroid-stimulating hormone (TSH); *lipotropin*, or lipotropic hormone (LPH); *growth hormone* (GH); *prolactin* (PRL); and two *gonadotropins*, or gonadotropic hormones (GTHs). The two GTHs are *follicle-stimulating hormone* (FSH) and *luteinizing hormone* (LH), both of which are named for their effects in female mammals. All of these hormones are polypeptides or proteins. These seven hormones, together with MSH from the pars intermedia, are referred to as the *tropic hormones* of the adenohypophysis. The cells responsible for secreting tropic hormones are called *tropes*. Hence a corticotrope is a cell that produces ACTH.

Cellular Types of the Adenohypophysis

Staining procedures and light microscopy

A variety of cellular types can be distinguished in the pars distalis by utilizing special dyes in particular staining combinations. This differential uptake of dyes is due to the differential affinity of cytoplasmic granules for these dyes. Some cytoplasmic granules bind acidic dyes, and cells with these granules are termed *acidophilic cells* or *acidophils* (*philos* love). Other granules bind basic dyes and the cells containing these granules are termed *basophils*. Cells containing granules that may bind acidic and basic dyes are said to be *amphophilic*. A basophil in the pars distalis is defined as a cell that is stained by the aniline blue dye of the Mallory trichrome staining method. However, since aniline blue is in reality an acidic dye, some investigators have preferred to use the term *cyanophil* (*cyanos* blue) to designate these cells. The endocrine literature is filled with the term basophil, so use of this term continues, although it is recognized to be technically inappropriate. Cells that do not contain stainable cytoplasmic granules are called *chromophobes*. Various types of basophils or acidophils may be distinguished from one another in terms of their specific affinities for other dyes. Some cells contain granules that stain with the periodic acid-Schiff(PAS) method for staining glycoproteins and mucopolysaccharides. Such cells are termed PAS(+).

Cellular types in the pars distalis

Generally speaking there is one cellular type responsible for synthesis and release of each tropic hormone. Only two strongly basophilic cells are found, however, and they are responsible for secretion of three glycoprotein hormones: TSH, LH and FSH. The traditional mammalian designation for these various cellular types is given with the corresponding generalized vertebrate cellular type indicated within the parentheses. The PAS(+) *beta basophil* (type 1 basophil) that secretes TSH can be distinguished from the gonadotropic PAS(+) *delta basophil* (type 2 basophil) by the affinity of the former for aldehyde fuchsin (AF[+]). It has not been possible to differentiate between LH- and FSH-secreting gonadotropes on the basis of stainability for light microscopy or with the aid of the electron microscope. In the past, immunological tests have not been able to distinguish separate cellular types, and it may be that both mammalian GTHs are produced by the same cell. Recent studies employing sensitive microscopic and immunocytochemical procedures suggest there may be separate cellular sources for gonadotropins in tetrapods. However, throughout this book the source for gonadotropic hormones will be referred to as the type 2 basophil.

Two cells staining with acidic dyes have been found to be sources for GH and PRL respectively. The *alpha acidophil* stains with orange G (orangeophilic) and is the source of GH in mammals. The *epsilon acidophil* exhibits a strong affinity for azocarmine (carminophilic) and secretes PRL. The PRL-secreting epsilon acidophil (type 1 acidophil)

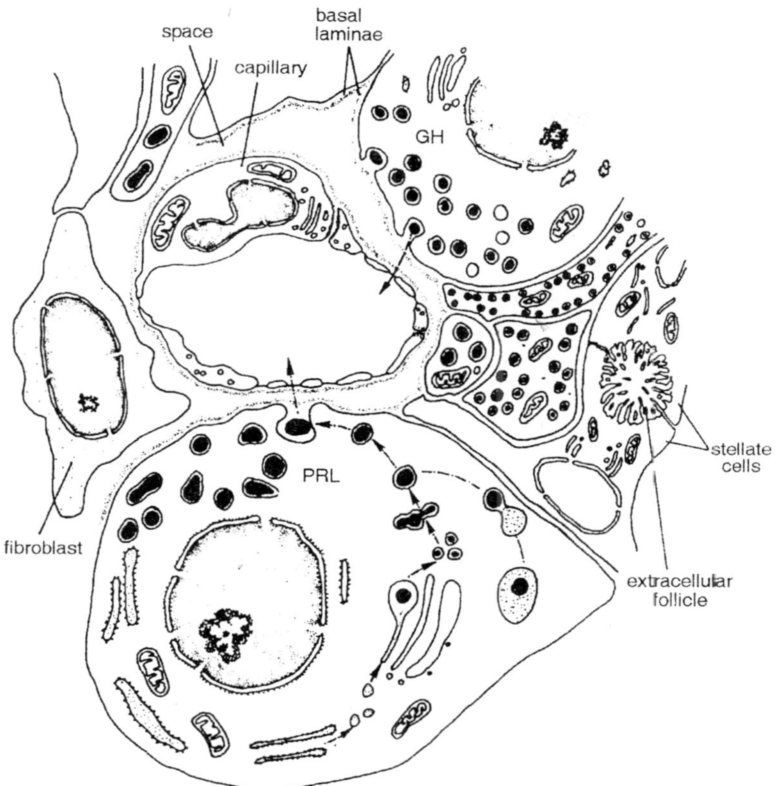

Fig. 3.4. Some cellular types characteristic of the mammalian adenophyophysis. GH, growth hormone; PRL, prolactin.

is called a lactotrope. The alpha acidophil (type 2 acidophil) is known as a somatotrope.

Corticotropin-secreting cells have proved especially difficult to identify, partly because of their similarity to other cellular types and the chemical similarity of ACTH to other pituitary hormones, MSH and LPH. Corticotropes are considered to be intermediate between chromophobes and basophils for they stain very weakly PAS(+) and AF(+). Corticotropes may be termed *type 3 basophils*. They have been identified at the light microscope level by means of immunofluorescent techniques coupled with adrenalectomy and administration of corticosteroids. Corticotropes are polyhedral with occasional long cytoplasmic processes. In many cytological respects, corticotropes are like thyrotropes. Both cells are PAS(+) and AF(+) and frequently respond to the same experimental manipulations. Furthermore, TSH and ACTH activities have been associated with the same size of small granules (35-150 nm) extractable from the pars distalis. The type 3 basophil may also be the source of lipotropin.

The sixth cellular type in the mammalian pars distalis is the chromophobe. Chromophobes may represent inactive, depleted or undifferentiated cells, and they may differentiate into either basophils or acidophils, depending upon the experimental manipulations applied. A special type of chromophobic cell, the *stellate* or *follicular* cell, has been observed in some mammals with the aid of the electron microscope. The processes of the stellate cell are very long and form a sort of network or reticulum between capillaries throughout the pars distalis. They are called follicular in some mammals because of the way these processes will sometimes surround or enclose small spaces or lumina. These stellate cells may perform a supportive or nutritional function. They are probably not the source of any tropic hormones, although some investigators have suggested them to be the source of ACTH. The term stellate cell, however, should not be applied to the ACTH-secreting cell.

The electron microscope has been used to characterize cellular types of the pars distalis on the basis of general cellular morphology and the size and shape of electron-dense cytoplasmic granules. A combination of ultrastructural, tinctorial and immunofluorescent techniques plus differential isolation and biological assay of granules leaves little doubt as to the origins of most of the tropic hormones. The source for LPH remains unresolved as well as whether one or two cellular types are responsible for production of GTHs (FSH and LH).

Cellular types in the pars intermedia

One type of epithelial cell occurs in the pars intermedia. This MSH-secreting cell characteristically stains with the lead hematoxylin procedure (PbH). Corticotropin and LPH activities have been identified in the pars intermedia, but they appear to be present within the MSH-secreting cells.

Cellular types in the pars tuberalis

Three cellular types have been identified in the primate pars tuberalis by means of light and electron microscopic techniques. Type I stains with PAS and exhibits small cytoplasmic granules (170 nm diameter). Type II cells stain with the dye alcian blue (AB) and contain somewhat larger granules (350 nm). The remaining cellular type is described as having occasional granules. Cells reacting specifically to antibody to pituitary LH and FSH are believed to be gonadotropic cells and probably correspond to type I. Thyroptropic cells that specifically bind antibody to TSH are thought to be the type II cells, although granule sizes are much larger than for pars distalis thyrotropes. Occasionally, rare cells are observed that bind antibody to ACTH and GH, but PRL cells are absent. Gonadotropin-containing cells have also been found in the pars tuberalis of rats. These data would suggest that the pars tuberalis may represent only a "fragment" of the pars distalis and that in some species it may function to secrete tropic hormones in response to either hypothalamic regulatory hormones or other substances in the portal blood or cerebrospinal fluid.

Neurohypophysis

The mammalian neurohypophysis consists of two distinct components. The *median eminence* is defined as the more anterior portion of the neurohypophysis that has a blood supply in common with the adenohypophysis, specifically the hypothalamo-hypophysial portal system. An abundant but separate blood supply characterizes the *pars nervosa*, which is that portion of the neurohypophysis in contact with the pars intermedia. Both the median eminence and the pars nervosa are neurohemal structures composed of axonal tips of NS neurons originating in hypothalamic nuclei, capillaries and special cells known as *pituicytes*.

The function of pituicytes is unknown, but they could play a role as supportive elements or may be involved actively in storage and release of neurohormones from the neurohypophysis. Pituicytes are thought to be derived from ependymal or glial cells. Ependymal cells are specialized neural cells that line the ventricles of the brain and the central canal of the spinal cord as well as form an outer protective layer around the brain and spinal cord. Glial cells are sometimes called the connective tissue of the nervous system although they may play a nutritive or some other role with respect to the neurons. There is some evidence that might suggest an influence of ependymal cells on tropic hormone release, but more work must be done in this area before a definitive role can be established.

Mammalian Hypothalamus

The hypothalamus contains many NS nuclei that produce neurohormones. The neurohormones associated with the median eminence and adenohypophysis are termed *hypothalamic hypophysiotropic hormones* and are identified as either *releasing hormones*, or *release-inhibiting hormones*, depending on whether they stimulate or inhibit tropic hormone release from the adenohypophysis. These regulating hormones are mostly small peptides, although at least one has been found to be an amine derived from an amino acid.

The neurohormones associated with the pars nervosa are all of very similar structure. Each is an *octapeptide neurohormone* consisting of nine amino acid residues. The use of "octa" has come about because the two cysteine residues form a disulfide bridge to become a single amino acid, cystine. Formation of the disulfide bond results in the characteristic five amino acid-ring structure with a side chain of three amino acids.

The region of the mammalian hypothalamus that controls hypophysial function consists primarily of the mediobasal hypothalamus containing a number of bilaterally paired NS nuclei, including the

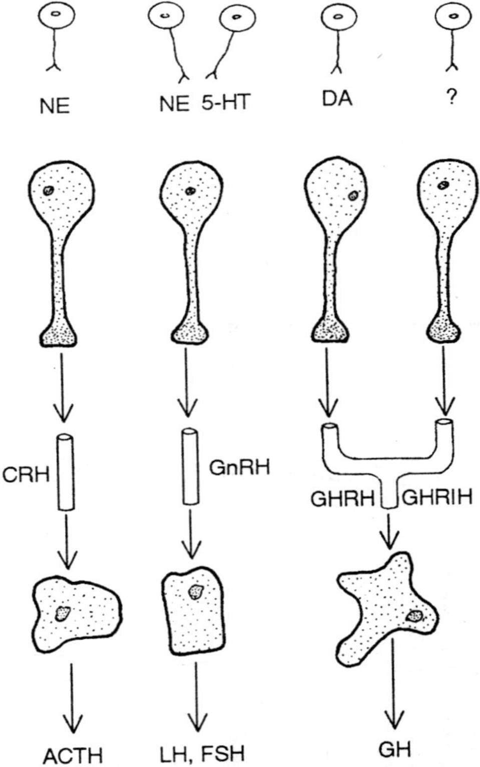

Fig. 3.5. Representation of relationships between the nervous system, hypothalamic neurosecretory neurons and release of tropic hormones.

anterior hypothalamic, *suprachiasmatic*, *ventromedial*, *dorsomedial*, *posterior hypothalamic* and *arcuate* nuclei. These nuclei are responsible for producing the hypothalamic hypophysiotropic neurohormones that regulate release of tropic hormones from the hypophysis. In addition there are two adjacent, bilaterally paired nuclei responsible for production of the octapeptide neurohormones of the pars nervosa. These are the *supraoptic* nuclei and, immediately posterior, the dorsoanteriorly located *paraventricular* nuclei.

Much of our knowledge about these hypothalamic centers has been accumulated from observing the effects of particular hypothalamic lesions or of localized electrical stimulation in the hypothalamus and adjacent regions or from studies involving implants of crystalline hormones into the hypothalamus (related to negative feedback effects). Some cautions of interpretation from such studies are in order, however. The use of disruptive lesions, for example, requires careful bilateral placement of comparable lesions and leaves some uncertainty as to exactly what was destroyed by the lesions. Alterations in pituitary function following placement of lesions might involve destruction of the NS neurons that elaborate a given hypothalamic hypophysiotropic hormone or may only disrupt a NS tract. The lesion might have damaged non-NS neurons that would normally modulate the activity of these NS neurons. Damage to vascular elements of the median eminence might also alter tropic hormone release patterns. Ideally, secretion of all tropic hormones should be monitored following placement of a particular lesion, yet, for practical reasons, this is rarely done. Usually only one or, at most, two tropic hormone systems are examined (such as, TSH or GTH), whereas others (PRL, ACTH and GH) are ignored.

In spite of such drawbacks the use of combinations of these approaches has helped establish the general localization of different NS centers. The thyrotropic center appears to be located anteriorly behind and dorsal to the optic chiasma (anterior hypothalamic and suprachiasmatic nuclei). Gonadotropic centers are located more posteriorly (mediobasal hypothalamus), and, in particular, the ventromedial nuclei appear to be responsible for controlling ovulation in some species. Corticotropin release is influenced most strongly by experimental alterations associated with the ventroanterior and ventromedial portions of the hypothalamic hypophysiotropic area. Release of GH appears to be localized in the ventral hypothalamic hypophysiotropic area, and the hypothalamic center for regulating PRL release seems to be located somewhere posterior to the optic chiasma in the ventral portion of the hypothalamic hypophysiotropic area.

Hypothalamic Hypophysiotropic (Regulating) Hormones

Numerous studies have confirmed that the hypothalamus exerts a direct influence over functioning of the adenohypophysis. Observations indicate the absence of neural connections between the hypothalamus and the adenohypophysis but the presence of the hypothalamo-hypophysial portal system led to the establishment of what is now termed the *neurovascular hypothesis*. Severing the portal connections or transplanting the pituitary to some avascular site elsewhere in the body (an *ectopic* transplant) causes marked changes in the secretory pattern of the tropic hormones. Generally such operations are followed by a marked reduction in circulating levels of TSH, FSH, LH, GH and ACTH, whereas PRL and MSH levels increase. These data led to the interpretation that release of the first group of tropic hormones is under stimulatory hypothalamic control (releasing hormones) and that MSH and PRL release is under inhibitory control. If the severed blood vessels of the hypothalamo-hypophysial portal system are allowed to regenerate so that blood may again flow from the median eminence to the adenohypophysis, the normal secretory patterns for the tropic hormones resume. These latter observations support strongly the neurovascular hypothesis of hypothalamic control over tropic hormone release.

In recent years as many as nine different hypothalamic regulatory hormones have been proposed. The chemical identities are established for most of these regulatory hormones. The others are frequently referred to as factors rather than as hormones. In this text a unified terminology is used for simplicity. Each regulatory hormone is named for the tropic hormone it influences and is designated according to whether it causes release (R) or is release inhibiting (RI). Thus the neurohormone that stimulates release of thyrotropin is termed TRH (thyrotropin-releasing hormone). Similarly the single neurohormone believed to cause release of both FSH and LH is termed the gonadotropin-releasing hormone (GnRH). Prolactin release is inhibited by PRIH (prolactin release-inhibiting hormone).

Thyrotropin-releasing hormone

The first hypothalamic regulatory hormone that was identified chemically was the tripeptide TRH. It is found in the hypothalamus, appears in the hypophysial blood following electrical stimulation of the hypothalamus and causes release of TSH in vivo and in vitro. It appears that this tripeptide is the endogenous hypothalamic hypophysiotropic TRH in mammals.

Extrahypothalamic TRH is present in the brain as well as in the spinal cord, the pineal gland and the neurohypophysis. The common occurrence of TRH outside the hypothalamus and its presence in extrahypothalamic regions of the nervous system of mammals, nonmammalian vertebrates and even invertebrates has led to the suggestion that TRH may also function as a neurotransmitter. It has been demonstrated that administration of synthetic TRH causes depression of firing in certain neurons.

The biological half-life for TRH in peripheral blood is very short (for example, 2 minutes in mice) apparently because peptidases in the blood rapidly inactivate TRH. Consequently it is very difficult to measure endogenous levels of circulating TRH in peripheral blood even with radioimmunoassay.

Gonadotropin-releasing hormone

Endogenous GnRH is a decapeptide originating in the arcuate and ventromedial NS nuclei of the hypothalamus. Apparently GnRH causes release of both GTHs, FSH and LH, from the adenohypophysis, although a greater amount of LH release is always observed following administration of synthetic GnRH. It is the marked effect of this regulatory hormone on LH release and subsequent induction of ovulation that accounts for its originally being named an LH-releasing hormone (LHRH or LRH). Although there is evidence for a separate FSHRH, it is generally accepted that endogenous GnRH causes release of both gonadotropins. Since GnRH was first synthesized, approximately 1400 analogues have been created, ranging from stimulatory to inhibitory on gonadotropin release.

Like TRH, GnRH has been identified immunologically in extrahypothalamic nervous tissue and in the pineal gland of some species. Similarly, GnRH, like TRH, can cause depression of neural function in the central nervous system. Thus GnRH may have a role as a neurotransmitter in addition to its involvement with gonadotropin release.

Gonadotropin-releasing hormone is rapidly degraded in peripheral plasma, and the success of several potent synthetic analogues appears to be related to their relative resistance to degradation. One super-releaser has approximately 150 times the activity of the native GnRH.

Growth hormone-releasing and release-inhibiting hormones

The release of GH by the hypothalamus is under both inhibitory (GHRIH) and stimulatory (GHRH) control. Also known as *somatostatin*, GHRIH is a very strong inhibitor of GH release. Somatostatin also blocks the action of synthetic TRH on TSH release from the adenohypophysis. It is a tetradecapeptide.

Somatostatin occurs in extrahypothalamic nervous tissue (both brain and spinal cord), and a possible neurotransmitter function has been suggested similar to that for TRH and GnRH. It is also present in the mucosa of the stomach and has been shown to inhibit release of the gastric hormone, gastrin. In addition, somatostatin is present in the pancreatic islets where it appears to be involved in the inhibition of both glucagon and insulin release.

The importance of stimulatory control over GH release is emphasized by the reduction in GH release following disruption of the portal vessels. There is evidence for the existence of hypothalamic GHRH associated with the ventromedial nucleus of the hypothalamus. It is possible that direct neural control of GH release occurs and may account for the stimulatory actions of various physiological stresses on increasing GH release. A potent releaser of GH has been isolated from a human pancreatic tumor. This peptide (40 amino acid residues) is related chemically to the glucagon family of peptides.

Prolactin-releasing and release-inhibiting hormones

The primary control over prolactin release, as mentioned earlier, is inhibitory in mammals. The mammalian PRIH appears to be dopamine released from hypothalamic neurons into the hypothalamo-hypophysial portal system. The inhibitory nature of hypothalamic control is evidenced by enhanced release of PRL following certain hypothalamic lesions, separation of the adenohypophysis from the hypothalamus by insertion of a physical barrier to blood flow or transplantation of the adenohypophysis to an ectopic site.

There is evidence for an endogenous factor that stimulates PRL release. This PRL-releasing hormone (PRH) may be the tripeptide already known as TRH since administration of synthetic TRH stimulates PRL release as well as TSH release. This action of TRH on PRL secretion is not influenced by somatostatin. It is not clear whether endogenous TRH has any physiologically important influence over PRL secretion since its effects would normally be counteracted by the presence of PRIH. Furthermore, the occurrence of enhanced PRL release in vitro in the absence of any hypothalamic factors tends to argue against any endogenous PRH. It remains to be demonstrated that factors which evoke TRH release with ensuing release of TSH are accompanied by increases in PRL release without a change in PRIH levels.

Corticotropin-releasing hormone

The amino acid sequence of CRH has been determined, and synthetic CRH is now available. It is a peptide of 41 amino acid residues and releases ACTH from the corticotrope. There is considerable homology (at 20 positions) to sauvagine, a 40-residue peptide isolated from skin of the frog, *Phylomedusa sauvagii*. There is also overlap with the structure of a hypotensive agent, urotensin I, isolated from teleosts.

Melanophore-stimulating hormone release-inhibiting hormone

Experimental studies similar to those described for PRL established that hypothalamic control of MSH release from the pars intermedia was primarily inhibitory. It is not clear whether MSH release is controlled by a neurohormone or by direct innervation. Several peptides have been shown to possess intrinsic MSHRIH activity, but the exact chemical identity of the endogenous neurohormone is not known. Extrahypothalamic MSHRIH activity has been demonstrated, and at least one of the synthetic releasers of MSH is an antidepressant when administered to humans. Numerous investigators believe that dopamine released into blood traveling to the pars intermedia is MSHRIH.

Role of Cyclic Nucleotides

Hypothalamic hypophysiotropic hormones produce several initial changes in adenohypophysial cells prior to release of tropic hormones. One such effect is an increase in the permeability of the membrane to calcium ions, causing an influx of these ions. The role of calcium ions is essential to tropic hormone release, and release does not occur in calcium-free medium in vitro. The specific involvement of calcium has not been settled but it appears to be necessary for the action of cyclic adenosine 3',5'-monophosphate (cAMP). Most of the hypothalamic regulating hormones have been shown to activate adenyl cyclase in the cell membrane, which causes an increase in cAMP production. Cyclic AMP operates through a cAMP-dependent protein kinase to effect hormone release.

Control of Hypothalamic Hypophysiotropic Hormone Release

Release of hypothalamic regulatory hormones is influenced by neural activity. A variety of neurotransmitters (for example, dopamine, norepinephrine, serotonin) and drugs that selectively block neuronal function (such as reserpine, which blocks adrenergic neurons by depleting them of neurotransmitters) affect tropic hormone release. The same agent may stimulate release of one tropic hormone and inhibit release

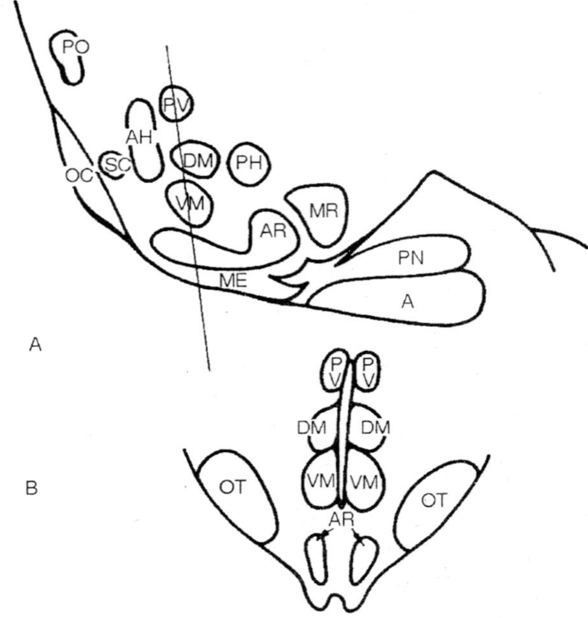

Fig. 3.6. The hypophysiotropic centers of the hypothalamus. A—Sagittal section; B—Cross section. OC, optic chiasm; MB, mammaillary body; A, adenohypophysis; PN, pars nervosa; ME, median eminence; OT, optic tract; AH, anterior hypothalamic nucleus; AR, arcuate nucleus; DM, dorsomedial nucleus; PH, posterior hypothalamic nucleus; PO, preoptic area; PV, paraventricular nucleus; SC, suprachiasmatic nucleus; VM, ventromedial nucleus.

of another. Presumably these neuronal agents influence release of the hypothalamic hypophysiotropic hormones.

Many pharmacological studies of nervous regulation over hypothalamic NS centers have been conducted employing different neurotransmitters or drugs that either mimic or block the activity of various known neurotransmitters. Studies of this type have led to identification of the kinds of neurons that regulate release of individual hypothalamo-hypophysiotropic hormones. For example, application of dopamine to cultured pituitary cells with and without cocultured hypothalamic tissue has made it possible to distinguish between the indirect stimulatory activity of dopamine on LH release and its inhibitory action on PRL release. Utilization of catecholamines and related drugs has also contributed much to the understanding of neuronal regulation of hormone release. Norepinephrine and epinephrine may both influence release of hypothalamo-hypophysiotropic hormones.

The response of NS cells to specific neurotransmitters is determined by the presence of receptors for these substances on the cell membranes of the NS neurons. The inhibitory action of dopamine on PRL release mentioned above is accomplished through the binding of dopamine to receptors in the plasmalemma of the PRL-secreting cells. The ergot alkaloids such as ergocornine and ergocryptine can mimic the action of dopamine by binding to another receptor on the PRL cell membrane termed an alpha receptor. Stimulation of this receptor might ordinarily evoke PRL release, but the use of these alpha receptor-blocking drugs (alpha blockers) can completely inhibit hormone release.

Oxytocin and vasopressin, two of the octapeptide neurohormones associated with the pars nervosa, as well as melatonin, a derivative of tryptophan secreted by the pineal gland, have been shown to alter tropic hormone release, probably at the hypothalamic level. These hormones may reach the hypothalamus either via the cerebrospinal fluid or through the general circulation following their release from the pars nervosa. It is also possible that neural fibers release octapeptide neurohormones in the vicinity of the median eminence and influence release of regulatory hormones into the portal vessels. Most of the supportive data for the action of octapeptide neurohormones and melatonin on tropic hormone release are based on studies applying large doses, and it may not be justified to extend such data to in-vivo control mechanisms. Another interesting observation is that two of the peptides have MSHRIH activities. are fragments of one of the neurohypophysial octapeptides, oxytocin. Furthermore, melatonin is structurally very similar to the neurotransmitter serotonin, an intermediate in the synthesis of melatonin from tryptophan. This structural similarity may account in part for some of its purported activities on hypothalamic neurons.

Other peptides have been identified in hypothalamic neurons, and some of these may influence tropic hormone release, at least indirectly. Two peptides, known as *neurotensin* and *substance P*, are probably neurotransmitters and thus could indirectly affect release of the hypothalamo-hypophysiotropic hormones. A family of peptides that mimic morphine effects in the central nervous system are also present in the median eminence. They are known as the *endorphins* and *enkephalins*, and they appear to be capable of stimulating GH and PRL release due to presynaptic effects on dopaminergic neurons rather than direct actions on cells of the pars distalis. Enkephalins and endorphins are extractable from the adenohypophysis, too.

Feedback and Tropic Hormone Release

The release of adenohypophysial tropic hormones can be influenced through negative feedback at the hypothalamus or at the pituitary. Thyroid hormones produced by the action of TSH on the thyroid gland can reduce circulating TSH by inhibiting release of either TRH or TSH or both through effects on the hypothalamus and the pituitary gland respectively. Similarly, corticosteroids produced by the adrenal cortex and gonadal steroids from ovaries or testes inhibit release of ACTH and GTHs respectively through similar action. Growth hormone release can be influenced by either circulating glucose or amino acid level in a similar negative feedback loop. Circulating glucose and amino acid levels are in part determined by the actions of GH on its target cells. Negative feedback of PRL and MSH release does not seem to be necessary because of the strong inhibitory influence of the hypothalamus over their release.

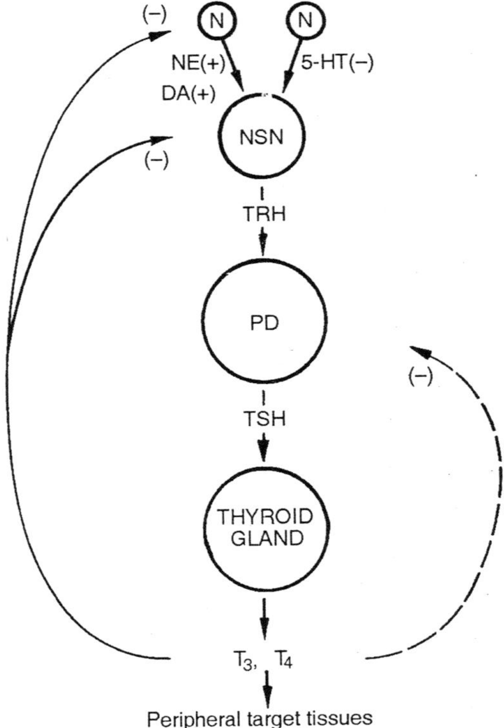

Fig. 3.7. The hypothalamo-hypophysial-thyroid axis.

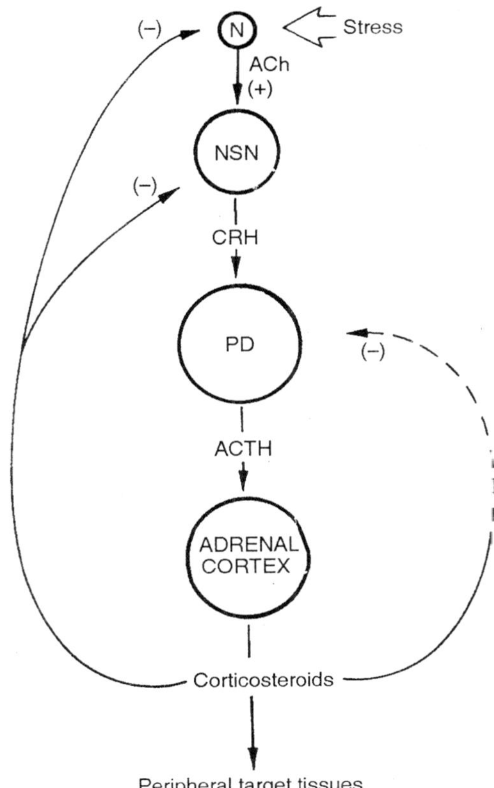

Fig. 3.8. The hypothalamo-hypophysial-adrenal axis.

In female mammals, gonadal steroids have differential effects on hypothalamic NS centers. There appears to be a "tonic center" for regulating gonadotropin release that is inhibited by either estrogens or progesterone and a "surge center" inhibited by progesterone but stimulated by estrogens. Consequently the increase in estrogens produced by the growing follicles in the ovary causes a massive release (surge) of gonadotropin that triggers ovulation. Estrogens have also been shown to stimulate release of PRL, but it is not clear whether it inhibits dopamine release (PRIH) or whether it stimulates TRH release (PRH?).

Hypothalamic Octapeptide Neurohormones

This second group of hypothalamic neurohormones, the so-called octapeptide hormones, are synthesized in the supraoptic and paraventricular hypothalamic nuclei. The octapeptides consist of nine

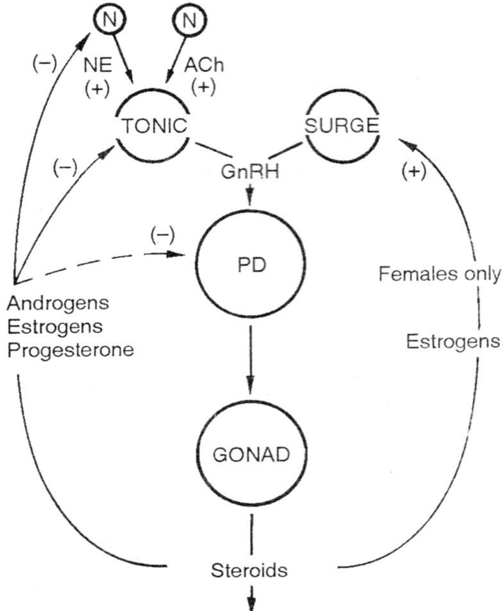

Fig. 3.9. The hypothalamo-hypophysial-gonadal axis.

amino acid residues, but two cysteine residues are joined by disulfide bonds, resulting in formation of cystine; hence, "octapeptide" rather than "nonapeptide." These hormones are stored in the pars nervosa until they are released into the general circulation. The targets for these hormones are located at considerable distances from the pars nervosa (for example, kidney, mammary gland, uterus).

Two neurohypophysial octapeptide hormones are present in the pars nervosa of most adult mammals. *Arginine vasopressin* (AVP) and *oxytocin* (OXY) are the common two octapeptide neurohormones. A variant of AVP known as *lysine vasopressin* (LVP) is produced by the Suina, which includes peccaries, domestic pigs and the hippopotamus. Most of the species in this group exhibit both AVP and LVP in addition to OXY. A unique vasopressin-like molecule, *phenypressin*, has been isolated from the pars nervosa of marsupials. Fetal mammals have been shown to produce what at first appears to be a hybrid of AVP and OXY, being composed of the side chain of AVP with the ring structure of OXY. This octapeptide is known as *arginine vasotocin* (AVT) and is characteristic of adult nonmammalian vertebrates. The pineal gland of at least some adult mammals also contains AVT.

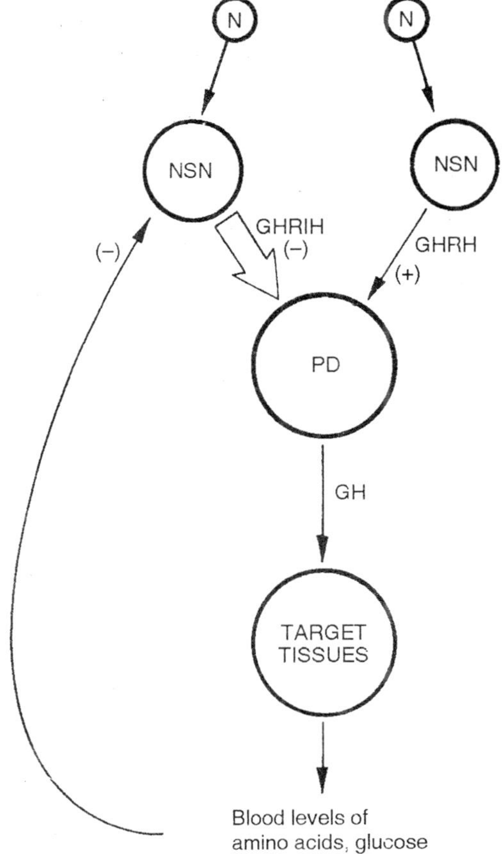

Fig. 3.10. Hypothalamic regulation of growth hormone secretion.

The vasopressins (AVP, LVP) function as antidiuretic agents affecting the ability of the kidneys to reabsorb water from the glomerular filtrate. At higher doses, vasopressins cause vasoconstriction and can elevate blood pressure (a pressor effect). This action may increase glomerular filtration and water excretion. Arginine vasotocin produces a similar type of action in nonmammals and has been suggested to play an osmoregulatory role in fetal mammals.

Oxytocin stimulates contraction of uterine smooth muscle and contraction of the myoepithelial cells lining the ducts of the mammary glands. The action of OXY on the uterus is related to the induction of labor and the birth process. Ejection of milk from the mammary gland

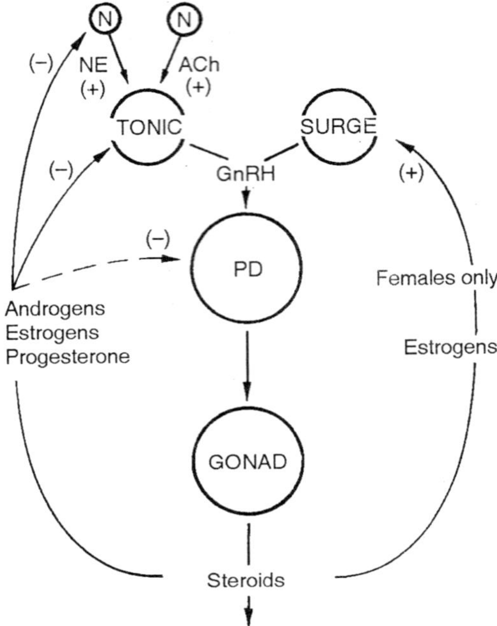

Fig. 3.11. Regulation of prolactin release. Dopamine (DA) is the prolactin release-inhibiting hormone.

is caused by OXY released from the pars nervosa in response to suckling at the breast by the young.

Clinical Aspects of the Hypothalamo-Hypophysial System

Hypothalamus

The most common clinical disorder of the hypothalamus is *diabetes insipidus*. The patient produces an abnormally large volume (3 to 8 liters/day) of dilute urine. This diuresis is due to the absence of AVP which normally controls water reabsorption in the kidney. Forty to 50% of such patients are *idiopathic* (denoting a disease of unknown cause) and exhibit no other evidence of neuroendocrine dysfunction. Some individuals have been identified at autopsy as having degeneration of the supraoptic and paraventricular nuclei. About 15% of the cases are related to the presence of tumors within the brain which indirectly affect production of AVP. Physical damage (such as a lesion) or infections (e.g., encephalitis) account for the remainder. Nephrogenic diabetes insipidus is the result of a failure of the kidney tubules to respond to normal levels of AVP.

The syndrome of inappropriate antidiuresis (SIAD) is caused by excessive AVP release. High levels of AVP result in reduced urine production. Drugs such as demeclocycline block the action of AVP on the kidney and are used to treat this condition. Such drugs induce nephrogenic diabetes insipidus. Medications that control AVP release from the pars nervosa often are not uniformly effective.

Tumors within the central nervous system are responsible for a number of other disorders. One of the more dramatic consequences is precocity. Precocity is much more common in males. About one-fourth of precocity cases are correlated with the presence of a pineal tumor, and 95% of these occur in males. A number of other cases are associated with hypothalamic tumors, most of which also are found in males. One type of tumor, associated with precocity, the *harmatoma*, occurs in the posterior hypothalamus. It consists of masses of partially disoriented glial and ganglion cells or of normal cells located in abnormal sites. Harmatomas may secrete GnRH which could explain their effects.

Several other pathologies have been identified including the following:

1. *Hyponatremia*: Low blood sodium may be correlated with a number of factors including brain carcinoma, basal skull fractures, meningitis and encephalitis.
2. *Hypernatremia*: Excessive levels of sodium may occur as a result of aneurysms in the brain, pineal tumors and so forth. This condition is not accompanied by fluid imbalance.

Adenohypophysis

Acromegaly is a spectacular disorder of GH regulation leading to gigantism. This is a rare disorder affecting from 3 to 40 individuals per million people in the United States each year. Acromegaly was the first disorder of the pituitary gland to be recognized. It is caused by overproduction of GH due either to the absence of adequate somatostatin to suppress GH release or by the absence of negative feedback to suppress release. Growth hormone-secreting tumors release GH autonomously, but such tumors are uncommon. Approximately half of acromegalic patients are deficient in one or more additional pituitary hormones, usually the gonadotropins. Not only do such patients exhibit excessive growth, but body proportions become distorted. Cartilage tends to proliferate in joints resulting in abnormally proportioned hands and elongate jaws. There are also marked effects on other systems.

For example, the skin exhibits excessive sweating and secretion of sebum, the heart is enlarged and hypertension may develop.

Pituitary *chromophobe adenomas* are the most common source of pituitary-related problems. (Any benign or noncarcinogenic glandular tumor can be termed an adenoma.) They rarely secrete any hormones (occasionally GH and seldom TSH or ACTH) and their effects are usually due to pressure on the brain or optic chiasm caused by growth of the adenoma. Most patients experience severe headaches and visual disturbances. Sometimes the production of one or more pituitary hormones may be reduced.

Partial or total *hypopituitarism* refers to selective or total absence of pituitary hormones. These defects may reside in the adenohypophysis itself (primary disorder) or be due to hypothalamic dysfunction (secondary disorder). There has been considerable interest in the role(s) of endogenous opiates on mental disturbances such as depression and schizophrenia. Clinical studies have yielded mixed results and it is not clear whether mental illness is associated with endogenous opiates.

Comparative Aspects of the Hypothalamo-Hypophysial System in Nonmammalian Vertebrates

Fishes

The piscine hypothalamo-hypophysial system is separable into the same major divisions as that of mammals: hypothalamus, neurohypophysis and adenohypophysis. Some marked differences occur as well. There is no pars tuberalis in fishes, although there is possibly a homologous structure in the elasmobranchs (Chondrichthyes). The pars distalis of the adenohypophysis, with the possible exception of the hagfishes (Agnatha), shares a common blood supply with the median eminence, the hypothalamo-hypophysial portal system. The pars distalis is differentiated into two subregions or zones. The pars intermedia is posterior to the par distalis and is intimately interdigitated with the pars nervosa of the neurohypophysis to form a *neurointermediate lobe*. This structure is characteristic of all but two groups of fishes, the hagfishes and the lungfishes (Dipnoi). Finally, posterior to the neurointermediate lobe in cartilaginous fishes and most bony fishes is a unique structure formed from the floor of the diencephalon. This structure is known as the *saccus vasculosus*. It is especially well developed in some groups, but its function is unknown.

Two different terminologies have been proposed for the subregions that are recognized in the piscine adenohypophysis. The nomenclature

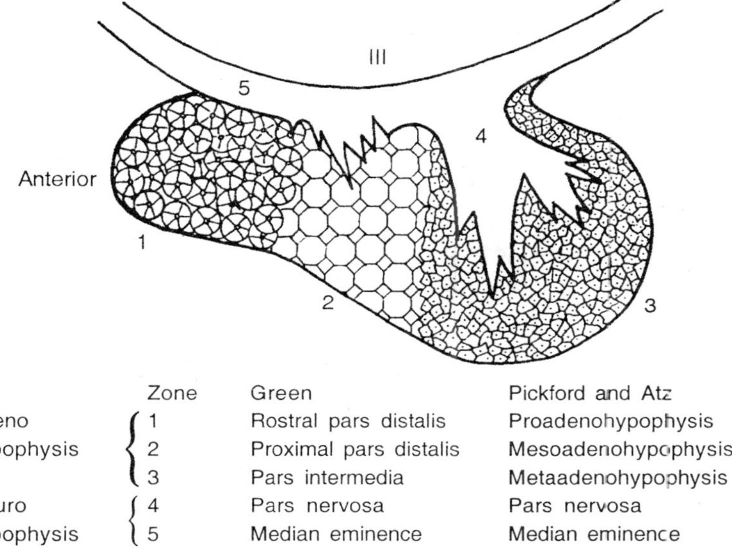

	Zone	Green	Pickford and Atz
Adeno hypophysis	1	Rostral pars distalis	Proadenohypophysis
	2	Proximal pars distalis	Mesoadenohypophysis
	3	Pars intermedia	Metaadenohypophysis
Neuro hypophysis	4	Pars nervosa	Pars nervosa
	5	Median eminence	Median eminence

Fig. 3.12. Terminology for designating regions of the piscine adenohypophysis.

proposed by Green will be used in favour of the alternative system proposed by Pickford and Atz because the Green system is more similar to mammalian terminologies. There are three distinct zones recognized by both schemes. The most anterior and rostral (dorsal) portion of the piscine adenohypophysis typically consists of follicles and is termed the *rostral pars distalis* (proadenohypophysis). The remainder of the pars distalis comprises the *proximal pars distalis* (mesoadenohypophysis). The *pars intermedia* is termed the metaadenohypophysis according to the Pickford and Atz nomenclature. Each of these regions of the adenohypophysis is readily distinguished cytologically because each contains different types that produce different tropic hormones. The cellular types found in each region and the hormones they are thought to produce are discussed for each taxonomic grouping in the following sections.

Class Agnatha

Family Myxinoidea (Hag fishes)

The Atlantic and Pacific hagfishes possess the most primitive hypothalamo-hypophysial system among the vertebrates. Consequently it is lacking many of the features characteristic of other piscine groups. Futhermore, they are much more primitive than their living agnathan

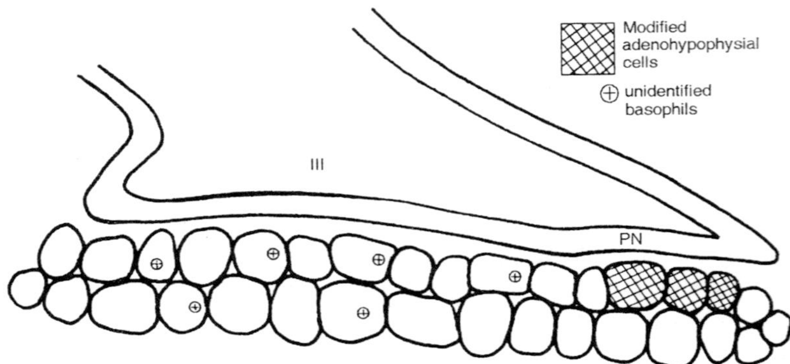

Fig. 3.13. The hagfish hypophysis.

relatives, the lampreys (Petromyzontidae). There is a single NS center in the hagfish hypothalamus, the *preoptic nucleus*, located dorsal to the optic chiasm at the anterior end of the hypothalamus. This nucleus appears to produce NS product that is stored in the neurohypophysis. There is no anterior neurohemal region in the Atlantic hagfish comparable to the median eminence, although a neurohemal area has been described for the Pacific hagfish and has been termed a median eminence. There is no evidence, however, to support homology to the mammalian median eminence. The origin of the adenohypophysis of hagfishes appears to be from endoderm rather than from ectoderm. If this observation proves to be correct, it represents an additional puzzle with respect to the origin of the pituitary. Furthermore, it raises the question of possible homology of the hagfish adenohypophysis to that of other vertebrates and supports the viewpoint that hagfishes are abberant vertebrates and are not on the mainline evolutionary pathway.

The hagfish adenohypophysis is not differentiated into subregions. It is composed primarily of chromophobic cells and rare PAS(+) basophils or an occasional acidophil. Electron micrographs of the hagfish adenohypophysis show rare granular cells with cytoplasmic granules of 100-200 nm diameter. These granular cells are believed to represent the two rare stainable cellular types identifiable with the light microscope. When hagfish adenohypophysial tissue is cultured in vitro, no observable changes take place in either granular or agranular cells.

Bioassays of hagfish pituitaries for PRL activity have proven negative, and antiovine PRL antibody does not bind to hagfish adenohypophysial cells. Furthermore, hypophysectomy of Pacific hagfish produces no convincing alterations in either thyroid or gonadal tissue.

Gonadotropin activity has been demonstrated in the pituitary of *Eptatretus burgeri*, a shallow water, seasonally breeding hagfish.

Certain cells of the myxinoid adenohypophysis exhibit cytological modifications where they make contact with the neurohypophysis. These altered cells are termed *modified adenohypophysial tissue*, and it has been proposed that this apparent induction by neurohypophysial tissue may represent phylogenetically the origin of the pars intermedia.

Family Petromyzontidae (Lampreys)

In the lampreys three regions are distinguished in the adenohypophysis as they are in jawed fishes. The more anterior rostral pars distalis is composed of basophils and chromophobes. These same cellular types as well as azocarmine(+) cells, carminophils, are found in the proximal pars distalis, which is located between the rostral pars distalis and the pars intermedia. Only one PbH(+) cell occurs in the pars intermedia.

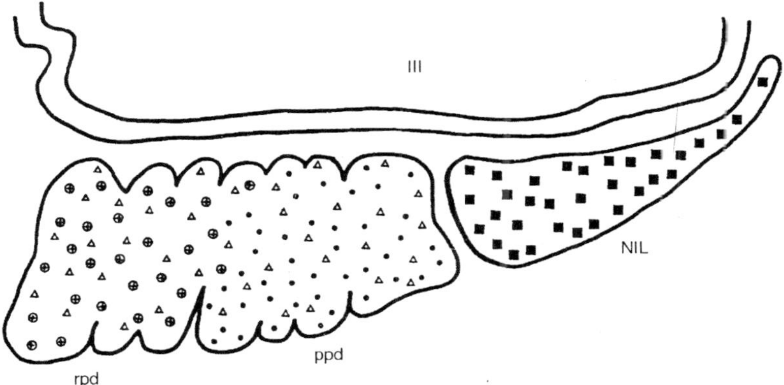

Fig. 3.14. The lamprey hypophysis. rpd, rostral pars distalis; ppd, proximal pars distalis; NIL, neurointermediate lobe; III, third ventricle.

The distinct pars nervosa and the pars intermedia form a well-developed neurointermediate lobe. Peptidergic neurons terminate in the pars nervosa where the single octapeptide neurohormone AVT is stored. The lamprey may possess a second neurohemal region associated with the pars distalis. Although this structure has been referred to as a median eminence, this interpretation has not been accepted universally.

Class Chondrichthyes

Selachii (Sharks, Rays, Skates)

The selachian (elasmobranch) hypothalamo-hypophysial system possesses two features not found in agnathan fishes. The *pars ventralis*

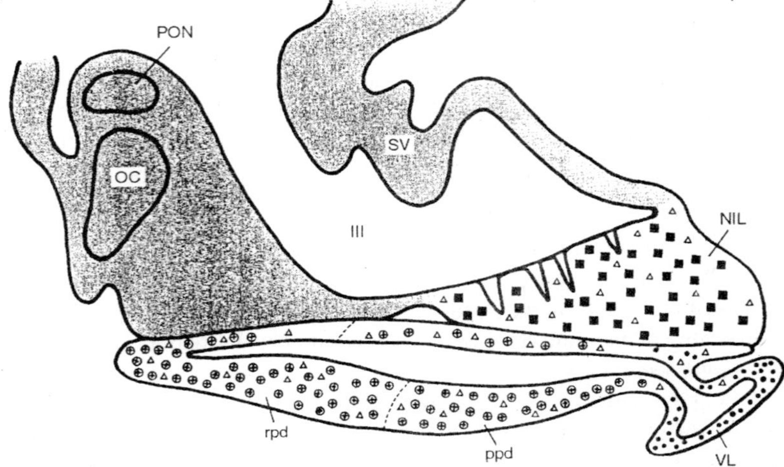

Fig. 3.15. The elasmobranch hypophysis. rpd, rostral pars distalis; ppd, proximal pars distalis; NIL, neurointermediate lobe; VL, ventral lobe; SV, saccus vasculosus; PON, preoptic nucleus; OC, optic chiasma; III, third ventricle.

represents a fourth subdivision of the adenohypophysis that is unique to selachians. This unique region is located ventral to the proximal pars distalis to which it is connected by a stalk. Some investigators have suggested that the pars ventralis is homologous to the pars tuberalis of the tetrapod adenohypophysis. Localization of GTH and TSH activity in the pars ventralis supports such a homology. The proximal pars distalis also has been found to have GTH activity. Prolactin activity and ACTH activity have been localized in the rostral pars distalis. Growth hormone activity has been demonstrated in the proximal pars distalis.

Seven cellular types have been identified in the pars distalis of the shark *Scyliorhinus caniculus*, on the basis of ultrastructural criteria. The rostral pars distalis possesses two cellular types believed to produce PRL and ACTH. Thyrotropes do not appear to be confined to any particular region of the pars distalis although most of the TSH activity demonstrated by bioassay resides in the ventral lobe. Possibly two types of gonadotrope have been identified in the proximal and ventral lobes respectively, but they may only be variants of a single cellular type. Somatotropic activity has been assigned to a single cellular type localized in the proximal pars distalis. A stellate chromophobe similar to the mammalian stellate cell is found throughout the pars distalis and it may play a similar role to its mammalian counterpart.

The terms rostral and proximal may be somewhat misleading when applied to the selachian pars distalis because of the presence of the ventral lobe homologous to part of the proximal pars distalis of other fishes or possibly to the pars tuberalis of tetrapods. Consequently these regions are sometimes designated as the rostral, median and ventral lobes of the pars distalis.

The second feature appearing for the first time in selachians is the *saccus vasculosus* derived from the hypothalamus and located immediately posterior to the neurointermediate lobe. It is not as well developed as its homologue in bony fishes, but it does possess the unique *coronet* cellular type characteristic of the saccus vasculosus of bony fishes.

The neurointermediate lobe is well developed. The pars nervosa portion contains unique octapeptide hormones in addition to AVT. These hormones are presumably produced in the preoptic nucleus. The pars intermedia contains only one cellular type associated with MSH activity.

In selachians there is a well-developed median eminence with a hypothalamo-hypophysial portal system connecting it to the pars distalis. The median eminence consists of anterior and posterior neurohemal areas. The posterior region receives both peptidergic and aminergic NS axons and appears to be linked by portal vessels to the proximal pars distalis and probably the ventral lobe. There is considerably less NS material in the anterior neurohemal area, which appears to be connected by capillaries to the rostral pars distalis. It is tempting to suggest that the median eminence has differentiated to increase the efficiency of delivering hypothalamic neurohormones to specific adenohypophysial cells.

The occurrence of a highly developed hypothalamo-hypophysial connection suggests that the median eminence and its regulatory relationship to the adenohypophysis probably evolved in the ancestral placoderm fishes. The presence of a similar system in bony fishes would support this view.

Holocephali (Ratfishes)

The hypothalamo-hypophysial system of ratfishes has been well studied structurally. The holocephalan adenohypophysis is readily subdivided cytologically into rostral pars distalis, proximal pars distalis and pars intermedia. Although several cellular types have been demonstrated with selective staining procedures, no experimental studies have determined which cells produce tropic hormones.

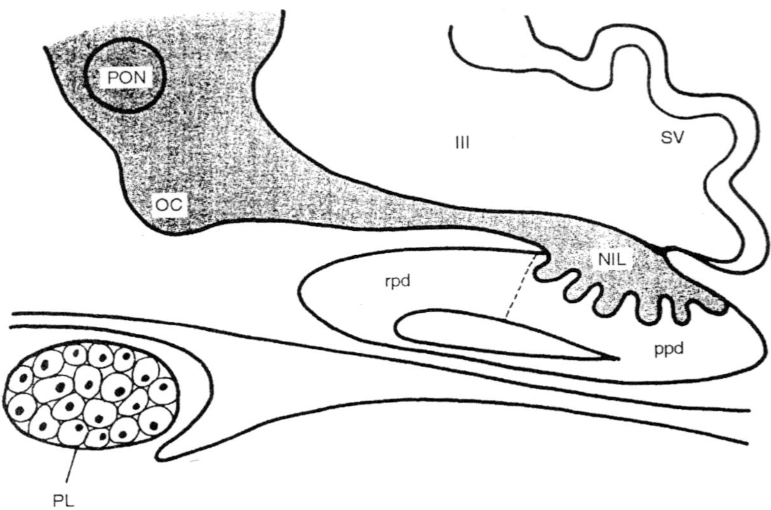

Fig. 3.16. The ratfish hypophysis. PL, pharyngeal lobe.

Holocephalans possess a unique region associated with the adenohypophysis called the *pharyngeal lobe*. This structure is located in the roof of the mouth outside the cranium as though a portion of Rathke's pouch had stayed behind and had not become incorporated into the adenohypophysis proper. The pharyngeal lobe consists of follicles, and may be homologous to the follicular rostral pars distalis of bony fishes. Their extreme dissimilarities ontogenetically, however, make this an unlikely homology. It may be homologous to the ventral lobe of the elasmobranch and the buccal lobe of the coelacanth pituitary.

The ratfish neurohypophysis includes a prominent median eminence connected to the rostral pars distalis and the proximal pars distalis by the hypothalamo-hypophysial portal system. Somatostatin, TRH and GnRH immunoreactivity is present in the hypothalamus. The pars nervosa is mingled with the pars intermedia of the adenohypophysis to form a typical neurointermediate lobe. A well-developed saccus vasculosus is present.

Class Osteichthyes: the primitive Actinopterygii

Polypteri (Polypterus and Erpetoichthyes)

The polypterid fishes have been examined structurally and possess all the typical piscine features. These primitive African fishes retain as adults a connection between the hypophysis and the mouth cavity, the *buccohypophysial canal* (*bucco* mouth), which is believed to be a

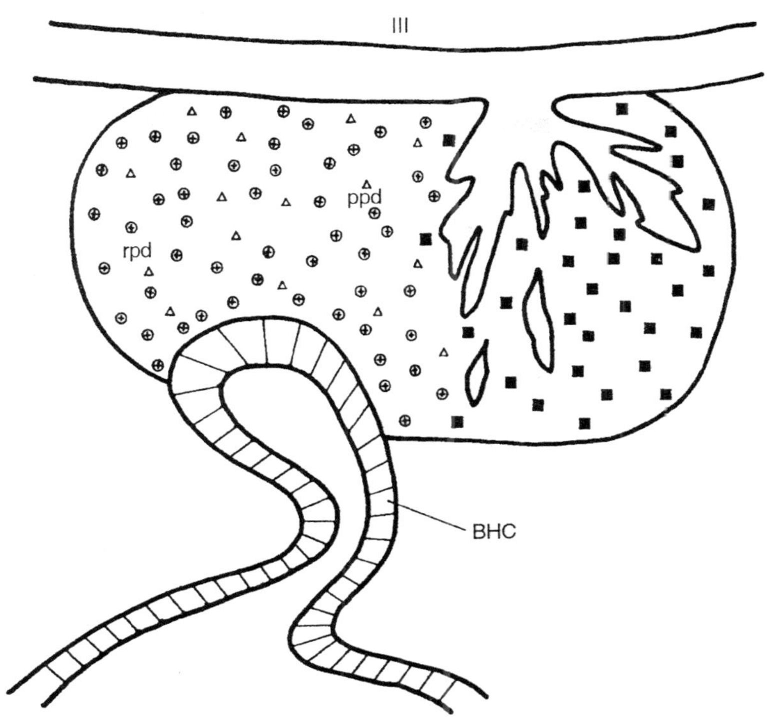

Fig. 3.17. The hypophysis of Polypterus. BHC, buccohypophysial canal.

remnant originally connecting Rathke's pouch to the oral cavity. The suggestion that Rathke's pouch may have originated from neural ectoderm rather than oral ectoderm should be carefully considered in these fishes. The buccohypophysial canal or duct is lined with chromophobic cells that have a small quantity of PAS(+) and AB(+) cytoplasm. This cellular type does not appear to be associated with production of any tropic hormones, however.

The pars distalis consists of separate rostral and proximal portions. A number of cellular types have been identified in these two areas, including various basophils acidophils and chromophobes. Experimental studies are needed, however, to identify these cells as sources for particular tropic hormones. Acidophilic cells in the rostral and proximal pars distalis react positively to antibodies against sheep PRL, suggesting that they are a source of PRL or a PRL-like molecule. Growth hormone activity has been demonstrated in the adenohypophysis of *Polypterus*, but no specific cellular type is implicated.

The pars intermedia is closely associated with the pars nervosa, producing a typical neurointermediate lobe. There is a single cellular type in the pars intermedia that is PAS(+) but PbH(-). This observation is curious since cells responsible for secreting MSH in other vertebrates are all PbH(+).

The pars nervosa contains AVT and another octapeptide neurohormone known as *isotocin* (IST). These two neurohormones are characteristic of the ray-finned fishes. The preoptic nucleus is believed to be the source for these octapeptide neurohormones. Peptidgeric fiber tracts travel from the preoptic nucleus via the median eminence to terminate in the pars nervosa. Aminergic neurons terminate in the median eminence although their source is not known. It is possible that peptidergic and aminergic fibers both originate in the preoptic nucleus. The hypothalamo-hypophysial portal system is well developed, and there are no neural fibers penetrating into the pars distalis. As will be seen, direct innervation of adenohypophysial cells becomes common in the more advanced bony fishes.

Chondrostei (Sturgeon, Spoonbill)

There is less information concerning the hypothalamo-hypophysial system of chondrostean fishes than for any other ray-finned bony fishes. There is no buccohypophysial canal in chondrosteans, but a hypophysial cavity remains to represent the space within Rathke's pouch. This cavity separates the pars distalis and pars intermedia and may be homologous to the buccohypophysial canal. The pars distalis consists of a rostral zone and a proximal zone (rostral and proximal pars distalis). Numerous follicles occur throughout the pars distalis, and their lumina are considered to be remnants of the hypophysial cavity. The lumina of these follicles are filled with a basophilic colloidal material. The entire pars distalis may be homologous to the proximal pars distalis of teleostean fishes, although the follicles themselves may be homologous to the follicles of the teleostean rostral pars distalis. If this interpretation proves correct the terms rostral pars distalis and proximal pars distalis may no longer be applicable to the two zones of the chondrostean pars distalis.

The cytology of the pars distalis is poorly understood. Acidophils in the rostral portion are thought to be the source for PRL. Gonadotropin activity has also been demonstrated and tentatively assigned to one of the basophilic cells distributed throughout the pars distalis, although the evidence is circumstantial. Growth hormone activity has been demonstrated by bioassay in the sturgeon pituitary, but the cellular source is unknown. The pars intermedia of the sturgeon is large and closely associated with the pars nervosa to produce a typical neurointermediate lobe. The pars nervosa is basically hollow and is similar to the saccus vasculosus. Both peptidergic and aminergic fibers have been reported in the pars nervosa.

The hypothalamus contains a well-developed preoptic nucleus that provides peptidergic fibers to the pars nervosa. The *nucleus lateralis tuberis* consists of peptidgeric, aminergic, and NS neurons, and it may have separated from the preoptic nucleus of the more primitive bony fishes, as suggested by the situation described for *Polypterus*. There is a well-developed median eminence consisting of aminergic axonal endings from the nucleus lateralis tuberis. The median eminence is separated from the pars distalis by a connective tissue sheath so that no neurons penetrate the pars distalis. A hypothalamo-hypophysial portal system possibly conducts neurohormones from the median eminence to the pars distalis, although experimental verification of this hypothesis is not available.

Holostei (Gars and the Bowfin)

Adult holostean fishes do not exhibit a buccohypophysial duct nor is there any hypophysial cleft to suggest a relationship to Rathke's pouch. A transient hypophysial cleft does occur during development of the pituitary, however. The adenohypophysis consists of a rostral and proximal pars distalis and a pars intermedia. The rostral pars distalis is follicular. The cellular types have been carefully described for the adenohypophysis of the bowfin *Amia calva*. The rostral pars distalis consists of erythrosinophilic cells (type 1 acidophils) that are believed to produce PRL. Bioassays have confirmed the presence of PRL in the adenohypophysis, but the specific cellular type responsible for its production has not been identified. The follicular lumina contain a colloidal solution that is PAS(+), AB(+) and AF(+). A second PbH(+) acidophil in the rostral pars distalis is thought to be the source for corticotropin, although there is no experimental evidence that ACTH is present.

At the interface of the rostral and proximal pars distalis occurs an amphophilic cell reputed to produce GTH. Another acidophilic type of cell located in the proximal pars distalis stains positively with orange G (OG[+]) and may be the source of GH that has been demonstrated by bioassay. Two basophilic cells have been differentiated in the proximal pars distalis. One is believed to be the source of TSH, but the role of the other is unknown.

The pars intermedia interdigitates with the pars nervosa to form a neurointermediate lobe. There are two distinguishable cellular types in the pars intermedia of *Amia*. The predominant cell is PbH(+) and presumably secretes MSH. The second cell is a rare PAS(+) cell of unknown significance.

The preoptic nucleus has separated into two distinct portions, the dorsal *pars magnocellularis* and the ventral *pars parvocellularis*. This differentiation of the preoptic nucleus had begun to some degree in *Polypterus*, but complete separation has occurred in the holosteans. The significance of this separation is not clear. Peptidergic fibers from the preoptic nucleus pass to the pars nervosa where octapeptide hormones (AVT and IST) are stored. Aminergic fibers also appear in the pars nervosa and are believed to come from the nucleus lateralis tuberis.

The median eminence is connected to the pars distalis by a well-developed portal system. In addition, there are a limited number of peptidergic and aminergic axons that penetrate the pars distalis. However, only aminergic fibers are associated with the median eminence of *Amia*. Here, in these near-relatives of the teleostean fishes, is the modest beginning of a direct neural innervation of pars distalis cells so well developed in teleosts. In the holostean fishes is the beginning of a shift from neurovascular control to neuroglandular control of the adenohypophysis directly by the hypothalamus.

Class Osteichthyes: Teleostei

The hypothalamo-hypophysial systems of several teleostean species have been studied in detail. In view of the vast adaptive radiation that teleosts have undergone, it is not surprising to find considerable variation in this system. Two major patterns are indicated for "lower" and "higher" teleosts. However, the major features of the systems in teleosts are rather similar, including cytological features of cells related to the different tropic hormones. Although only a few species have been examined, they have been examined in greater detail than other piscine groups, and experimental efforts have been made to determine the function of the various cellular types. The tendency to localize cellular types in particular regions of the adenohypophysis is very strong in teleosts and has aided their identification.

In general, the teleostean rostral pars distalis consists of follicles filled with PAS(+) colloidal material. Two cellular types have been identified in the rostral pars distalis, and they have been linked experimentally to secretion of two tropic hormones. The *eta cell* (a

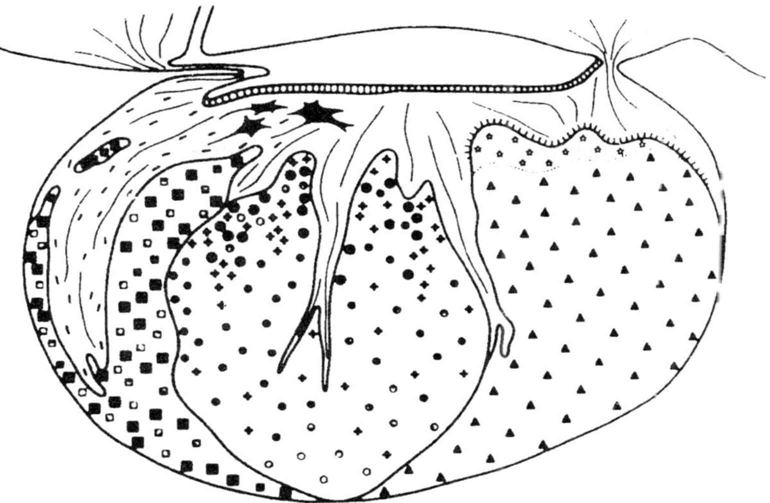

Fig. 3.18. The hypophysis of a "higher" teleost.

carminophil or erythrosinophil; type 1 acidophil) produces PRL. These cells are best developed in freshwater fishes. The activity of these cells increases when euryhaline fishes are held in fresh water, a correlation with the osmoregulatory role for PRL in freshwater fishes. Treatment of fishes with the drug metyrapone interferes with the synthesis of corticosteroids in the teleostean homologue of the adrenal cortex, the interrenal gland. A reduction in circulating corticosteroids results in stimulation of corticotropic cells and hence identifies the epsilon cells (type 3 basophils) of the rostral pars distalis as the source of ACTH.

Thyrotropin-secreting cells (delta cells) have been identified in the proximal pars distalis of several species by using thyroid inhibitors and thyroidectomy. Two kinds of GTH-secreting cells have been demonstrated with selective staining in the proximal pars distalis of chinook salmon and the European eel, and it has been suggested they are the cells responsible for secretion of LH and FSH respectively. However, experimental studies support the existence of only one GTH that is LH-like in activity. A non-glycoprotein factor is present in teleostean pituitaries that stimulates uptake of yolk precursors by the ovary. Growth hormone activity is associated with the OG(+) alpha cell of the proximal pars distalis.

There is a strong tendency to regionalize the various cellular types as unitype clusters within the proximal pars distalis. For example, in *Xiphophorus* the cellular types of the proximal pars distalis are segregated. Thyrotropes and somatotropes are restricted to the dorsal portion of the proximal pars distalis, and the gonadotropes occupy the ventral portion.

The pars intermedia is intimately associated with the pars nervosa of the neurohypophysis to form a neurointermediate lobe. Only one

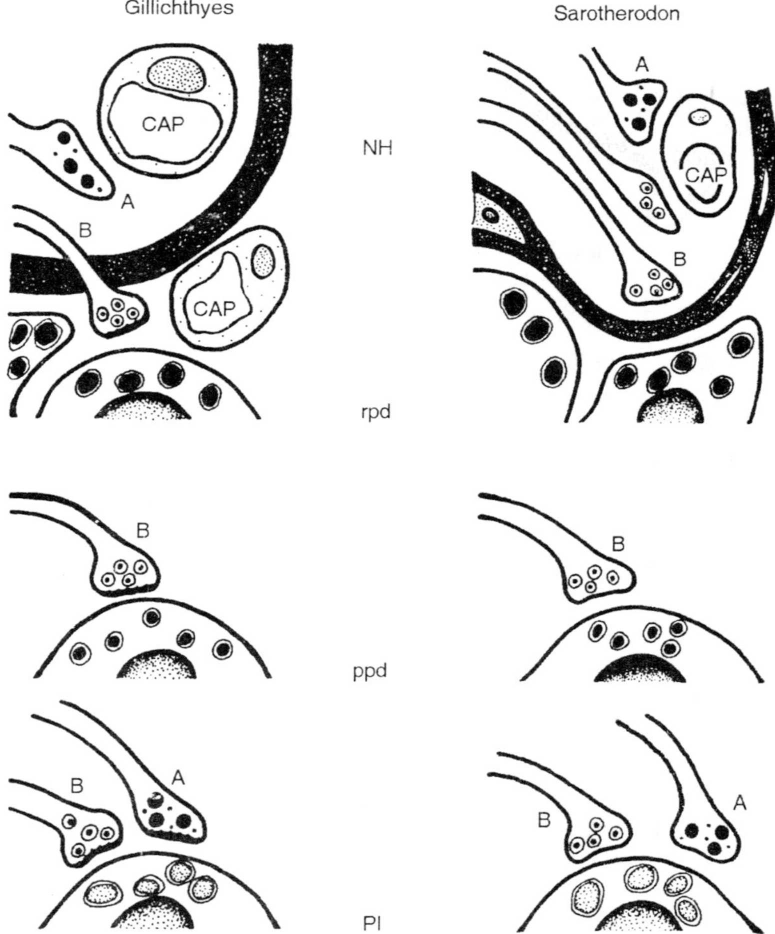

Fig. 3.19. Patterns of direct innervation of adenohypophysial cells in teleostean fishes. A, type A neurosecretory fibers; B, type B neurosecretory fibers; CAP, capillary; NH, neurohypophysis; rpd, rostral pars distalis; ppd, proximal pars distalis; PI, pars intermedia.

cellular type is found in the pars intermedia of some species, for example, salmonids, but two cellular types are found in other species. Type 1 cells are PbH(+), and they are found in all species examined. This cellular type contains granules of 250-400 nm diameter and is believed to secrete MSH. Type 2 pars intermedia cells contain PAS(+) granules (120-200 nm), but their function is unknown.

Neurosecretory centers for regulating secretion of TSH, ACTH, GTH and PRL appear to be localized in the nucleus lateralis tuberis. Release of both TSH and PRL is under inhibitory hypothalamic control, whereas release of ACTH and GTH is under stimulatory control. In contrast, stimulatory control of GH release would seem to reside in the nucleus anterior tuberis located near the preoptic nucleus. The nucleus lateralis tuberis is composed of peptidergic and aminergic neurons. Some of these neurons terminate in the median eminence, but others terminate (synapse) directly on pars distalis cells, providing cytological evidence for direct neural control over adenohypophysial function. The preoptic nucleus in the teleostean hypothalamus consists of peptidergic cells and is responsible for the synthesis of AVT and IST that are stored in the pars nervosa.

Class Osteichthyes: Subclass Sarcopterygii

The hypothalamo-hypophysial system of the lungfishes (Dipnoi) is basically like that of tetrapods, especially the amphibians, although a pars tuberalis is absent in lungfishes. It is not like actinopterygian fishes in that it lacks many of the features that characterize the piscine system. There is less regionalization of cellular types in the lungfish adenohypophysis than in other bony fishes. Furthermore, both the neurointermediate lobe and the saccus vasculosus are absent in lungfishes.

The coelacanth, *Latimeria chalumnae*, which is considered to be closer to the ancestral line of the tetrapods than are the lungfishes, is much more piscinelike than it is tetrapodlike. The anatomy of the adenohypophysis is much like that of the elasmobranch fishes. There is a buccal lobe in the pars distalis which may be homologous to the ventral lobe of elasmobranchs. The pars distalis may be subdivided into rostral and proximal portions with stainable cellular types appearing that are similar to those of teleosts. No experimental studies have been performed, however, to link stainable cellular types with specific tropic functions. Growth hormone is present in the coelacanth adenohypophysis according to bioassay and immunological tests, but this activity has not been assigned to either region of the pars distalis.

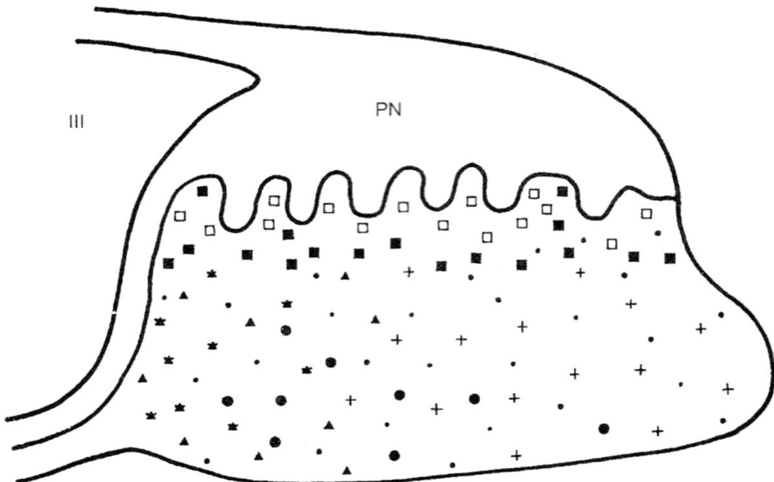

Fig. 3.20. The lungfish pituitary. PN, pars nervosa.

Only one stainable cellular type has been demonstrated in the pars intermedia, which is probably the source of MSH. The pars intermedia forms a typical neurointermediate lobe characteristic of fishes. The median eminence of the neurohypophysis is connected by a well-developed hypothalamo-hypophysial portal system to the adenohypophysis, particularly the proximal region. Neurosecretory axons appear to penetrate the proximal pars distalis similar to the situation described for the holostean fish *Amia calva*.

Although the sarcopterygian fishes represent the closest living relatives to the tetrapods, extant members of this group provide no transitional stages with respect to changes in the hypothalamo-hypophysial system. The living coelacanth appears to be completely fish-like, whereas the lungfishes are like the tetrapods, with the exception of the missing pars tuberalis. Perhaps some embryological studies of pituitary development in lungfishes might reveal a transient homologue to the pars tuberalis.

Tetrapod Vertebrates

Class Amphibia

The amphibian hypothalamo-hypophysial axis has most of the features characteristic of the tetrapod system, including the presence of a pars tuberalis. The prominent preoptic nucleus is a major bipartite NS center although other NS nuclei are important as well. At least two kinds of peptidergic fibers originate in the preoptic nucleus, but

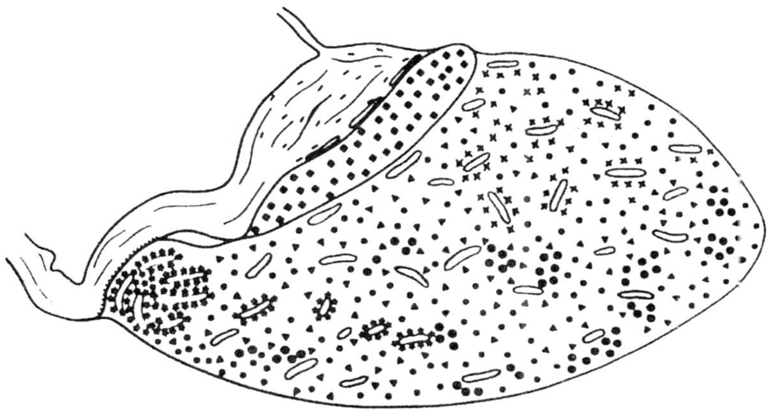

Fig. 3.21. The anuran hypophysis (Amphibia).

the significance of this finding is not clear. The *infundibular nucleus* located in the basal hypothalamus supplies aminergic and peptidergic fibers to the median eminence. This nucleus is homologous to at least part of the major hypophysiotropic region of the mammalian hypothalamus, although its relationship to the nucleus lateralis tuberis of fishes is uncertain. A *gonadotropic center* has been identified that contributes aminergic and peptidergic axons to the median eminence. The pars nervosa of the neurohypophysis receives peptidergic fibers originating in the preoptic nucleus. At least two octapeptide hormones are stored in this well-developed neurohemal organ. There is no tendency

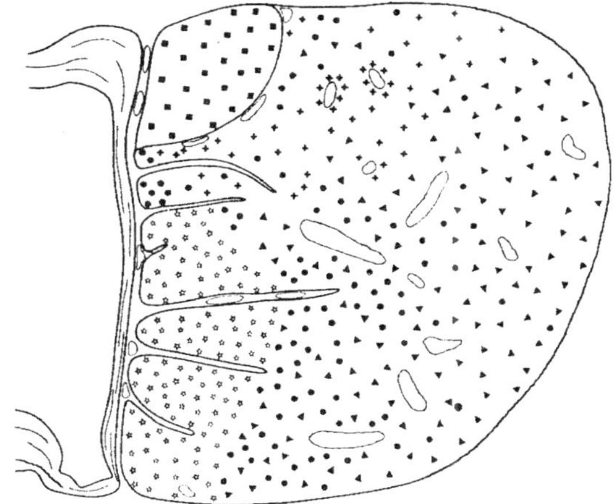

Fig. 3.22. The hypophysis of a caudate amphibian.

for development of a neurointermediate lobe, nor is a saccus vasculosus or any comparable structure found in amphibians.

The adenohypophysis consists of pars tuberalis, pars intermedia and pars distalis. There is no known function for the pars tuberalis. Based upon ultrastructural comparison of cytoplasmic granules and other features, what appear to be two separate cellular types have been observed in the anuran pars tuberalis. There appears to be a neural pathway extending from the ependymal lining of the third ventricle to the pars tuberalis, but no functional correlations have been reported.

The pars intermedia has a poor vascular supply, but it is directly innervated by aminergic neurons thought to originate in the *caudal aminergic nuclei* of the hypothalamus. Release of MSH appears to be under direct neural control.

The pars distalis is not separable into discrete regions although there is a tendency for some regionalization of cellular types. Urodele amphibians exhibit greater regionalization than do anuran species. Careful cytological studies have been performed on many amphibian species, often coupled with experimental manipulations, bioassays or correlations with specific life history events. Unfortunately there has been considerable disagreement among researchers in this area with respect to establishment of specific sources for the various tropic hormones. The following account of the cytology of *Rana temporaria* will serve as the basic amphibian type for purposes of this text.

Five stainable cellular types can be distinguished in the pars distalis of *R. temporaria*. There are three basophilic cells and two acidophils plus a number of chromophobic cells.

Type 1 basophils possess granules that stain positively with PAS, AF and AB procedures and are considered to be thyrotropes. Type 2 basophils stain rather weakly with the same procedures but contain cytoplasmic granules that are stainable with OG. This cell is believed to be the source of GTHs. Type 3 basophils are similar to type 1, except they do not stain with AB. This cell is thought to be the source of ACTH.

Acidophils of type 1 are large erythrosinophilic and orangeophilic cells. These cells bind antibody to sheep PRL and presumably are the source for PRL in amphibians. Type 2 acidophils are OG(+) and have been suggested to be the source of a GH. There is no evidence, however, to support this contention. Furthermore, there is considerable evidence to support a role for PRL as a larval GH.

Class Reptilia

In reptiles the preoptic nucleus has become two separate nuclei: the *supraoptic nucleus* and the *paraventricular nucleus*. These nuclei consist of peptidergic NS neurons that terminate in the pars nervosa. They produce the octapeptide neurohormones that are stored in the pars nervosa. A separate hypophysiotropic region, the *infundibular nucleus*, is considered to be homologous to the nucleus of the same name in amphibians and birds as well as to the hypophysiotropic area of the mammalian hypothalamus. Aminergic and peptidergic NS fibers originate in this nucleus and terminate in the median eminence.

The adenohypophysis is well developed in most reptiles, and it consists of a pars distalis, a pars tuberalis and a pars intermedia. The primitive reptilian condition probably is best represented by the Rhynchocephalia (*Sphenodon*), Chelonia (turtles) and Crocodilia. The pars distalis appears as two distinct regions reminiscent of the condition in bony fishes. There is a rostral or *cephalic lobe* and a *caudal lobe*.

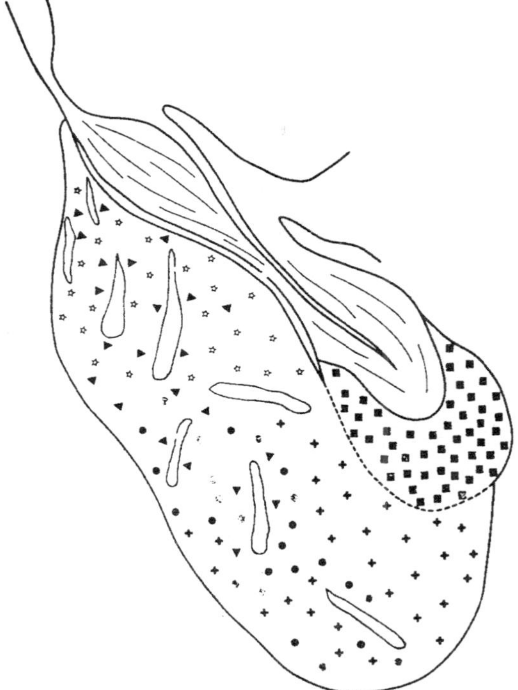

Fig. 3.23. General organization of the reptilian hypophysis.

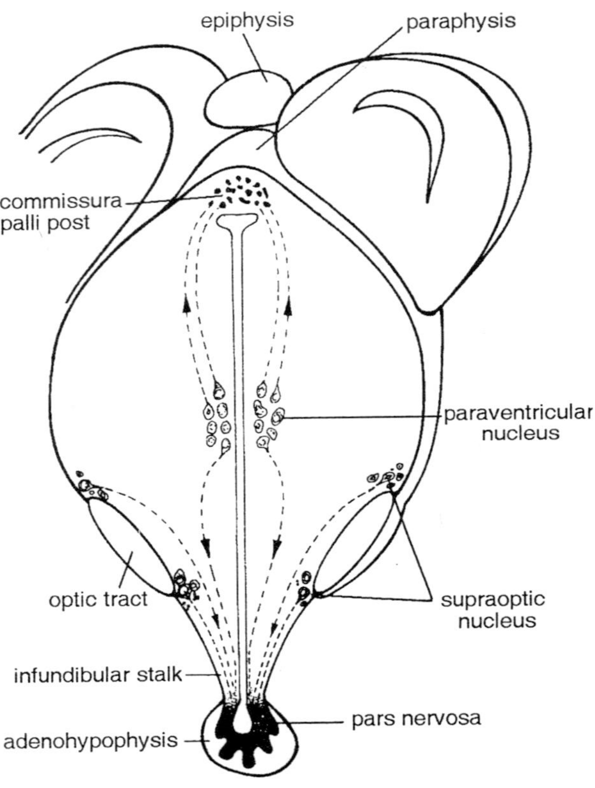

Fig. 3.24. Neurosecretory connections between reptilian (snake) hypothalamic nuclei and the epithalamus and hypophysis.

The distribution of cellular types in these lobes of the pars distalis is similar to that described for fishes. The pars tuberalis is well developed in the Rhynchocephalia, Chelonia and Crocodilia, but it is greatly reduced and sometimes absent in lizards. In adult snakes, the pars tuberalis is completely absent. Reptiles have the best developed pars intermedia of any vertebrate group. It is especially elaborate in chelonians, crocodilians, snakes and many lizards.

There are five stainable cellular types in the reptilian pars distalis similar to those described for the amphibians. More than one scheme has been proposed for their nomenclature, but a system similar to that adopted for the Amphibia is used here.

Type 1 acidophils are carminophilic cells located in the cephalic lobe that presumably secrete PRL. Bioassays of the two portions of the pars distalis indicate that PRL activity resides in the cephalic lobe. Type 2 acidophils are located in the caudal lobe. These cells are OG(+) in most reptiles and are believed to secrete GH. Bioassays of the caudal lobe indicate GH activity is present.

Type 1 basophils are somewhat scattered throughout the pars distalis but are more concentrated in the caudal lobe. These cells stain similarly to the thyroptropic type 1 basophils of amphibians and probably secrete TSH. Cytologically these cells are sensitive to thyroidectomy. Bioassay of the separate lobes of the pars distalis of a lizard, *Anolis carolinensis*, reveals that TSH activity is in the caudal lobe where the majority of type 1 basophils are found in this species.

Type 2 basophils tend to be scattered throughout the pars distalis. These cells are localized ventrally toward the midline in most snakes and some lizards but are limited, to the cephalic lobe of other lizards. Gonadotropes as well as some thyrotropes occur in the pars tuberalis of turtles. Type 2 basophils vary greatly in size among different species, but all stain like the gonadotropes of amphibians. Furthermore, these cells respond to castration or treatment with gonadal steroids, making it likely that they secrete GTHs.

The last basophilic type consists of amphophilic cells located in the cephalic lobe. They exhibit similar staining properties to the corticotropes described for amphibians. Bioassays have localized ACTH activity in the cephalic lobe, and the type 3 basophils respond to experimental manipulations such as administration of metyrapone, which blocks corticosteroid synthesis in the adrenal.

The reptilian pars distalis contains chromophobic cells in addition to the stainable cellular types. Reptilian chromophobes appear as agranular, stellate cells and may function as a transport system carrying nutrients from the blood to the hormone-producing cells.

The pars intermedia contains only one stainable cellular type, which is the presumed source for MSH. The skin of many reptiles is sensitive to MSH, and the American chameleon *Anolis carolinensis* is used for a sensitive MSH bioassay. This lizard can change from green to brown under the influence of MSH, which allows it to colour adapt to different backgrounds in its environment. Colour changes in the true chameleon, *Chamaeleo* spp., are under direct neural control, however, and MSH is not involved.

Class Aves

The avian hypothalamo-hypophysial system differs from that of other tetrapods (and fishes) in that the pars intermedia is absent in all species. A well-developed pars tuberalis is present, and the pars distalis consists of cephalic and caudal lobes homologous to those described for reptiles.

The median eminence differs from mammals in that the primary capillaries of the portal system lie superficially or in grooves on the surface rather than penetrating to form the complex vascular bed seen in mammals. This anatomical difference allows an investigator to sever the portal connection with the pars distalis without interrupting fiber tracts. The hypophysiotropic region of the hypothalamus supplies aminergic and peptidergic fibers to the median eminence. The supraoptic nucleus and the paraventricular nucleus are responsible for secretion of the octapeptide neurohormones.

In some birds, i.e., the pigeon, the Japanese quail and the white-crowned sparrow, the median eminence is separable into an anterior neurohemal area and a more posterior neurohemal area similar to that described for elasmobranchs. Each of these areas has is own portal connection to the pars distalis, which consists of a cephalic and a caudal lobe. It is not certain how widespread this phenomenon of two median eminences is among other avian species; however, such potential regionalization of both the median eminence and the pars distalis could represent a mechanism to increase the efficiency of delivery of hypothalamic neurohormones to cellular types regionalized in various

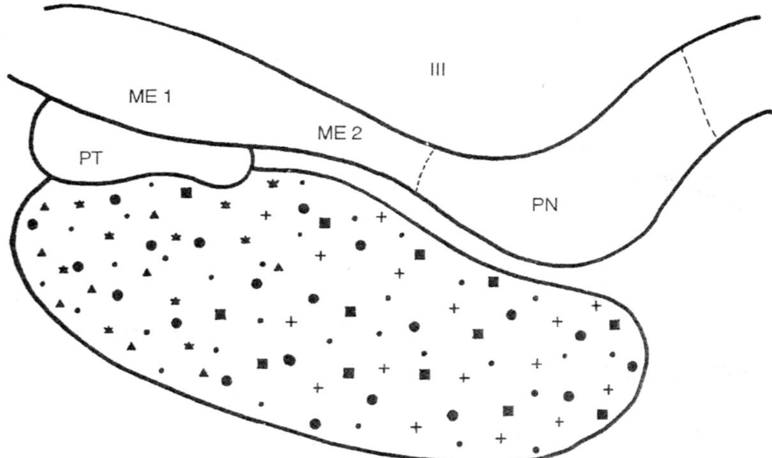

Fig. 3.25. The avian hypophysis. Note the presence of two separate neurohemal areas associated with the pars distalis (ME 1, ME 2) and the absence of a pars intermedia. III, third ventricle; PT, pars tuberalis; PN, pars nervosa.

parts of the pars distalis. This neurovascular specialization is analogous to the system of direct innervation of pars distalis cells that was observed in teleosts.

The nomenclature for the cytological features of the avian adenohypophysis is somewhat different from the schemes used for other vertebrates. The avian pars distalis contains seven stainable cellular types as well as chromophobic cells. The stainable cells have been grouped into three categories: *serous cells*, *glycoprotein-containing cells* and *PbH(+) cells*.

The serous cells include the OG(+) alpha cell (type 2 acidophil) located in the caudal lobe and possibly responsible for GH secretion. The second serous cellular type is the erythrosinophilic eta cell (type 1 acidophil) located in the periphery of the cephalic lobe. The eta cell is synonymous with the so-called *broody cell* of egg-incubating birds and is the source of PRL.

Three types of glycoprotein-containing cells can be distinguished in the avian pars distalis. The beta cell of the cephalic lobe is strongly PAS(+) and supposedly produces FSH. The gamma cell is found in the caudal lobe. It is acidophilic and PAS(+) and has been claimed to be the probable source of LH. Both of these cellular types respond in the same manner to experimental manipulations such as castration and administration of synthetic GnRH. If separate gonadotropic cells for LH and FSH do exist in birds, this would be the only vertebrate group in which one cellular type did not appear to produce both GHs. Future studies may reveal that these beta and gamma cells of birds are only variants of a single type of cell comparable to the type 2 basophil of other vertebrates.

The delta cell appears to be a type 1 basophil on the basis of its stainability features and is thought to be the source of TSH. This cellular type has been reported to occur throughout the pars distalis; however, only those delta cells of the cephalic lobe of the Japanese quail show cytological changes following thyroidectomy. Bioassay data obtained for this same species indicate that TSH activity is restricted to the cephalic lobe. These data suggest that the actual distribution of thyrotropes may be highly dependent upon the species.

There are two PbH(+) cells in the avian pars distalis: the kappa cell and the epsilon cell. The kappa cell presumably secretes MSH. Cytologically the kappa cell is similar to the MSH-secreting PbH(+) cells of other vertebrates, including the PbH(+) cell in the human pars distalis that is also believed to produce MSH. Bioassays for

MSH activity support the proposal that MSH-secreting cells are present in the avian pars distalis, but since ACTH also has MSH activity in these bioassays this is not conclusive proof. The epsilon cell located in the cephalic lobe has stainability characteristics like the type 3 basophils of other vertebrates. This cell responds to experimental manipulations such as adrenalectomy and metyrapone treatment, indicating that it secretes ACTH.

Some Comparative Aspects of Hypothalamic Control of Adenohypophysial Function in Nonmammals

Since some hypothalamic hypophysiotropic substances of nonmammalian vertebrates have not been identified many researchers refer to them only as factors. For consistency, however, they will be referred to as hormones throughout this text.

Fishes

Immunoreactive GnRH has been demonstrated in the hagfish, lamprey and shark. Somatostatin is present in the brains of hagfish and sharks and TRH has been found in lampreys and sharks. There is no evidence of any functional roles for these peptides, however.

Numerous studies of releasing hormones have been conducted on teleosts. Teleostean GnRH differs structurally from mammalian GnRH, but is effective in sheep bioassays for mammalian GnRH. Similarly, mammalian GnRH induces ovulation in teleosts. Thyroid function appears to be under inhibitory hypothalamic control, and administration of mammalian TRH depresses thyroid activity. Dopamine blocks PRL release, and TRH stimulates release. Somatostatin is present and can block GH release. Dopamine also blocks GH release. There is some evidence to suggest presence of CRH.

Tetrapod Vertebrates

Class Amphibia

Hypothalamic control over TSH release in amphibians may be of a stimulatory nature as in mammals, or it may have no influence as demonstrated for some adult amphibians. Molecules identical to mammalian TRH are present in relatively large amounts in the amphibian hypothalamus, and even greater amounts are found in the skin. However, synthetic mammalian TRH has proven ineffective as a stimulator of thyroid function in a number of species. Intravenous injection of TRH did elevate plasma levels of the thyroid hormones in one species, *Rana ridibunda*, but a similar approach failed for the Mexican axolotl.

Release of PRL and MSH is considered to be under inhibitory control in amphibians. Release of MSH appears to be under direct aminergic neural control in anurans, but the situation for PRL release is less clear. The similarity in regulatory mechanisms for MSH and PRL release is reflected also by the actions of certain pharmacological agents on their release. Drugs that affect PRL or MSH release in mammals have similar actions in amphibians. For example, ergot derivatives such as ergocornine or ergocryptine inhibit release of both PRL and MSH. These agents are alpha-adrenergic blocking agents and support the hypothesis that aminergic substances are involved in the release of both PRL and MSH.

There is also evidence for a PRL-releasing hormone (PRH). Extracts prepared from the hypothalamus of *Xenopus laevis*, the clawed frog, stimulate PRL release from chicken, rat and anuran pituitaries in vitro. Hypothalamic content of both PRH and PRIH activities shows a seasonal variation in *Rana temporaria*, and such variations if common to other species could explain conflicting results obtained by various researchers. It is not clear whether any of the reported effects of hypothalamic activities influencing PRL release are due to their TRH content or to substances such as aminergic neurotransmitters present in these extracts.

In mammals TRH causes release of PRL, which is a known antagonist of thyroid hormones in amphibians. No effect of synthetic TRH on PRL release could be demonstrated in the red-spotted newt *Notophthalmus viridescens*, but PRL release was stimulated following injection into bullfrogs and by addition of synthetic TRH to pituitaries of *Xenopus laevis* cultured in vitro. Synthetic TRH also causes release of MSH in frogs.

Gonadotropin release is under positive hypothalamic control. Amphibian GnRH appears to be similar to mammalian GnRH which is effective in both anurans and urodeles. Seasonal changes in hypothalamic GnRH reflect changes in gonadal function. Growth hormone release is autonomous in urodeles but is under stimulatory hypothalamic control in anurans. Immunoreactive somatostatin is present in amphibian brains, but its role in regulating GH release is not clear.

There is evidence for a CRH in the anuran preoptic nucleus, but release of ACTH in urodeles seems to be autonomous.

Class Reptilia

No generalization can be formulated about hypothalamic control in reptiles because very little research has been conducted in this area. Corticotropin release is probably under a mammalian pattern of control. The regulatory pattern for PRL is not clear as some conflicting data have been reported. Release of MSH is under inhibitory control. Thyrotropin release has not been examined but is presumed to be under stimulatory control. Synthetic mammalian TRH does not stimulate the thyroid of the turtle.

Gonadotropin release is under stimulatory hypothalamic control. Immunoreactive GnRH is present in the preoptic area but is not identical to mammalian GnRH. Synthetic mammalian GnRH induces ovulation in some species. Immunoreactive somatostatin is present in turtles and lizards, but its role is not clear. Release of GH is primarily under stimulatory control.

Lesions in the infundibular nucleus of the lizard *Sceloporus cyanogenys* result in reduced adrenal weight, probably due to reduction in ACTH levels. Corticosteroid administration produces similar reductions in adrenal weight, presumably due to negative feedback on ACTH release via the hypothalamus.

It is not clear whether PRL release in turtles is under stimulatory hypothalamic control as in birds or under inhibitory control as in most vertebrates. Cultured pituitaries of *Malaclemys terrapin* release PRL, but extracts prepared from the median eminence of female *Pseudemys scripta* or male rats stimulate PRL release from cultured *Pseudemys* pituitaries. Hypothalamic extracts from another turtle, *Chrysemys picta*, similarly evoke PRL release from either avian or mammalian pituitaries. Cultured pituitaries of a lizard, *Dipsosaurus dorsalis*, also release PRL following addition of an extract of rat median eminence. These extracts may have been rich in a legitimate PRH or some other substance such as a neurotransmitter or even the TRH tripeptide. Synthetic mammalian TRH does stimulate PRL release from *C. picta* pituitaries, supporting the interpretation that these effects are due to presence of TRH in the extracts. Whether or not endogenous TRH tripeptide is a physiological regulator of PRL release remains to be seen.

Class Aves

Some of the first studies of hypothalamic control in birds involved lesions in the hypothalamus paralleling mammalian work. Placement of lesions in the hypothalamus produces testicular atrophy in the duck. Similar experiments have resulted in localization of hypothalamic centers regulating release of GTHs, TSH and ACTH.

Release of adenohypophysial hormones appears to be influenced by a mammalian pattern of hypothalamic control except for PRL. The pattern of PRL release in birds is under direct stimulatory control by the hypothalamus, which is unlike the condition in other vertebrates. The avian PRH might be similar if not identical to mammalian TRH since TRH causes PRL release from the bird pituitary in vivo and in vitro.

Thyroid glands of the lesser snow goose, *Anser caerulescens caerulescens*, and the duck, *Anas platyrhynchos* are activated within 20 minutes following injection of synthetic mammalian TRH. Injections of GnRH were without effect on thyroid hormone secretion.

Ovulation can be induced in the Japanese quail with synthetic mammalian GnRH, and mammalian GnRH activates gonadotropes in the avian pars distalis. Although some correlations have been noted between amine levels in the avian hypothalamus and tropic hormone release, the role of aminergic substances in hormone release is poorly understood.

Possible Role for the Epiphysial Complex in Hypothalamic Function

The epiphysial complex includes the pineal gland, which is believed to secrete its products into the blood and possibly into the cerebrospinal fluid. The pineal gland may secrete the amine melatonin or small peptides, including AVT, or both. Numerous studies have documented antigonadal actions of pineal extracts, melatonin, or both in fishes, amphibians, reptiles, birds and mammals. In addition, effects on thyroid function have been noted in some of these groups. It is possible that these pineal actions are mediated via effects on the hypothalamus, or the specific sites of action may be at the target endocrine glands themselves.

4

Pars Intermedia

Many vertebrates show changes in pigmentation or pigmentary patterns that are correlated with environmental factors or particular events in their life histories. Rapid responses are under direct neural control, and the slower responses are generally the result of endocrine control or a combination of neural and endocrine control. Pigmentation patterns mediated through melanotropin (MSH) from the pars intermedia are emphasized in this chapter with some minor reference to other endocrine factors and neural innervation. Vertebrates possess a variety of specialized, pigmented cells referred to as *chromatophores*. These cellular types singly or in special combinations are the bases for colour patterns associated with the integument. *Physiological colour changes* involving displacement of pigments within a pigment cell are characteristic only of fishes, amphibians and reptiles. All vertebrates exhibit increases in the number of chromatophores or increases in the amount of pigment contained within the chromatophores or both. This type of colour change is termed *morphological colour change*.

The terminologies employed here are from Bagnara and Hadley. These authors do not recognize the term melanocyte as adopted by the Nomenclature Committee of the Sixth International Pigment Cell Conference in 1966. Mammalian researchers refer to the melanin-containing cells of the skin and other organs as melanocytes. The epidermal melanophore can be considered equivalent to the mammalian melanocyte.

Chromatophores

The types of chromatophores found among the vertebrates include the *melanophores*, usually containing black or brown pigments;

Melanocytes
A cell that synthesizes the pigment melanin

Melanoblast
Precursor of melanocyte; contains premelanosomes but does not synthesize melanin

Melanosome
Melanin-containing organelle in which melanization is complete; no tyrosinase activity present

Premelanosome
Active in melanin synthesis; tyrosinase activity present

Melanophore
A dermal or epidermal cell that participates with other cells in rapid colour changes by intracellular displacement (migration) of melanosomes

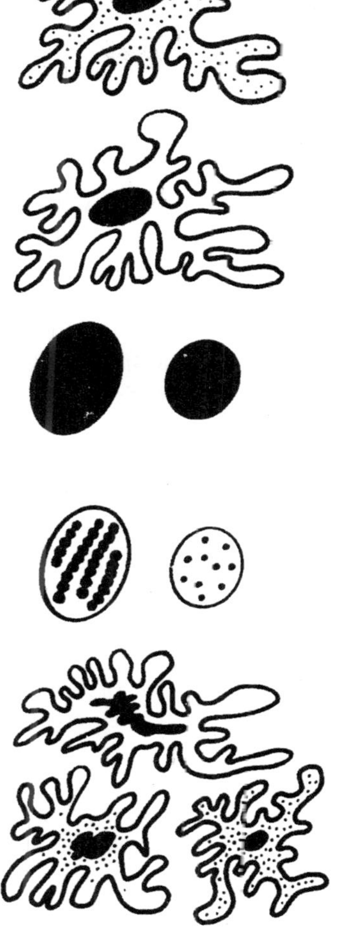

Fig. 4.1. Terminology of vertebrate melanin-containing cells.

iridophores (guanophores), appearing silvery or golden when viewed with reflected light; *xanthophores*, containing yellow or orange pigments; and *erythrophores* with red or orange pigments. In many cases a combination of these different chromatophore types occurs in the integument to provide a particular colour or to allow for changes in colouration according to environmental conditions.

Structural colouration is due to interference, diffraction or *Tyndall scattering* of light waves. The Tyndall effect results from scattering of light by the reflecting platelets in iridophores. This scattering often imparts a blue colouration to amphibians and reptiles. If these iridophores are overlayed by xanthophores containing yellow pigments, the integument will appear green. Combinations of iridophores and other chromatophores may result in shades of blue, green, yellow, orange, red, bronze or gold.

Epidermal Melanophores

Epidermal melanophores are common in reptiles and occur at some stage in the life history of amphibians (embryo, larva or adult). They are also present in the hair of mammals and the feathers of birds. Fishes do not prominently exhibit epidermal melanophores. This type of chromatophore typically is fusiform with long, slender processes that have few branches. Epidermal melanophores synthesize the pigment *melanin* and often deposit it extracellularly. These cells are involved in morphological colour changes only and do not play a role in physiological colour changes.

Dermal Melanophores

Dermal melanophores are found in the dermis, in various organs and on the surfaces of certain blood vessels and nerves. These cells are stellate and are principally involved in physiological colour changes in lower vertebrates. Dermal melanophores contain the pigment *melanin* localized in organelles called *premelanosomes* or *melanosomes*. Premelanosomes are organelles that are still synthesizing melanin from its basic precursor, the amino acid tyrosine. These organelles exhibit high tyrosinase activity, the enzyme responsible for melanin synthesis. Melanosomes have ceased melanin synthesis and exhibit no tyrosinase activity. When examined with the aid of the electron microscope, melanosomes appear as uniformly electron-dense bodies in the cytoplasm, whereas the premelanosomes may show varying degrees of density even within a given organelle.

Dispersion of melanosomes into the stellate processes of the melanophore causes a darkening of the integument. Aggregation of melanosomes near the nucleus of the melanophores will cause the integument to lighten. Migration of melanosomes is the principal mechanism employed whereby fishes, amphibians and reptiles adapt to light or dark backgrounds. These melanosome migrations may be under neural or endocrine control or both.

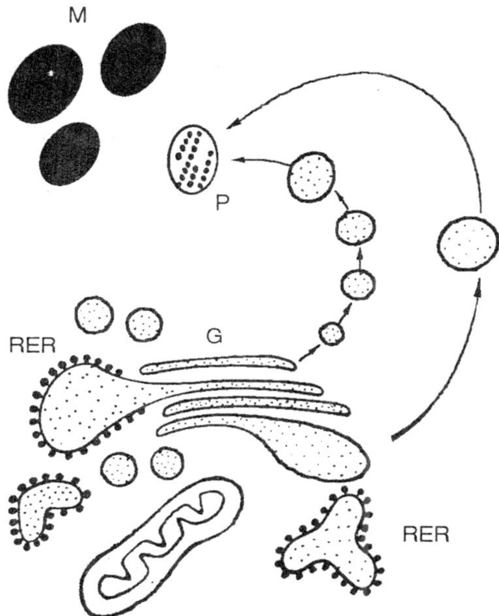

Fig. 4.2. Formation of premelanosomes and melanosomes in melanophore (melanocyte). RER, rough endoplasmic reticulum; G, Golgi apparatus; P, premelanosome; M, melanosomes.

The ability to adapt to dark or light backgrounds is used as an index for endocrine state with respect to MSH. The degree of dispersion of melanosomes within a given melanophore may be quantified on an arbitrary scale from 1 to 5 where 1 indicates complete aggregation and 5 indicates maximum dispersion. This index of melanophore

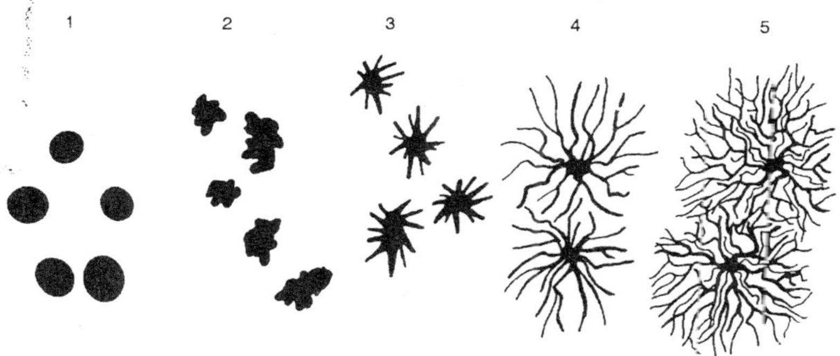

Fig. 4.3. The melanophore index for determining the degree of dispersion of the melanosomes within the stellate dermal melanophore of amphibian larvae.

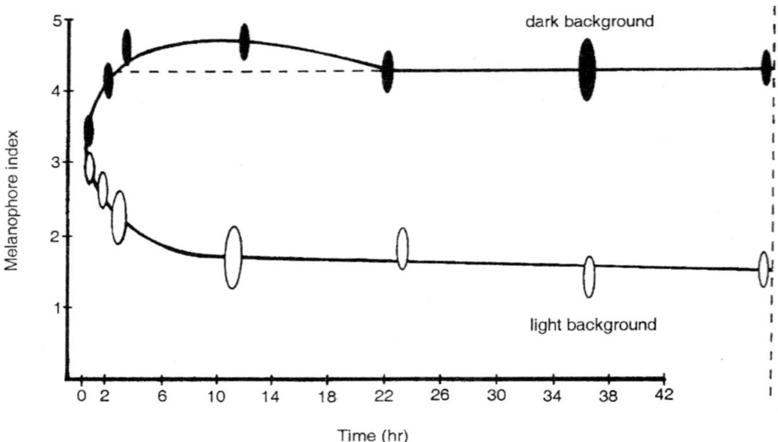

Fig. 4.4. Melanophore index for animal transferred from darkness into light and placed on a black or white background.

expansion is referred to as the *Hogben or melanophore index*. This index is useful in examining intact animals when it is not feasible to utilize the MSH bioassays.

Iridophores

Iridophores are found in the dermis, and they are highly variable in general appearance in different vertebrate groups. They reflect light because of the presence of *reflecting platelets* in their cytoplasm. These flat, reflecting platelets contain *purines*, especially guanine, although hypoxanthine, adenine or uric acid may be employed. Reflecting platelets are arranged in oriented stacks. Iridophores are sometimes involved with morphological and physiological colour changes as well as with structural colour patterns, but their regulation is not fully understood.

Xanthophores and Erythrophores

Xanthophores typically contain yellow pigments, and erythrophores contain red pigments, although either of them may possess orange pigments. These chromatophores are dermal, although they are occasionally located in the epidermis, as in the red spots of the red-spotted newt *Notophthalmus viridescens*.

The pigments contained within these chromatophores are usually *carotenoids* but may be *pteridines*. Pteridines are found in distinct cytoplasmic organelles called *pterinosomes* that are derived presumably from smooth endoplasmic reticulum. These organelles are spherical or ellipsoidal (about 0.5 mm diameter) and are composed of a series of concentric lamellae. Pterinosomes are evenly distributed throughout the cytoplasm of certain xanthophores and erythrophores of fishes, amphibians and reptiles. The *carotenoid vesicles* are organelles containing carotenoid pigments. These organelles are much smaller than pterinosomes and are not lamellar in structure.

Overlapping of Characteristics in Chromatophores

Melanophores are sometimes observed that contain reflecting platelets (characteristic of iridophores) or pterinosomes (containing pteridine pigments as found in xanthophores and erythrophores). Either xanthophores or iridophores may contain melanosomes in addition to their characteristic organelles. Some chromatophores have been described that contain both pterinosomes and reflecting platelets. Obviously the descriptive categories outlined above are not mutually exclusive. These categories, however, are useful in characterizing the majority of chromatophores and in discussing their neural and endocrine regulation.

Dermal Melanophore Unit

Dermal melanophores appear in some amphibians in a precise arrangement with xanthophores and iridophores to form a functional entity that has been termed the *dermal melanophore unit*. Similar associations of chromatophores may occur in lizards as well. This unit consists of an iridophore surrounded by extensions of a more basal dermal melanophore. Both of these cells are overlain by a xanthophore. The interactions of xanthophore pigments and Tyndall scattering caused by reflecting platelets of the iridophore can be altered by the movement of melanosomes in the dermal melanophore. Such alterations in position of the melanosomes can result in gradations of colour.

Pigmentation Patterns

Colouration patterns in some vertebrates may be determined by the distribution of types of chromatophores and by the relationship of dermal chromatophore units to other chromatophores. For example, the dark spots on the skin of the frog *Rana pipiens* are due to concentrations of epidermal melanophores and extracellular deposition of melanin, whereas adjacent regions of the skin that may vary from light green to black are occupied exclusively by dermal chromatophore units.

Regulation of Chromatophores by Melanotropin

The pars intermedia is the primary endocrine regulator of pigmentary responses in vertebrates through elaboration of MSH. Recall that MSH release is under inhibitory hypothalamic control and that MSH belongs to the family of adenohypophysial polypeptides that includes lipotropin, corticotropin (ACTH) and the endorphins.

The major target for MSH is the dermal melanophore, and in a few cases, other chromatophores may be affected.

Adenyl cyclase activity is enhanced by MSH in melanin-synthesizing cells of fish, amphibians and mammals. Melanotropin activates adenyl cyclase in the melanophore plasmalemma, which in turn produces cyclic adenosine 3´, 5´-monophosphate (cAMP). Melanosome dispersion is effected by cAMP and is enhanced by caffeine through its inhibition of the cAMP-degrading enzyme phosphodiesterase. Corticotropin also causes

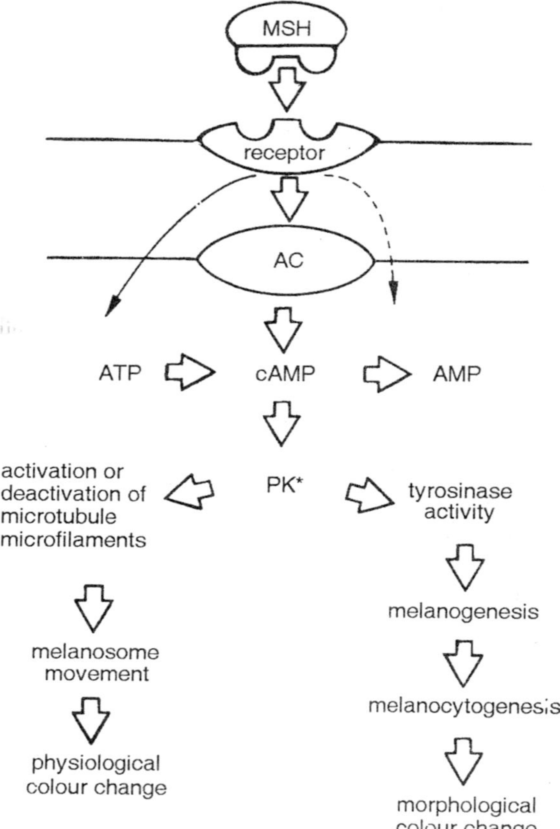

Fig. 4.5. Mechanism of action by MSH on the melanophore. AC, adenyl cyclase; PK, activated or phosphorylated protein kinase.*

melanosome dispersion by activating adenyl cyclase. Cyclic guanosine 3′, 5′-monophosphate (cGMP) however, is without effect. Melanophore regulation by nervous elements may operate by inhibiting the action of MSH on cAMP formation. For example, norepinephrine, the neurotransmitter of postganglionic sympathetic neurons, inhibits the MSH-induced increase in cAMP.

The actual mechanism of how melanosomes migrate in and out of the stellate processes of the melanophore is unknown. It appears that microtubules are essential for aggregation of the melanosomes, and these may be activated by cAMP. Any experimental treatment known to interfere with microtubule formation, such as application of the drug colchicine, blocks melanosome aggregation. Ultrastructural observations, however, in some amphibian melanophores do not support the presence of microtubules oriented properly to cause melanosomes to disperse, and the role of microtubules needs further clarification.

Class Agnatha

Hagfishes (*Myxine glutinosa*) show no ability to adapt to different backgrounds, suggesting no role for MSH with respect to background adaptation. However, hypophysectomy of lampreys (Petromyzontidae) produces permanent paling, presumably due to loss of MSH. Exogenous MSH causes melanosome dispersion in lampreys. Nothing is known about factors controlling MSH release in cyclostomes.

Class Chondrichthyes

Sharks do adapt to background but very slowly, requiring up to 100 hours to achieve maximum adaptation. Hypophysectomy abolishes this ability to "background adapt," whereas ectopic transplants of the neurointermediate lobe cause permanent darkening regardless of the background upon which the shark is placed.

The dogfish shark pituitary contains α-MSH, and sharks darken following injection of either α- or β-MSH. Melanophore responses to MSH are paralleled by similar responses in xanthophores.

Both aminergic and peptidergic fibers penetrate into the dogfish pars intermedia. Possibly peptidergic fibers control synthesis of MSH by the pars intermedia cells, and aminergic fibers are responsible for controlling MSH release. The pars intermedia of the holocephalan *Hydrolagus colliei* also exhibits NS innervation, although the type of innervation has not been determined. The vascular arrangement between the hypothalamus and the pars intermedia of the dogfish shark implies neurovascular control rather than direct neural innervation or neurohumoral control. It is not clear what mechanism predominates in controlling release of MSH in these fishes.

Class Osteichthyes: Teleostei

The pigmentary responses of fishes are under control of direct aminergic innervation of melanophores and other chromatophores and MSH does not play a major role in physiological colour changes. The melanophores of most species do not respond to either hypophysectomy or exogenous MSH, whereas others show only a limited response to MSH following denervation of the melanophores. Dispersion of pigments in melanophores of goldfish *Carassius auratus* and erythrophores of the European minnow *Phoxinus phoxinus* occurs following treatment with MSH. Melanophore dispersion can be induced only in denervated melanophores of the killfish *Fundulus heteroclitus* because of the presence of overriding neural stimuli in intact fish.

MSH may play some role in morphological colour changes in teleosts. Treatment of xanthic goldfish (goldfish normally lacking any melanophores) with MSH stimulates melanophore differentiation and melanogenesis. Hypophysectomy of *Fundulus heteroclitus* causes a reduction in the number of melanophores, and treatment with MSH results in an increase in the number of melanophores. A conflicting report suggests melanophore differentiation in xanthic goldfish is caused by ACTH and not by MSH, but this difference has not been resolved.

Class Amphibia

Melanophores

Direct control of amphibian melanophores is under endocrine regulation, and there is no evidence for innervation of amphibian melanophores with the possible exception of limited neural control in *Rana pipiens*. There is, however, evidence for neural control of MSH release. It was once claimed that there were two hormones controlling pigmentation in *Xenopus laevis* a B or darkening substance and a W or lightening substance. The B substance is MSH, and the existence of the W substance believed to be produced by the pars tuberalis has been disproved.

There have been reports of dual "innervation" of the pars intermedia in amphibians of a different type than in the cartilaginous fishes. Adrenergic fibers have been shown to penetrate the pars intermedia of *Rana temporaria* and *Bufo arenarum*, and two types of adrenergic fibers synapse with pars intermedia cells of *Rana pipiens*. No innervation of the pars intermedia is observed in the African clawed hog *Xenopus laevis*, however. Dual control of MSH release as suggested for some anurans may operate through the presence of alpha and beta adrenergic receptors in the plasmalemma of the pars intermedia cells. The catecholamines, dopamine, norepinephrine, and epinephrine, as well as alpha receptor agonists such as phenylephrine all inhibit MSH release from the pars intermedia. Isoproterenol stimulates MSH release by activating beta adrenergic receptors. Known antagonists of alpha receptors (for example, dibenamine and dihydroergotamine) block the inhibitory actions of the catecholamines. Finally, antagonists of beta receptors, such as propranolol, block the action of isoproterenol.

Certain cautions must be employed when one is attempting to interpret the actions of various drugs on MSH release from the pars intermedia. Epinephrine, for example, inhibits MSH release when applied at higher doses (10^{-5} M) but stimulates MSH release when applied at lower doses (10^{-6} or 10^{-7}). Apparently at higher doses epinephrine saturates both alpha and beta receptors, but the alpha effect predominates. At lower doses, epinephrine is more readily bound to the beta receptors so that the beta effect predominates, and MSH is released. The observation that a certain compound stimulates release of MSH does not necessarily imply a normal in-vivo role for that substance or for the type of innervation it might suggest.

Dopamine is the most potent inhibitor of MSH release, and several investigators have proposed that it is the hypothalamic MSH release-inhibiting hormone (MSHRIH) in amphibians as in mammals. The actions of the other catecholamines under experimental conditions may only reflect an influence on the normal mechanism whereby dopamine controls MSH release.

Other amphibian chromatophores

Pigment organelles of xanthophores are normally in an expanded state in most amphibians. In the tree frog *Hyla arenicolor*, however, the xanthophores are normally in an aggregated condition, and they may be dispersed by the application of MSH. An endogenous role for MSH on xanthophore pigment organelles has not been confirmed.

The aggregation of reflecting platelets in the iridophores of *Rana pipiens* skin is stimulated by cAMP or MSH. These iridophores possess alpha and beta adrenergic receptors, and stimulation of the alpha receptors produces dispersion of the reflecting platelets. These data suggest a possible role for catecholamines in iridophore regulation.

Class Reptilia

Reptiles exhibit a variety of mechanisms for regulating dermal pigment cells, including neural and endocrine mechanisms. Unlike the well-established pattern of adrenergic innervation in most amphibians, no innervation of the pars intermedia has been observed at either the light or electron microscopic level in several lizard species. Even the primitive tuatara of New Zealand, *Sphenodon punctatus*, exhibits no innervation of the pars intermedia. Innervation of pigment cells does occur in some reptiles, and pigmentary control of dermal melanophores may be neural, endocrine or both.

Colour changes in the true chameleon *Chameleo pumilis* are under direct neural control, and MSH has no effect on the skin chromatophores. Melanophores of *Chameleo jacksoni*, however, respond to both MSH and ACTH. The opposite extreme is found in the American chameleon *Anolis carolinensis*, in which there is no neural control over melanophore responses, and melanosome dispersion can be readily induced in vitro by application of MSH or cAMP to isolated pieces of *Anolis skin*. Pretreatment with alpha adrenergic-blocking agents obliterates the response of *Anolis* melanophores to MSH. Horned lizards, *Phrysonoma* spp., exhibit both neural and hormonal regulation of melanophores.

Neither xanthophores nor iridophores of *A. carolinensis* show any response to MSH. It is not known whether any non-melanin-containing chromatophores of other reptiles show any regulatory control.

Class Aves

Feather pigments (including melanin) are under the control of gonadal, thyroidal and gonadotropic hormones. The only reported action for MSH in birds is related to a developmental action. Embryonic implants of chicken pituitaries cause formation of black feathers where normally only white feathers would develop. This effect can be mimicked by treatment with either α-MSH or ACTH.

Class Mammalia

The first report for an action of MSH in mammals was the observation that injection of a highly purified MSH preparation into humans caused a darkening of skin and nevi (a nevus is a congenital mole). This action was due both to dispersion of melanosomes in melanophores and an increase in free melanin contained within surrounding keratinocytes.

Hypophysectomy of the short-tailed weasel results in regrowth of white hair characteristic of the winter coat even in animals possessing summer pelage (brown) at the time of hypophysectomy. Treatment of hypophysectomized weasels with MSH or ACTH causes regrowth of brown pelage. Ectopically transplanted pituitaries also result in production of brown pelage in hypophysectomized weasels. Similar actions of MSH on pelage of mice have been reported.

Alpha MSH cooperates with testosterone to stimulate lipogenesis and the synthesis of wax esters in the sebaceous glands of rats. This observation suggests that other roles for MSH should be sought in mammals in addition to actions on melanophores.

Influence of the Epiphysial Complex on Regulation of Chromatophores

The epiphysial complex consists of one or more dorsal evaginations from the epithalamic portion of the brain. These epithalamic structures include the pineal gland or pineal organ, the *epiphysis cerebri*. In

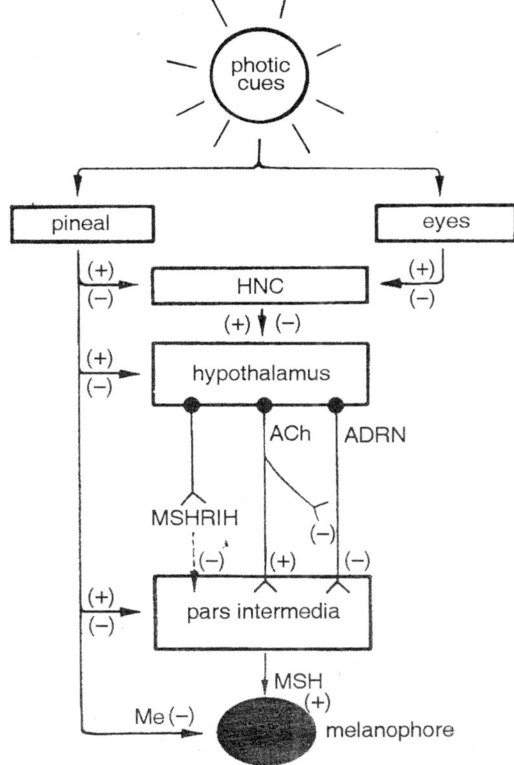

Fig. 4.6. Influence of photoreceptors (pineal, eyes) on release of MSH from the pars distalis. HNC, higher neural centers; Me, melatonin; ACh, acetylcholine; ADRN, adrenergic transmitters.

fishes and amphibians the epiphysial complex includes a photoreceptor as well as a glandular component. In amniotes there is a progressive tendency to reduce the photoreceptive function and emphasize the endocrine function of the pineal. Avian and mammalian pineal organs completely lack any photoreceptive structures.

Many studies have demonstrated an influence of the epiphysial complex on pigmentary responses. The blanching that had been observed in amphibian larvae maintained in the dark is due to melatonin produced in the pineal gland. Melatonin may inhibit MSH release or directly influence the distribution of melanosomes in the melanophores, causing the animals to lighten, or it may do both.

Fishes

There is little evidence to support a role for the pineal of fishes in pigmentary responses. Pinealectomy does abolish the diurnal rhythm of colour change in lampreys but the effects of melatonin and pineal implants are inconsistent in different species. No role for the pineal has been reported for members of the class Chondrichthyes. Some teleosts exhibit differential day and night colouration patterns, and melatonin can influence dispersion in some melanophores and aggregation in others to produce the typical "night pattern." In general, no physiological role for the pineal in colour changes has been established in teleosts.

Class Amphibia

Melatonin does cause blanching of dark-adapted frogs and is equally effective in hypophysectomized frogs treated with MSH. Pinealectomized larvae do not blanch when placed in the dark as do intact larvae, and it is concluded that the influences of light on pigment dispersion and aggregation in amphibian melanophores is mediated through the pineal gland.

Amniotes

There is no evidence to support a role for epithalamic structures as regulators of pigmentary responses in either reptiles or birds. In mammals, however, several examples of colour change have been investigated and a role for pineal influence established. The production of the white winter coat in the weasel *Mustela erminea bangsi* can be induced in the spring molt with melatonin treatment. This action of melatonin is believed to occur at the hypothalamus where it influences the output of hypothalamic MSHRIH. Normally the production of brown pigment in the hair of this mammal is dependent upon pituitary MSH.

Influence of Other Hormones on Pigmentation

Prolactin (PRL) preparations cause marked expansion of the xanthophores of the teleost *Gillichthyes mirabilis* and are the basis for the *Gillichthyes* xanthophore-expanding bioassay. In addition, prolonged treatment with mammalian PRL has been shown to cause darkening of the integument of fishes and amphibians. Expansion of the xanthophores of *Gillichthyes* can also be induced with MSH and ACTH, and some reports of pigmentary changes with crude preparations of PRL on melanophores may be attributable to contamination with either or both of these peptides. Purified tilapia PRL is ineffective in the *Gillichthyes* xanthophore-expanding bioassay, but the assay responds to PRL obtained from eels.

Gonadotropins and gonadal steroids have been reported to influence pigmentation in fishes, amphibians, birds and mammals. Follicle-stimulating hormone preparations induce integumentary darkening in tiger salamander larvae and in the American chameleon.

Thyroid hormones stimulate the deposition of guanine in the scales of teleosts and induce silvering in radiothyroidectomized and intact juvenile trout. The increased darkening observed in radiothyroidectomized trout and salmon may be due in part to stimulation of melanophores following their concentration of radioactive iodide used to destroy the thyroid tissue.

5

HORMONES OF PITUITARY

The adenohypophysial tropic hormones are separable into three distinct categories. The hormones within each category exhibit considerable overlap in chemical structures (that is, amino acid sequences) and in some cases overlap in biological activities as well. Category 1 includes the glycoprotein hormones: thyrotropin (TSH), follicle-stimulating hormone (FSH) and luteinizing hormone (LH). Each of these hormones is comprised of two polypeptide subunits, which each contain specific carbohydrate moieties. Growth hormone (GH) and prolactin (PRL) constitute the category 2 tropic hormones. Both PRL and GH are fairly large, single polypeptide chains, and they exhibit considerable structural and some functional overlap. Category 3 includes the smallest adenohypophysial peptides: corticotropin (ACTH), melanotropin (MSH), lipotropin (LPH) and the endorphins. These molecules are similar chemically, and there is overlap in some of their biological activities.

In addition to the three categories of pituitary tropic hormones, certain tropic hormones of similar chemical structure and biological activity are produced in the placental mammals. As many as five tropic-like hormones are produced by the chorionic (fetal) portion of the placenta, including *chorionic gonadotropin* (CG), which is LH-like in both structure and function, and *chorionic somatomammotropin* (CS), which has some GH but mostly PRL-like activity. Both a *chorionic thyrotropin* (CT) and a *chorionic corticotropin* (CC) have been isolated from human placentas and may occur in other mammals as well. Pregnant mares produce large quantities of a placental gonadotropin that has both FSH-like and LH-like properties. This glycoprotein hormone is termed *pregnant mare serum gonadotropin*.

An additional type of gonadotropin has been obtained from postmenopausal women, *menopausal gonadotropin* (MG). Human MG is basically FSH-like and is produced by the postmenopausal adenohypophysis in large amounts because of the failure of the ovaries to produce adequate levels of estrogens; that is, the normal negative feedback loop is absent. Menopausal gonadotropin has been employed in both clinical and experimental studies, but the advent of synthetic gonadotropins and synthetic gonadotropin-releasing hormone (GnRH) may soon make its use obsolete.

Initially the activities of the various tropic hormones were determined by bioassays. The most commonly employed bioassays are described below. These biological approaches are still used in the biochemical isolation and characterization of tropic hormones, especially in nonmammals. Once highly purified hormones became available, radioimmunoassays (RIA) were developed for several of the mammalian tropic hormones and are routinely employed to measure circulating levels. There are, however, a number of drawbacks to widely employing RIA for measurement of circulating hormone levels. Production of antibodies against hormones purified from pituitary glands may result in an antibody against a prohormone or some portion of the prohormone rather than against the circulating biologically active form. Use of such an antibody might yield results that do not correlate with biological bioassay data. The close similarities in structure among the various tropic hormones of a given category (for example, PRL, GH and CS all have gross structural similarities) may result in cross-reactivities to the antibody produced against only one hormone. The specificity of the antibody for the structure of the purified hormone antigen makes it difficult to use one antibody prepared against sheep FSH to estimate circulating levels of FSH in another species. For example, the endogenous gonadotropins of the sheep and the alligator may be sufficiently different in structure that it is impossible to know whether one is measuring FSH or LH activity or both when assessing the presence of some immunologically active substance in the blood. These problems of structural similarities make it absolutely essential that any RIA be validated against bioassay, especially if the species used for the antibody preparation is not phylogenetically close to the species in which it is being used.

Greater specificity in immunoassay of hormones may be provided by the newer radioreceptor assays. In these assays, hormone receptors are extracted from target tissues removed from animals of the same species. These receptors theoretically should bind preferentially the endogenous circulating hormone with a much higher affinity than for nonhomologous hormones (that is, hormones of other species). Extracted receptors are used in place of the antihormone antibody of the RIA. A radioreceptor assay can readily distinguish between GH and PRL because the receptors are extracted from tissues that will bind only one of the two hormones. Such procedures may have great potential for studies in species in which it is not yet practical to obtain purified hormones for developing a RIA. Because of technical problems, however, it may be some time before this procedure achieves its full potential.

Much of our initial knowledge concerning structures and functions of tropic hormones has come about as a result of the availability of pituitary glands and placentas from slaughtered domestic livestock. Since many differences may occur in the structure of polypeptides with similar but not identical biological activity, it has become increasingly important to designate the source of the hormone used in experimental studies. This is especially true when using mammalian hormones in nonmammals where two different molecules may function similarly in mammals but might provide different results in a nonmammal. Investigators who study mammalian tropic hormones usually designate the source of the hormone such as bovine (cow), ovine (sheep), porcine (pig), equine (horse). An additional lower case letter preceding the abbreviation of a tropic hormone designates the species source. For example, bovine growth hormone would be designated bGH, whereas growth hormone prepared from humans would be hGH. This system is useful only so long as there remain relatively few purified hormones.

Category 1 Tropic Hormones

All of the mammalian glycoprotein tropic hormones examined to date are composed of two peptide subunits with an assortment of carbohydrate moieties attached. Molecular weights for these glycoproteins are about 32,000 daltons. Each hormone consists of an *alpha subunit* that is identical in all three adenohypophysial glycoproteins as well as chorionic gonadotropin. The *beta subunit* is specific to each hormone and is responsible for its unique biological activity. Each subunit is synthesized as a separate pre-subunit, modified and eventually coupled. There is considerable overlap between beta subunits of LH and CG, which have very similar biological activity.

The carbohydrate components of the glycoprotein hormones also slow considerable specificity. Follicle-stimulating hormones contain larger quantities of sialic acid than do the others, and the sialic acid

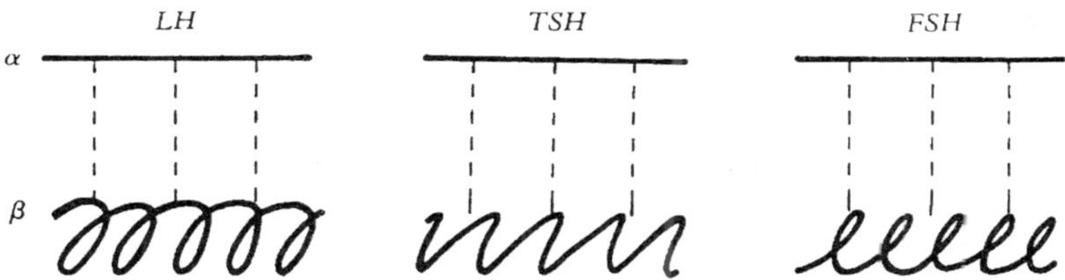

Fig. 5.1. Schematic representation of the structures for the three pituitary glycoprotein hormones.

is largely associated with the beta subunit. Sialic acid protects FSH from rapid degradation by the liver. Treatment of FSH with the enzyme *neuraminidase* selectively removes sialic acid and allows it to be degraded rapidly.

It is relatively easy to dissociate these glycoproteins into their respective subunits. These subunits have little if any biological activity by themselves when administered to mammals. It is possible to recombine the dissociated subunits and restore full biological activity. Any alpha subunit combined with any beta subunit results in a fully active hormone characteristic of the source of the beta subunit. Thus, an alpha subunit isolated from TSH when combined with a beta subunit from FSH yields a glycoprotein with FSH activity and not TSH activity.

Bioassays for FSH

In male mammals FSH typically stimulates spermatogenesis in the testis. In females FSH stimulates follicular growth in the ovary. Follicle-stimulating hormone may also stimulate steroidogenesis in the testis although there is little if any contribution to circulating levels.

The major bioassay for FSH prior to development of RIAs was the increase in testis or ovarian weight in hypophysectomized rats. Another specific bioassay for FSH is the maintenance of testis weight in male lizards *Anolls carolinensis* following hypophysectomy. The lizard testis is very unresponsive to LH, and this bioassay is very specific for small quantities of highly purified FSH.

Bioassays for LH

In contrast to FSH, LH stimulates circulating androgens and estrogens in males and females and causes ovulation and subsequent formation of the corpus luteum in females. Luteinizing hormone may also be responsible for progesterone synthesis and release by the corpus luteum. In males, LH causes release of mature spermatozoa from the testes (spermiation). Major actions for LH are the stimulation of steroid secretion and gamete release, whereas FSH is primarily involved with gamete preparation (follicle development in females and spermatogenesis in males). Since three pituitary tropic hormones are named for their actions in females (PRL, FSH and LH), it seems reasonable that at least one of these, LH, might be renamed for its action on the androgen-producing cells that occur between the seminiferous tubules of the testis rather than for inducing corpus luteum formation (luteinization) in the female. Hence it was suggested that LH be named the *interstitial cell-stimulating hormone* for its action on the steroidogenic *interstitial cell* (Leydig cell) of the testis. Equality of the sexes has not been realized, however, and the feminist name of LH has remained.

Numerous bioassays have been developed for quantitatively measuring LH activity. One of the first bioassays was the spermiation response observed following injection of pituitary extracts or purified LH into frogs or toads. At one time this bioassay was widely employed to detect the presence of CG in urine of women. The structural and functional similarity of the hCG and hLH to amphibian gonadotropins is the basis for this bioassay. A spermiation test, however, responds to highly purified FSH and cannot be considered a specific bioassay for LH-like hormones. Pregnancy is now determined from urine samples with a simple immunoassay involving an antibody produced against hCG.

A popular bioassay for LH is the *ovarian ascorbic acid depletion test* (OAAD), which is based upon a quantitative reduction of ascorbic acid in ovarian tissue following administration of LH. The degree of depletion is proportional to the dose of LH, and this bioassay is not affected by FSH. The relationship of ascorbic acid to hormone secretion is not understood, however. The pigment response to LH by feathers of the African weaver finch *Euplectes franciscanus* is also specific for LH or CG, but it is not employed routinely.

The specific response of amphibian ovaries in vitro to LH or CG (ovulation) has resulted in development of several similar bioassays. Administration of pituitary LH or progesterone is the most potent stimulator of ovulation in isolated fragments of anuran or urodele ovaries. Luteinizing hormone causes progesterone synthesis in follicular cells surrounding the oocyte, and it is progesterone that induces the ovulatory event. The system is responsive to some other steroids but is unresponsive to other tropic hormones.

Bioassays for TSH

Stimulation of thyroid gland function can be quantified cytologically by means of the TSH dose-dependent increase in epithelial height of the thyroid follicles in hypophysectomized animals. A more rapid technique involves measurement of the rate at which an injected dose of radioactive iodide (radioiodide) is accumulated by thyroid follicles. This bioassay can also be performed in vitro so that other variables can be eliminated. These parameters indicate only the degree of TSH-stimulation proportional to circulating TSH levels but provide no consistent information concerning rates of thyroid hormone synthesis and release from the stimulated thyroid gland.

The human has been shown to produce variant TSHs, one of which is associated with a pathological condition known as Graves' disease. Normal TSH has a biological half-life of about 0.25 hours, and maximum radioiodide uptake in thyroids of hypophysectomized rats is observed 4 hours after TSH administration. The abnormal TSH has a biological half-life of 7.5 hours and causes maximum radioiodide uptake 12 hours after administration. This so-called *long-acting thyroid-stimulator* is frequently employed in mammalian studies of thyroid function.

Category 2 Tropic Hormones

Two pituitary tropic hormones, GH and PRL, plus one placental tropic hormone, CS, comprise category 2. Prolactin and GH are large single polypeptide hormones of similar structure and molecular weight (about 23,000). Human CS is extremely similar to both hGH and PRL. There is an 85% homology between hGH and hCS as well as a 48% overlap with ovine PRL. Both GH and PRL produce a number of common effects on osmoregulation (renal function, intestinal fluid absorption), selective tissue growth (prostate gland, sebaceous gland), lactation and other processes. Ovine PRL and hCS have only weak effects on body growth, however. The human placenta also produces PRL which accumulates in amniotic fluid. This placental PRL is indistinguishable from pituitary PRL.

Growth Hormone

Growth hormone is often termed a protein anabolic hormone because it stimulates incorporation of amino acids into proteins. It is assayed biologically by means of the somewhat cumbersome measurement of the thickness of the epiphysial cartilage in the tibia of the hypophysectomized rat or mouse. Currently RIA techniques are employed routinely to measure blood and pituitary content of GH in mammals.

Growth hormone represents about one half of the total hormone content of the human adenohypophysis. It has been characterized as a protein composed of 191 amino acids (MW = 21,500) having a biological half life in blood of 20 to 40 minutes. Human GH has been synthesized in the laboratory. Crude GH preparations consist of a collection of protein isohormones. Each form is thought to have its own actions, and collectively they produce all the effects normally attributed to pituitary GH activity. The gene responsible for synthesis of the 21,500 dalton form of GH has been constructed and inserted successfully into the genome of mice. Growth rates of mice with the inserted genes and growth rates of their offspring are about twice that of normal mice.

Growth hormone secretion can be increased by rising levels of certain amino acids (for example, arginine), and GH in turn lowers blood levels of amino acids through stimulation of protein synthesis. Secretion of GH is also stimulated by low blood glucose levels and is inhibited by high ones. Growth hormone cooperates with insulin to channel utilization of these nutrients following a meal. Furthermore, GH becomes an important regulator of blood glucose and amino acid utilization during short-term and long-term starvation.

Circulating levels of hGH are highest during the period of maximum growth (ages 2 to 17 years). A daily secretory rhythm becomes established at about 4 years of age and continues throughout adult life. This pattern of GH secretion is both irregular and spontaneous, depending upon the physiological state of the individual, but episodes of GH release are frequently correlated with the onset of deep sleep.

Optimum growth-promoting actions of GH are obtained in hypophysectomized animals only when thyroid hormones are administered together with GH. This relationship between thyroid hormones and GHs has been described as a synergism, that is, the growth response elicited by combined therapy with thyroid hormones and GH is greater than predicted by adding together the responses obtained with each hormone administered alone. Either thyroid hormones or GH will reinitiate some growth in hypophysectomized animals, but complete resumption of normal growth requires combined therapy. Furthermore, animals that exhibit thyroid deficiencies grow slowly and abnormally.

Thyroid hormones may influence growth in two ways. They stimulate synthesis of GH in intact rats, but their action pertinent to the effect in hypophysectomized animals is a peripheral one. Thyroid hormones maintain a "responsive state" in the target cells so that they are more sensitive to GH. Androgens and estrogens to a lesser extent can increase the responsiveness of human tissues to hGH, but the mechanism is not understood.

Another growth-related action of GH is the liberation of small peptides (mol wt = 6000-9000) from circulating plasma proteins. These peptides were first called sulfation factors because of effects on incorporation of sulfate into cartilage. They are now known as *somatomedins*. Several peptides have been isolated that exhibit somatomedin activities. Two of these (somatomedins A and C) stimulate formation of cartilage, whereas another (somatomedin B) stimulates incorporation of ^{3}H-thymidine by glial cells and fibroblasts (that is, stimulates new DNA synthesis and cellular division). Two of these growth stimulators are structurally related to insulin and have some insulin activity in addition to their growth-promoting actions. They are known as insulinlike growth factors I and II (IGF I, IGF II). Thus, GH controls proliferation of cartilage that serves as the matrix for bone formation via release of plasma-bound somatomedin. A bioassay for somatomedin has been developed using costal cartilage from pigs in vitro. Somatomedin activity has been demonstrated in plasma from three primates, including man, as well as from domestic pigs, cows, sheep, goats, horses, donkeys, dogs, rabbits, guinea pigs and rats.

Prolactin

Prolactin also occurs in multiple isohormones and has been shown to produce a variety of distinctive actions in animals, including effects associated with reproduction, growth, osmoregulation and the integument. In addition, PRL may produce synergistic actions with ovarian, testicular, thyroid and adrenal hormones. The best-known action for PRL is the lactogenic effect on the mammary gland of female animals for which the hormone was named. Prolactin stimulates the synthesis of milk proteins by the mammary gland. A similar effect is produced by CS from the placenta. In some species PRL may also influence the synthesis of progesterone by the corpus luteum. There is evidence in male mammals for effects of PRL on certain sex accessory structures.

Prolactin bioassays

Several specific bioassays for PRL have been reported utilizing animals representing four major vertebrate classes (teleosts, amphibians, birds and mammals):

1. The sodium-retaining bioassay in the cichlid teleost, tilapia, *Sarotherodon (Tilapia) mossambicus*.

2. The xanthophore-expanding bioassay in the teleost *Gillichthyes mirabilis*.
3. The red-eft water drive performed in the newt *Notophthalmus viridescens*.
4. The crop-sac assay performed in the domestic pigeon.
5. The in-vitro mouse mammary gland assay.

The extremely similar structures of GHs and PRLs within and among species makes specific RIA for either PRL or GH difficult, and extreme caution should be used when interpreting RIA data, especially the use of mammalian PRL antibodies in nonmammals. This point will be emphasized by the discussions of these hormones in nonmammals that follow.

Sodium-retaining bioassay in teleosts

This bioassay stresses the osmoregulatory actions of PRL in teleostean fishes, and particularly its effect on sodium uptake across the gill and resultant alterations in plasma sodium levels. Only some pituitaries from teleostean species, however, will produce a measurable response in this bioassay, and all other piscine and other nonmammalian preparations are inactive. Purified mammalian PRLs, curiously, work very well in this bioassay. A similar bioassay was developed prior to this one that employed hypophysectomized guppies, *Poecilia latipinna*. The intact tilapia bioassay is simpler because of the larger size of the assay animal and because hypophysectomy is unnecessary. Adaptation of these fishes to seawater almost eliminates PRL from the circulation.

Xanthophore-expanding bioassay in Gillichthyes

This bioassay is performed in the gobiid fish *Gillichthyes mirabilis*, the long-jawed mud sucker. It is an all-or-none bioassay for determination of the smallest dose of purified hormone or the greatest dilution of pituitary homogenate or extract that will cause local yellowing in 50% of the fish tested following injection of the test material beneath the preopercular skin. The bioassay does have the distinct advantage, however, of being extremely sensitive to piscine (including agnathan) and amphibian pituitary homogenates. For example, teleostean pituitaries are more than 100,000 times more effective in the assay than is ovine PRL that is used as a standard. As little as 1/100th of a tiger salamander pituitary gives a positive response in the xanthophore-expanding bioassay, whereas an entire pituitary of the same size is required to produce a minimal response in the pigeon crop-sac bioassay. The specificity of this bioassay is questioned by the failure of purified tilapia (teleost) PRL to produce a typical response, suggesting that the bioassay may be measuring some contaminant rather than PRL.

Red-eft water-drive bioassay

Prolactin induces a second metamorphosis or "water drive" in newts that is characterized by migration of the juvenile terrestrial form, the eft, back to water where breeding will occur. Water-drive behaviour is accompanied by a series of physiological and morphological changes as well, but these are not related to the bioassay per se. Following injection of PRL the efts will leave the dry areas of their laboratory containers and submerge themselves in water. No other hormone has been found to induce water-drive behaviour. This bioassay has the disadvantage of being an all-or-none response. Pituitary preparations for all classes of vertebrates except the class Agnatha, possess water-drive-inducing activity.

Pigeon crop-sac bioassay

This bioassay has the advantage over the previously described assays of being a quantitative, dose-related bioassay. Consequently a standard curve may be prepared and relative activities of unknown preparations quantitatively assessed. The disadvantages of the crop-sac bioassay are the time involved in performing it and the fact that most piscine PRLs will produce only an atypical response or no response at all. Positive results have been obtained with lungfish pituitaries as well as pituitaries from all tetrapod groups.

Mammary gland in-vitro bioassay

Mammary gland explants from pseudopregnant or midpregnant mice (or rabbits) are cultured in a precise medium that includes some other hormones. The addition of PRL or pituitary homogenates to the culture medium causes cytological changes that can be quantified according to a numerical index. These cytological changes are correlated with the ability of PRL to stimulate milk synthesis. Mammotropic (lactogenic) activity is present in the pituitaries of all tetrapod species, but piscine pituitary preparations yield minimal responses, if any. Like the other bioassays the mammary gland response is not a rapid assessment.

An electrophoretic assay

An assay for PRLs has been developed involving electrophoresis of hormone-containing extracts on polyacrylamide gels followed by densitometric quantification. Prolactin molecules migrate as a distinguishable band of protein in the gel, and the position of the PRL band in a gel prepared from an "unknown" source can be compared to purified PRL in another gel. Migration to identical positions is evidence of the presence of PRL in the unknown. The protein bands in the gel may be stained with a general protein stain (for example, amido Schwartz). The intensity of staining, which is proportional to the quantity of PRL in the band, can be determined with the aid of a densitometer. Growth hormone can be readily distinguished from PRL because the latter migrates through the gel well ahead of GH. This technique is utilized to prepare relatively pure preparations of tropic hormones as well. Initially, however, it was essential to locate and quantitate PRL activity and GH activity by bioassaying the banded proteins.

Category 3 Tropic Hormones

Corticotropin and lipotropin are associated with the pars distalis components of the adenohypophysis, whereas MSH is found primarily in the pars intermedia. All of these hormones are similar in structure and overlap considerably with respect to biological activities.

Corticotropin

Mammalian ACTH has been purified from several mammalian sources (bovine, porcine, ovine, human). It has 39 amino acids in a single peptide chain with a molecular weight of about 4500. Amino acids 1-23 of ACTH have full biological activity, 1-19 have 80% of full activity, but fragment 1-16 has very little biological activity. Amino acids 24-39 are obviously outside of that region of the molecule responsible for its biological activity. A synthetic peptide identical to amino acids 1-24 has been prepared; it is known as *beta corticotropin*.

The classical bioassay for ACTH has been the *adrenal ascorbic acid depletion* test (AAAD), which is similar to the OAAD bioassay for LH. The lowest effective dose of purified ACTH in this bioassay is 0.2 mU or about 1.2×10^{-9}g (about 1 ng). The adrenal cortex secretes certain corticosteroids in response to ACTH stimulation, but the relationship of ascorbic acid to steroid secretion is not understood. The AAAD test is rather cumbersome, however, and it has generally been replaced by measurement of either circulating ACTH or corticosteroids with RIA.

Melanotropin

The name of this tropic hormone stems actually from its actions in amphibians and reptiles where it causes dispersion of melanin pigment granules contained within a specialized cell, the melanophore, located in the skin. Dispersal of these melanin granules causes the animal to appear darker. Consequently the standard bioassays for MSH have been developed in amphibians and reptiles utilizing their ability to adapt to dark or light backgrounds. One bioassay involves measurement of the quantity of light reflected (using a reflectometer) from an isolated piece of skin removed from a light-adapted frog. A similar bioassay employs isolated pieces of skin from the light-background-adapted *Anolis carolinensis*, the American chameleon. A rapid all-or-none response (darkening) occurs in vitro.

The adenohypophyses of all vertebrates tested possess MSH activity, including birds that lack a pars intermedia. However, physiological roles for MSH have not been shown in birds or in most mammals. A physiological role has not been established for MSH in fishes, but their pigmentary changes apparently are under dual sympathetic-parasympathetic innervation, allowing for rapid responses.

Three mammalian MSH molecules have been identified, and a given mammal typically possesses two of these. *Alpha MSH* consists of only 13 amino acids, and *beta MSH* has about 22. Beta-Ser-MSH has been isolated from sheep.

Lipotropin

Lipotropin is believed to be an adenohypophysial hormone that stimulates lipolysis (that is, hydrolysis of fats to free fatty acids and glycerol) in adipose tissue. The standard bioassay for LPH is to culture mouse or rabbit epididymal (testis) fat pads and measure the release of glycerol or free fatty acids or both into the culture medium. Another technique employs measurement of the inhibition of incorporation of ^{14}C-labeled acetate into lipid following addition of LPH to the culture medium. This is, in effect, a measurement of lipogenesis, which is inversely related to lipolysis.

Two different LPHs have been characterized. The first peptide consists of 58 amino acids and has been termed gamma LPH. The larger peptide is known as beta LPH and consists of the 58 amino acids as gamma LPH plus 33 additional ones. Lipotropin, presumably of pituitary origin, has been identified in the systemic circulation, and its release can be stimulated by arginine vasopressin. Aldosterone release may be stimulated by LPH, but circulating LPH has not been linked to changes in lipid metabolism, leaving open the question of the physiological role for LPH.

Endorphins and Enkephalins

A search for endogenous compounds that produce analgesic opiate-like (morphine) effects on the central nervous system has resulted in identification and chemical characterization of two groups of biologically active peptides. The four larger peptides are known as the *endorphins*, so named because they are endogenous and produce morphine-like effects. Two smaller peptides with opioid activity were identified by other workers as pentapeptides. These latter opioids were named *enkephalins*, meaning "in the head." It was soon observed that the endorphins have a chemical structure identical to a large fragment of ovine beta LPH and that one of the enkephalins is identical to residues 61-65 of LPH. The other enkephalin is identical except for the substitution of leucine for methionine at position 65. Hence these enkephalins are known as Metenkephalin and Leu-enkephalin respectively. This same sequence of five amino acids in Metenkephalin is found in the amine terminus of all four endorphin molecules, suggesting that its presence in brain extracts may be an artifact of partial hydrolysis of an endorphin molecule. The distribution of endorphins in the central nervous system parallels that observed for ACTH and LPH. The enkephalins, however, are localized in different neurons, suggesting that their structural relationship to LPH and the endorphins is coincidence and not an artifact of preparation. Another potent endorphin, called *dynorphin*, has been isolated. Dynorphin contains a peptide sequence identical to Leu-enkephalin in its amino terminal end.

The endorphins have also been localized in cells of the pars intermedia and the pars distalis that react positively to antibodies prepared against both ACTH and LPH. The distribution of these same compounds as well as of the enkephalins within the central nervous system is not affected by hypophysectomy, and it is assumed that the pituitary is not the source for these neural peptides. Endorphins as well as ACTH and LPH have been identified in both cerebrospinal fluid and in the blood. Factors such as arginine vasopressin that increase ACTH and beta-LPH blood levels do not enhance endorphin release, however.

Painful stimuli elevate levels of endorphins and enkephalins in the cerebrospinal fluid, and they appear to exhibit the features required for endogenous opiate-like agents. There is little doubt that these compounds probably function as neurotransmitters within the central nervous system related to their morphine-like actions. The action of morphine, a non-peptide, is blocked by closely related molecules such as *naloxone*. The effects of endorphins also are blocked by naloxone implying closeness in mechanisms of action for morphine and the endorphins. These endogenous opioids can influence release of GH and PRL, an effect also induced by morphine. Endogenous opioids may be involved in stress responses by mediating stress-induced eating. The possible roles for endorphins and enkephalins as regulators of behaviour, especially as related to painful stimuli, and their possible endocrine implications represent one of the most exciting areas of neuroendocrinology to appear since the discovery of the first hypothalamo-hypophysiotropic hormone.

Most of the category 3 peptides are produced by post-translational processing of a large prohormone known as *pro-opiomelanocortin*, POMC. Dynorphin and the enkephalins apparently have separate prohormones. The pattern of the subsequent cleavage of POMC depends on the type of cell. For example, in corticotropes the cleavage fragments include a 16K fragment, ACTH and LPH. The 16K fragment includes the sequence of γ-MSH which probably is released only as a consequence of extraction procedures used to isolate the peptides. In melanotropes, ACTH is cleaved further to produce α-MSH and the fragment of ACTH known as CLIP (corticotropin-like peptide).

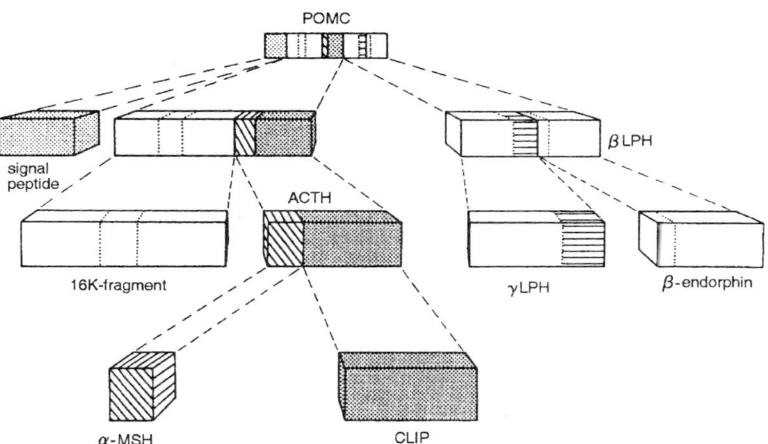

Fig. 5.2. Pro-opiomelanocortin (POMC) is a preprohormone from which various cells can produce biologically active peptides by enzymatically cleaving the parent molecule at predetermined sites.

Functional Overlap of Category 3 Hormones

There is a common heptapeptide core (seven amino acids) in α-MSH, ACTH and LPH. This similarity in structure is reflected clearly by overlapping biological activity. Corticotropin has considerable MSH activity and some lipolytic activity. Lipotropin has rather low ACTH activity but strong MSH action. Melanotropin has both weak ACTH and LPH activity. These overlaps in function affect interpretation of bioassayable data on MSH activity. Which peptide is the bioassay measuring, MSH or ACTH?

Comparative Aspects of Tropic Hormones in Nonmammalian Vertebrates

Gonadotropins

Some purified GTHs have been isolated from nonmammalian species, and they have been partially characterized. Although more GTHs are being isolated each year, no clear phylogenetic pattern has yet emerged.

Mammalian GTHs have been examined extensively for activity in lower vertebrates, and many recent reviews of this work are available for fishes, amphibians, reptiles and birds. Generally the hormonal control of reproduction involves two GTHs with LH-like and FSH-like activities. The apparent failure to find two GTHs in some vertebrate groups (squamate reptiles, teleostean fishes and some birds) may be due to derived secondary reliance on only one GTH with repression of synthesis of the other. Specific activity for both mammalian FSH and LH may be related to biological half-life differences between the two hormones when they are examined in vivo and emphasizes the need for good in-vitro bioassays in a variety of vertebrates. The question concerning the number of GTHs in all groups must be left incompletely answered at this time, however.

It has been proposed that the primitive glycoprotein hormone was an LH-like molecule that became modified into a TSH-like hormone. Follicle-stimulating hormone presumably diverged later from TSH. This scheme was suggested from studies of purified subunit structure. Part of this proposal resides on the presence of only an LH-like hormone in teleosts. Certain contradictory data must be resolved, however, such as the belief that both FSH and LH are produced by the same cell but TSH is produced by another cellular type.

Fishes

Class Chondrichthyes

Gonadotropin activity is present in both the proximal pars distalis and the ventral lobe of selachians. Antibody to mammalian GTHs binds to cells of the ventral lobe. Bioassay of the proximal pars distalis reveals the presence of an LH-like GTH that will stimulate oocyte maturation in the clawed frog *Xenopus laevis*.

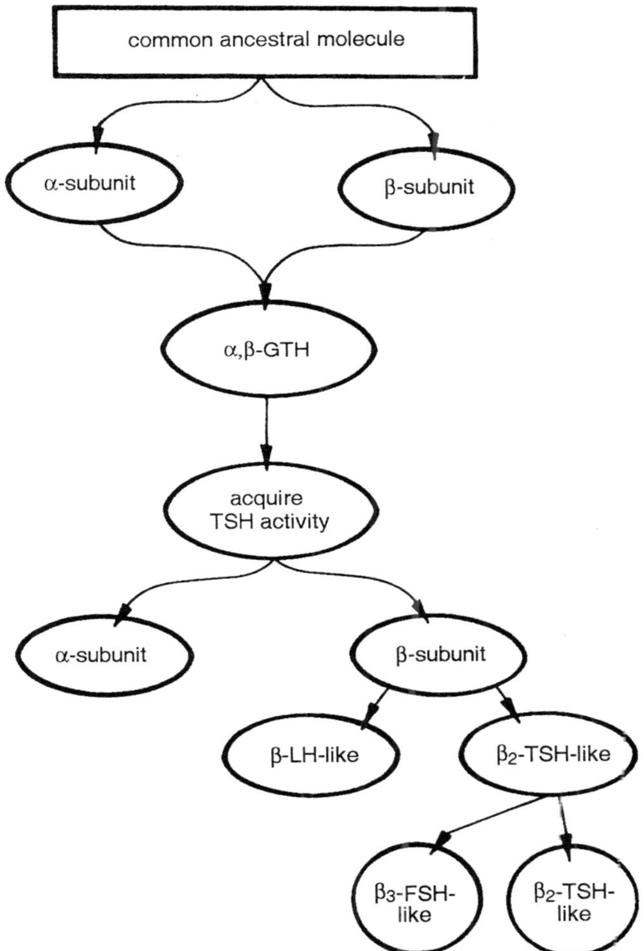

Fig. 5.3. Proposed scheme for glycoprotein hormone evolution based upon analysis of subunit peptide sequences for teleostean and mammalian gonadotropins.

Class Osteichthyes

The first observation suggesting a single LH-like GTH in teleosts was that mammalian FSH appeared to be inactive in teleosts and that mammalian LH could support the entire reproductive process. Gonadotropins purified from salmon, carp and rainbow trout have high LH-like activity when tested with the frog spermiation bioassay, as do pituitary extracts prepared from lungfishes. The spermiation reaction, however, is not as specific for LH as was previously supposed and will respond to highly purified FSH. Sialic acid is not reported to be essential for the biological activity of teleostean GTH, but salmon GTH lost much of its activity following removal of sialic acid prior to its assay in the lizard testicular bioassay. Crude extracts of salmon pituitaries, however, had no effect on chicken Sertoli cells, which are highly sensitive to FSH. The observation that antibody prepared against mammalian LH was bound by salmon gonadotropic basophils but not antibody against mammalian FSH lends further support to the contention that the gonadotropin of teleosts and possibly of lungfishes is LH-like.

Further extraction of certain teleostean GTH preparations yields a nonglycoprotein fraction that stimulates uptake from the blood of protein precursors utilized in yolk synthesis by growing oocytes. The official status of this "nonglycoprotein-gonadotropin" awaits further study.

Highly purified LH has been reported to have intrinsic TSH activity in teleosts, supporting the suggestion that an LH-like hormone was the primitive hormone that later gave rise to TSH.

Class Amphibia

Two distinct GTHs have been isolated and purified from the bullfrog, the tiger salamander *Ambystoma tigrinum*, and the leopard frog *Rana pipiens*. Purified bullfrog LH and FSH both stimulate spermatogenesis and spermiation in bullfrogs, but only bullfrog LH elevated plasma levels. *Ambystoma* LH was not as effective as bullfrog LH in raising plasma androgen levels in bullfrogs. Conversely, *Ambystoma* LH was more effective than bullfrog LH in *Ambystoma*, although the latter was considerably more effective than *Ambystoma* FSH. Bullfrog LH is more effective than ovine LH in stimulating progesterone synthesis in amphibians. These data strongly support the existence of two structurally and functionally separate GTHs in amphibians. Bullfrog LH also stimulates reptilian and avian thyroids emphasizing the closeness of LH and TSH.

Class Reptilia

Studies employing injections of mammalian hormones into squamate reptiles have emphasized a role for FSH-like GTHs but suggest no role for LH-like hormones. Earlier data might be interpreted as a consequence of a very short biological half-life for injected mammalian LH or that mammalian FSH might be similar enough to both a squamate FSH and LH to possess both activities. Purification of gonadotropic activity from squamate reptiles, however, has shown only one FSH-like molecule, although both FSH-like and LH-like GTHs have been isolated from chelonians and crocodilians. Presence of only one molecular species of GTH in squamates does not rule out the possibility that this molecule has intrinsic FSH-like and LH-like activities. Recent studies suggest snake GTH may not be homologous to either FSH or LH. Since both GTHs are found in other reptiles the squamate condition is probably secondarily derived and does not represent a primitive condition.

There is great variability in the responses of reptilian tissues to GTHs purified from different vertebrate groups. Testicular weight maintenance in hypophysectomized *Anolis carolinensis* is readily maintained by FSH-like hormones purified from mammals, birds, reptiles or amphibians, but this testicular parameter is very insensitive to LH-like gonadotropins. Nevertheless, plasma androgen levels in hypophysectomized *A. carolinensis* may be elevated by either FSH-like or LH-like GTHs from almost any source. Similar effects on plasma levels of androgens or on androgen synthesis by minced testes in vitro have been observed for crocodilians. The situation in chelonians is not clear as testes of some species do not respond to LH-like hormones in vitro, whereas at least one species (*Chrysemys picta*) does respond to in-vivo injections of mammalian LH but not as well as to FSH. This observation may be related to contamination of the LH preparation with FSH, or it could be due to species differences. Purified salmon GTH, which is LH-like in most bioassays, had no effect on circulating androgen levels in *C. picta*, however, supporting the possibility of contamination of the mammalian LH used.

The reptilian ovary, like the testis, exhibits broad sensitivity to FSH and LH. Follicle-stimulating hormones are generally the most effective, although there are some conflicting reports. In *C. picta* both partially purified ovine LH and chicken LH stimulated synthesis of estrogens and progesterone by preovulatory and postovulatory follicles respectively. Only minimal response to ovine or chicken LH was observed. In contrast, other studies have found that turtle ovaries respond best to FSH-like GTHs but that highly purified oLH is ineffective.

Obviously the story of the reptilian GTHs is incomplete. Although the squamates appear to possess only one FSH-like GTH, crocodilians and chelonians clearly possess two. The apparent lack of specificity by reptilian tissues for nonhomologous GTHs emphasizes the importance of working with purified GTHs from the species being studied or from very closely related species.

Class Aves

Birds are believed to conform to the typical mammalian pattern of two GTHs, LH and FSH, although avian species are rather insensitive to mammalian FSH and LH. Two separate GTHs have been extracted and partially purified from domestic galliform birds. Preliminary studies of the duck support existence of only one GTH. More species need to be examined.

Chicken LH stimulates interstitial cells in the testis of chickens or Japanese quail, and the seminiferous tubules are stimulated by chicken FSH. Mammalian LH and avian LH are reported to be more effective than FSHs in stimulating androgen production by chicken interstitial cells in vitro. Some conflicting reports suggest that avian FSH and LH are more nearly equal in their actions on minced pigeon or chicken testes, however. Reptilian LH is more effective than reptilian FSH in stimulating the avian testis. Purified amphibian GTHs (both LH-like and FSH-like) are equally effective, although their overall activity in birds is low. Purification of additional avian GTHs and development of specific RIAs for avian GTHs will provide more precise answers about the involvement of GTHs in avian reproduction.

Thyrotropins

Mammalian TSHs have been shown to stimulate thyroid function in all vertebrates examined with the usual exception of the hagfishes. Similarly pituitaries from most nonmammals exhibit TSH-like activity when tested in mammals.

Although little is known about the structures of nonmammalian TSHs, it is apparent that there is great similarity to those of mammals. Purified bullfrog TSH is thyrotropic in both anurans and urodeles but is ineffective on thyroids of reptiles and birds. Curiously, bullfrog LH exhibits thyrotropic activity in reptiles and birds. Purified LHs of tetrapods have also been shown to exert a TSH-like action in teleosts. This overlap in function has been used as support for the origin of TSH from an LH-like primitive glycoprotein hormone.

Growth Hormones and Prolactins

Prolactin

Prolactin has so many different actions in vertebrates that it is almost impossible to describe them all. Indeed, people have suggested a variety of names for this hormone that might be more descriptive, such as versitilin, ubiquitin, panaceanin and miscellanin. Many of the reported actions of PRL, however, may be grouped into a few general types: 52% of the reported actions are related to growth phenomena; 38% are related to reproduction; 25% involve water and electrolyte balance; 26% are concerned with integumentary structures; and 35% concern actions synergistic with other hormones. The most primitive role for PRL may be related to osmotic-ionic balance, with the other actions being acquired during vertebrate evolution. Certainly the major role for PRL in teleostean fishes is related to osmotic regulation in fresh water, and even in man PRL release can be induced following alterations in blood osmotic pressure. Regardless of the multiplicity of roles or even the establishment that it may be an osmoregulatory hormone, the term prolactin is here to stay.

Prolactin provides an excellent example for the evolution of endocrine systems from a variety of viewpoints. There have been evolutionary changes in the structure of the PRL molecule as evidenced by the failure of fish PRLs to work in mammalian bioassays, although mammalian PRLs retain piscine activity. The antigenic portion of the piscine PRLs is similar to mammals as evidenced by the binding of mammalian PRL antibodies to the lactotropes of the piscine rostral pars distalis. Furthermore, new target tissues have evolved (for example, the crop sac of birds and the mammary glands of mammals) that possess receptors specific for the "newer" portions of the molecules. This is evidenced by the failure of the piscine PRLs to activate responses in avian and mammalian tissues. It is unfortunate that there are no bioassays employing reptilian tissues and that purified reptilian PRLs are not available to fill in this part of the evolutionary changes that appear to be supported by studies with the bioassays of the other major groups.

Class Agnatha

Immunochemical studies suggest that the pituitaries of hagfish contain a molecule that binds antibody prepared against mammalian PRL, but only negative bioassays have been reported. Some cyclostome pituitaries, however, do contain bioassayable PRL, based on responses observed for lamprey pituitaries in the xanthophore-expanding *Gillichthyes* bioassay.

Class Chondrichthyes

The xanthophores of *Gillichthyes* expand following application of pituitary extracts from selachians, indicating the possible presence of PRL activity in the rostral pars distalis. No bioassays of holocephalan pituitaries have been reported.

Class Osteichthyes

As mentioned previously, PRL plays an important role in osmotic regulation in freshwater teleosts. Teleostean PRLs have little biological activity in amniote bioassays, but they work well in amphibians and teleosts. Purified PRLs have been prepared from five species of teleosts, and all chemically resemble mammalian GH more closely than mammalian PRL. However, both highly purified tilapia GH and PRL can stimulate synthesis of milk proteins in cultured rabbit mammary gland cells. Some investigators have proposed that since the structures of teleostean and other piscine PRLs are different from those of amphibians and the amniotes, they should be given a separate name. Thus, *paralactin* is sometimes used to refer to the piscine PRL-like substance in the rostral portion of the pars distalis.

Class Amphibia

Prolactins from most amphibians cross-react with antibody to rat growth hormone, and PRL is thought by some to be the larval GH of amphibians. Tiger salamanders were the only animals tested whose PRL did not cross-react with rat GH antibody. In addition to possible influences on larval growth, PRL is antimetamorphic in both anurans and urodeles, that is, it blocks metamorphosis from the aquatic larva to the semiterrestrial or terrestrial juvenile form. Prolactin induces water-drive behaviour in newts and possibly in salamanders and influences secondary sexual characters associated with breeding. Integumentary effects of PRL have been observed on the skin of newts and salamanders. Prolactin also affects water balance in anurans and urodeles as it does in fishes.

Class Reptilia

The reptiles are the only major vertebrate groups in which a good bioassay has not been developed, and little work has been done with PRL in reptiles. Prolactin stimulates growth in juvenile snapping turtles *Chelydra serpentina* and lizards *Lacerta s. sicula*. Appetite is also stimulated in *Lacerta* by PRL. A possible effect on water balance has been reported in turtles in which PRL influenced glomerular filtration. Reptilian PRLs give positive responses in other vertebrates, and this PRL activity appears to be confined to the cephalic (rostral) lobe of the pars distalis. Considerable research remains to be done with respect to PRL in reptiles.

Class Aves

Prolactin plays several essential roles in avian reproduction. In some cases PRL synergizes with other hormones, as in formation of the broodpatch for incubating eggs, fat deposition and induction of migratory restlessness. This last behaviour appears just prior to the actual migration of birds.

Prolactin has been purified from chickens. Chicken PRL is similar in amino acid composition and electrophoretic properties to mammalian PRLs and appears to be distinct from chicken GH.

Growth hormone

Compared to PRL, there are few comparative data on structure and function of vertebrate growth hormones. Mammalian GHs are effective in most nonmammals, and most nonmammalian preparations exhibit GH activity in mammals. Immunological studies of GHs prepared from different vertebrates have not clarified any relationships. Comparative studies are hampered by the similarities of GH and PRL, especially in the amphibians in which prolactin seems to be the larval growth factor.

Class Chondrichthyes

Growth hormone activity can be demonstrated in the selachian pars distalis with the rodent tibia bioassay. Somatomedin activity has also been demonstrated for two sharks, *Squalus acanthias* and *Mustelus canis*, although no definite link to GH has been established. No data on holocephalans are available.

Class Osteichthyes

Polypterid, chondrostean, holostean and teleostean pituitaries exhibit GH activity in the tibia bioassay. Furthermore, by means of immunochemical techniques employing anti-rat GH antibody, GH activity has been demonstrated in the pituitaries of chondrostean, holostean and teleostean fishes. Mammalian GHs are very active in promoting growth of fishes. Growth hormone isolated from tilapia is structurally unlike other GHs, but GH from sturgeon is mammalianlike.

There is a marked influence of thyroid hormones on growth in teleosts. Thyroid hormones accelerate growth in fishes and thyroid hormone treatment restores growth rates to normal in radio-thyroidectomized rainbow trout. Although it has not been conclusively demonstrated it is reasonable to assume a synergistic relationship of thyroid hormones and GHs similar to that reported for mammals. Androgens also produce a positive effect on growth of salmonid fishes, suggesting a protein anabolic action for these steroids that is probably independent of the action of GH.

The presence of somatomedin activity is reported in two teleosts examined by means of the porcine costal cartilage bioassay. No relationship between somatomedin and GH has been indicated, however.

Class Amphibia

The amphibians are complicated by the observation that PRL may be a larval growth hormone, whereas GH per se may operate only in transformed or metamorphosed individuals. Both GH and PRL have been isolated from frogs, *Rana catesbeiana*, and *Rana pipiens*. Purified frog GHs are not as effective as bovine GH in the rat tibia bioassay, however. The large number of similarities in the physical properties between amphibian and mammalian GHs and the similarities in amino acid composition suggest that there has been considerable conservatism expressed in the evolution of tetrapod GHs.

Cross-reaction occurs between antibody prepared against rat GH and PRLs prepared from most amphibian species, and only the tiger salamander of those species examined appears to produce GH distinctly different from mammalian PRL. Nevertheless, antibody prepared against bullfrog PRL does not cross-react immunologically with bullfrog GH. This observation supports the interpretation that inability to distinguish a distinct tropic hormone, such as PRL, in a nonmammal with antibody prepared against some mammalian tropic hormone, such as GH, does not mean the endogenous hormones (PRL and GH) are not separate and distinct hormones.

Somatomedin activity has been demonstrated in *Xenopus laevis*, but no link to either PRL or GH has been proposed.

Class Reptilia

Mammalian GH, like PRL, stimulated growth in juvenile snapping turtles and in the lizard *Lacerta*, although the sites of action for GH and PRL in the lizard appear to be different. Growth hormone stimulated appetite in *Lacerta*, as reported for PRL, and also produced marked increase in growth of the digestive tract (splanchnomegaly), an action reported for PRL in other vertebrate groups.

Purified GH has been prepared from the caudal lobe of adult snapping turtle pituitaries and sea turtles. Many characteristics of these GHs are similar to those of other tetrapods. Turtle GHs are very effective in the rat tibia bioassay.

Class Aves

Growth hormone activity has been demonstrated in pituitaries of chickens and turkeys by means of the mouse tibia bioassay. Antibovine

GH antibody apparently does not cross-react with chicken pituitary extracts. Purified GH from duck pituitaries does cross-react with rat GH antibody, however, and the apparent failure to observe cross-reactivity between chicken and bovine GH should be reexamined.

Corticotropins, Lipotropin and Melanotropin

Little comparative literature about the structures of ACTH and MSH exist for nonmammalian hormones. Peptides appear to be present in the pituitaries of all vertebrates (except hagfishes) that possess ACTH-like and MSH-like activities. Purified MSH has been prepared from the pituitary of the dogfish shark, but this is the only case of verified structure for nonmammalian category 3 tropic hormones.

The functional roles for mammalian ACTH and homologous pituitary extracts or homogenates in various vertebrate. Adrenal production of corticosteroids appears to be under control of a corticotropic factor in nonmammals. Corticotropins prepared from mammals are sometimes used to investigate melanophore responses in nonmammals, especially amphibians, since it has strong MSH activity.

Opiate activity (endorphins?) has been demonstrated in both the neurointermediate lobe and pars distalis of rainbow trout in amounts comparable to the pars intermedia of guinea pigs. Presumably data for other nonmammals will be available soon.

6

NEUROHORMONES

In this chapter the specific octapeptide neurohormones stored in the pars nervosa of the major vertebrate groups will be considered.

BIOLOGICAL ACTIVITIES ASSOCIATED WITH THE MAMMALIAN PARS NERVOSA

Five different biological activities have been associated traditionally with the pars nervosa of the neurohypophysis including a *pressor* effect (increased blood pressure), an *antidiuretic* action (reduced urine volume), *uterotonic* effect (induction of uterine contractions), a *milk-ejection* effect (ejection of milk from mammary glands) and a *depressor* effect (reduction in blood pressure). Diuresis or diuretic effect (increased urine production) has been reported but is believed to be simply a consequence of the pressor effect on renal vessels.

The pressor effect my be more primitive than the antidiuretic action, occurring primarily on the peripheral vasculature. Later the emphasis of octapeptide action shifted to the preglomerular renal vasculature allowing for reduced filtration and thus antidiuresis. Finally, the mechanisms for tubular reabsorption of water evolved and became the predominant antidiuretic mechanism of mammals. The uterotonic effect first appears among the chondrichthyian fishes and has been demonstrated in all gnathostomes. Milk-ejecting activity is possibly a modification of the uterotonic action acquired after the evolution of the mammary gland.

The name *pituitrin* was applied to the first crude preparation from the mammalian pars nervosa that contained these activities. Soon this preparation was separated into *pitocin*, which possessed uterotonic, milk-ejecting, and depressor activity, and *pitressin*, which contained

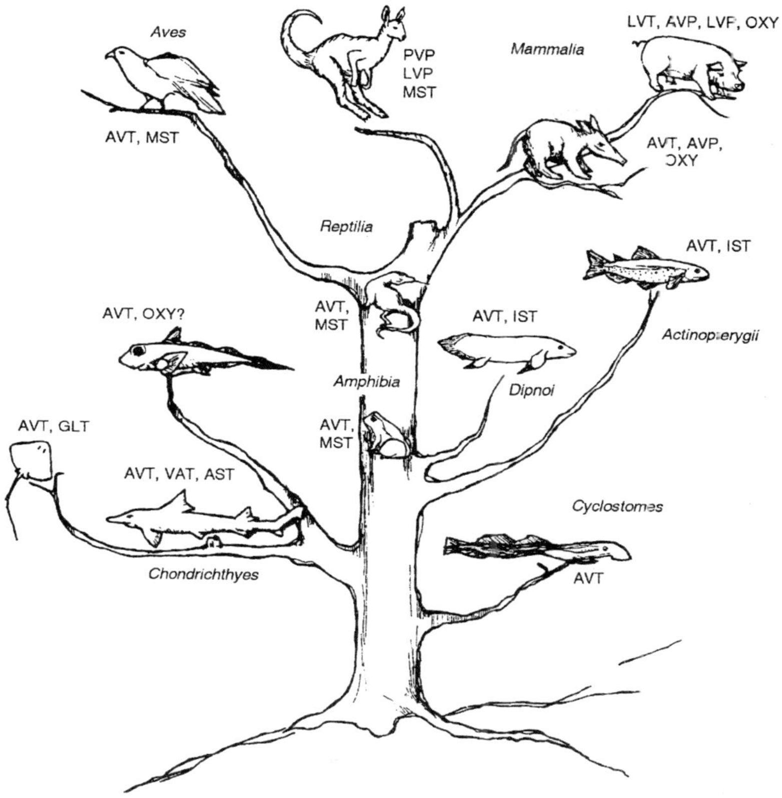

Fig. 6.1. Distribution of neurohypophysial octapeptide neurohormones among vertebrates. AST, aspargtocin; AVP, arginine vasopressin; AVT, arginine vasotocin; GLT, glumitocin; IST, isotocin, LVP, lysine vasopressin; LVT, lysine vasotocin; MST, mesotocin; OXY, oxytocin; VAT, valitocin; PVT, phenypressin.

the pressor and antidiuretic activities. *Oxytocin* (OXY) was eventually found to be the active principle of pitocin, whereas *vasopressin* was present in pitressin. Later it was discovered that there are three different vasopressins in mammals: 8-arginine vasopressin (AVP), 8-lysine vasopressin (LVP) and 2-phenylalanine vasopressin or phenypressin. Most mammals possess AVP. In the Suiformes (family Tayassuidae, peccaries; family Hippopotamidae, hippopotami) both AVP and LVP are present in the pars nervosa. The domestic pigs (family Suidae) appear to have only LVP. One strain of mice has been shown to produce only LVP. Marsupials produce the unique variant of vasopressin, *phenypressin* (PVP), as well as LVP and mesotocin (MST).

Pressor Effect

The first demonstration of biological activity in pars nervosa extracts was made by G. Oliver and E.A. Shafer in 1895. A rapid rise in blood pressure was observed following administration of extracts prepared from whole pituitary glands to anesthetized mammals. Later this activity was localized by W.H. Howell in the pars nervosa. Diuresis (excessive urine production and elimination) was observed by several investigators following administration of pars nervosa extracts. This diuresis was due primarily to increased blood pressure causing an increase in glomerular filtration rate (GFR) in the kidney. Eventually, this principle became known as *vasopressin*. The pressor activity of vasopressins is usually assayed by monitoring blood pressure changes in rats.

Antidiuretic Effect

The antidiuretic properties of pars nervosa preparations in man were independently reported by F. Farini and R. van den Velden in 1913. A correlation was observed between lesions in the pituitary gland and the occurrence of the clinical syndrome of abnormally high production of dilute urine, *diabetes insipidus*. Administration of pars nervosa extracts decreases urine output (antidiuretic) and increases the specific gravity of urine produced by sufferers of diabetes insipidus. (Increased specific gravity of the urine indicates a less dilute or more concentrated urine.) This antidiuretic effect can be demonstrated in the isolated kidney and is due to increased water reabsorption and not to changes in GFR. The pars nervosa principle responsible for this activity is also vasopressin. The permeability of the cells in the collecting duct of each nephron in the kidney is increased in the presence of vasopressin. This increased permeability to water results in a new diffusion of water from the lumen of the collecting duct into the interstitial spaces and eventually back into the blood vascular system. Artificially increasing the blood osmotic pressure causes release of vasopressin from the neurohypophysis, and urine output is decreased. An increase in blood volume or the presence of ethyl alcohol in the bloodstream reduces vasopressin release, and excess water is added to the urine with a resultant increase in urine volume and considerable dilution of the urine (decrease in specific gravity). The double-barreled action of consuming large quantities of beer on vasopressin release is well known to many individuals.

Antidiuretic activity is often assayed in rats, which are very sensitive to doses that are too low to cause any pressor effect. Obviously any pressor effect would cause diuresis and invalidate the assay. An additional bioassay for the vasopressins is the water-balance effect on urinary bladders of anuran amphibians.

Uterotonic Effect

Pars nervosa extracts were first observed by Dale in 1906 to stimulate uterine contractions, and this effect became known as the uterotonic effect. This initial observation eventually led to the clinical use of OXY to induce uterine contractions in labor and to reduce postpartum hemorrhage. The uterotonic activity of OXY is measured routinely with an in-vitro bioassay technique. For example, the contraction of an isolated rat uterus or strips of uterus can be easily quantified following application of purified OXY, and the uterotonic activity of unknown samples or other known substances can be compared quantitatively.

Bioassays for OXY all involve female target tissues, and one may ask what the role, if any, is for OXY in male mammals. Oxytocin has been shown to induce rhythmic contractions in the vas deferens and epididymis of rams. These results are related to earlier observations of increased semen production and semen containing greater numbers of spermatozoa in ejaculates of rams, bulls and rabbits treated with OXY prior to ejaculation. Based upon the actions of OXY on both the female and male genital tracts, some investigators have suggested that OXY release may be responsible for the induction of the rhythmic contractions of these structures associated in humans with the sensation of orgasm.

Milk Ejection

Application of pars nervosa extracts to a cannulated mammary nipple in a goat results in milk ejection. This action was later reported in cows as well although it was also shown that the extract had no effect on the total yield of milk (that is, synthesis of milk). These observations led to development of a sensitive in-vitro bioassay for OXY, the milk-ejecting principle of mammals. One such bioassay utilizes mammary glands dissected from pregnant mice.

Depressor Activity

Occasionally experimental studies of pars nervosa extracts demonstrated the presence of depressor activity or reduction in blood pressure. This depressor effect is caused by OXY and is usually assayed in the chicken.

Metabolic Actions

Regulation of certain metabolic events may prove to be a physiological action of neurohypophysial octapeptide hormones. The vasopressins depress circulating levels of free fatty acids in most mammals that have been examined, employing inhibitory effects on either lipolysis (hydrolysis of fats) or on release of lipolytic hormones such as growth hormone or epinephrine. They might stimulate the metabolism of free fatty acids by muscle cells although there is no evidence to directly support this suggestion. In contrast, blood glucose is elevated by the vasopressins as well as by OXY. The importance of these metabolic effects of neurohypophysial octapeptides in overall metabolic regulation in vertebrates has not been established, and additional research is needed to verify a physiological role for these hormones in metabolism.

Release of Tropic Hormones

Numerous studies of the effect of exogenous octapeptides on tropic hormone release have been reported, but their physiological significance as potential mediators of tropic hormone release has been debated. The demonstration of elevated levels of vasopressin and its specific "carrier" protein in hypothalamo-hypophysial portal vessels of rhesus monkeys, however, does support a possible role for vasopressin as a tropic hormone-releasing hormone. Thus the possibility of a physiologically important releasing role for pars nervosa octapeptides remains.

Neural Functions

The distribution of neurohypophysial hormones in the brain suggests that vasopressins and OXY may function as peptide neurotransmitters. Octapeptides have been shown to produce behavioural effects in animals and even to enhance learning in humans.

Characteristics of Mammalian Neurohypophysial Octapeptide Hormones

Chemistry

Each octapeptide molecule consists of a five-membered amino acid ring, including cystine, and a side chain of three amino acids. Although a neutral molecule, OXY is very similar in chemical structure to the basic vasopressins. Oxytocin has little activity in the biological assays for vasopressins, however. Similarly the vasopressins are relatively inactive in OXY-specific bioassays. Apparently the differences in biological activity reside primarily at position 3 in these molecules.

The presence of the amino acid isoleucine (Ile) at position 3 imparts OXY-like activity to the peptide, whereas phenylalanine (Phe) at the same position imparts strong pressor and antidiuretic properties to the molecule. Arginine or lysine at position 8 is responsible for the basic properties of the vasopressins.

Source

The production of pars nervosa hormones is accomplished by the supraoptic and paraventricular neurosecretory (NS) nuclei of the hypothalamus, and the pars nervosa acts only as a storage area for these hormones until they are released into the general circulation. Oxytocin is produced primarily by the NS neurons of the paraventricular nucleus, and the vasopressins are synthesized primarily in the supraoptic nucleus. Within the NS cells, neurohypophysial octapeptides are associated with proteins known as *neurophysins*, which have a molecular weight of about 12,000 daltons. In the rat, different neurophysins have been identified in association with OXY and AVP respectively.

The neurophysins are often termed carrier proteins and are transported and stored with the octapeptide hormones in the NS granules. Oxytocin and vasopressin are synthesized as parts of two larger peptides of between 20,000 and 25,000 daltons. As the secretion granules travel through the axons to the pars nervosa, the prohormones are cleaved into the specific hormones and associated neurophysins as well as another large peptide. The respective prohormones for vasopressin and OXY are *propressophysin* and *prooxyphysin*. Other studies suggest there are even larger precursors (80,000 daltons) for propressophysin and prooxyphysin which include the sequence identified as proopiomelanocortin.

Blood levels of OXY and AVP vary with the physiological state of the animal. There are no predictable rhythmic patterns of secretion. However, secretion of OXY and AVP into the cerebrospinal fluid shows a definite diurnal rhythm in cats and monkeys. Daytime levels are elevated implying an important role for octapeptides and demonstrating the use of cerebrospinal fluid as a medium to transport these hormones from one brain region to another.

Mechanism of Action

Arginine vasopressin produces its effects on collecting duct permeability to water in the kidney through stimulation of adenyl cyclase activity and an increase in cyclic adenosine 3´,5´-monophosphate (cAMP). Microtubule and microfilaments may be affected by increased cAMP levels and bring about permeability changes in the responsive cellular membrane.

One might assume OXY operates via a cAMP dependent mechanism as well. However, in uterine muscle OXY depolarizes the plasmalemma whereas agents known to operate via cAMP production (epinephrine, isoproteronol) cause hyperpolarization (decreased contraction). Such observations suggest a different mode of operation for OXY.

Biological Half-Life

The biological half-life for OXY and for AVP in mammals is about 1-5 minutes, although it may be longer in large mammals. Between 4 and 20 minutes are generally required to clear 90% of an injected dose through the activities of the kidneys and the liver. Pregnant mammals produce an enzyme, oxytocinase, that circulates in the blood. The presence of this enzyme in the blood may represent a mechanism to prevent OXY from acting on the uterus should its release be stimulated prior to the normal time of parturition. The dramatic increase in OXY seen in some mammals just prior to birth may be due in part to reduction of oxytocinase.

In the toad (*Bufo marinus*) the biological half-life for AVT and OXY is 32 and 10 minutes respectively. Unexpectedly the biological half-life for these hormones is very similar in the chicken (20 and 10 minutes respectively). Considering the lower body temperature for the toad, it would seem that clearance of these hormones in the toad through the actions of the liver may be actually much more rapid than in birds. This slow inactivation time in birds as compared to toads and mammals deserves closer attention.

Comparative Aspects of Neurohypophysial Octapeptide Hormones

The elucidation of an evolutionary pattern for the origin of the neurohypophysial octapeptides has been exciting for comparative endocrinologists. Comparative studies were initiated as long ago as 1908, when pressor effects were identified in extracts of the avian pars nervosa and the teleostean neurointermediate lobe. Following these initial observations an intricate story unfolded involving the evolution of not only functional aspects, related primarily to water balance and reproduction, but also structural alterations in the hormones themselves. Amino acid substitutions producing changes in molecular structure and function have been traced to single base changes (point mutations) in the sequences of bases in the DNA responsible for directing the synthesis of these neurohormones.

Molecular Evolution

Nine different neurohypophysial octapeptide hormones have been positively identified in vertebrates. Scrutiny of these molecules reveals two distinct molecular groupings, the basic *arginine vasotocin* (AVT) and a family of neutral OXY-like octapeptides. The majority of the vertebrates possess the neutral octapeptide known as *mesotocin* (MST) or 8-isoleucine OXY, and some groups exhibit more than one neutral octapeptide. Oxytocin may occur only in eutherian mammals. Arginine vasotocin is present in all nonmammalian vertebrates. This molecule is structurally similar to both the mammalian vasopressins and to OXY. Arginine occurs in position 8 as it does in the vasopressins, but Ile appears in the ring structure at position 3, causing the ring to be the same as for OXY. Consequently AVT has both OXY-like and vasopressin-like activities related to the ring structure and the basic side chain respectively. Although AVT might at first be assumed to be a hybrid molecule, its distribution among the vertebrates would argue that it may be the most primitive neurohypophysial octapeptide and hence "ancestral" to both the vasopressins and OXY as well as to other neutral octapeptides. The appearance of AVT in the pars nervosa of fetal mammals and its later disappearance after birth argues for AVT as an evolutionary precursor. Arginine vasotocin, however, may be present in the pineal glands of some adult mammals, suggesting the assumption of a new role.

Distribution and Genetics of Neurohypophysial Octapeptides

As previously mentioned, AVT is present in all nonmammalian neurohypophyses and occurs in the neurohypophysis of fetal mammals as well. Cyclostomes exhibit only AVT and no neutral octapeptide, supporting the hypothesis that AVT represents the most primitive octapeptide. Considerable variation is found with respect to the neutral octapeptides. Elasmobranchs exhibit three different neutral octapeptides: *aspargtocin* (AST), *valitocin* (VAT), *glumitocin* (GLT). The teleostean fishes produce a unique neutral octapeptide, *isotocin* (ichthyotocin, IST), but the lungfishes exhibit MST, as do all of the nonmammalian tetrapods. The vasopressins appear for the first time in the mammals.

A number of investigators have noted the similarities in structures of the various neurohypophysial octapeptides and have proposed evolutionary schemes based upon single base changes in the nuclear DNA (cistron) responsible for directing the synthesis of each octapeptide. The entire phylogeny of these neurohormones, with only a few exceptions, can be accounted for by a single base change in the DNA

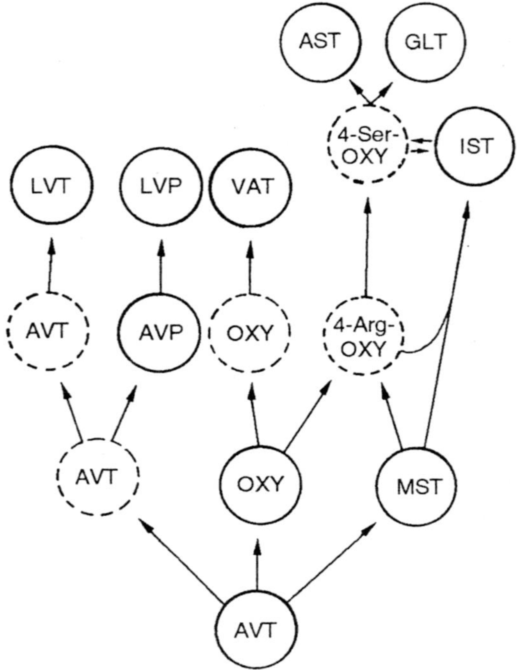

Fig. 6.2. Possible schemes for genetic changes leading to evolution of neurohypophysial octapeptide neurohormones.

responsible for determining the amino acid to be placed in position 3 or position 8 of the peptide. The exceptions require two steps, suggesting the existence of an intermediate peptide that either is still unknown or has disappeared with the extinction of some ancestor.

Class Agnatha: Cyclostomata

Larval and adult cyclostomes produce only one neurohypophysial octapeptide, AVT. The presence of only AVT in the cyclostomes is considered strong support for believing that AVT is the most primitive octapeptide from which the others have been derived through duplication, mutation and subsequent selection of the gene responsible for directing its synthesis.

A pressor effect was first reported for neurohypophysial extracts from the hagfish *Myxine glutinosa* in 1913. Since that time extracts prepared from lampreys and hagfishes have been shown to possess milk-ejection activity, antidiuretic effects, uterotonic activity and frog water-balance activity.

Purified AVT has not been shown to produce any effects on water balance or urine production in lampreys, but it does seem to influence renal sodium metabolism (*natriureferic* activity). Suggestive effects of AVT on sodium and water balance in hagfishes has been reported.

Class Chondrichthyes

Among these fishes there are several neutral octapeptides present in addition to the basic AVT. Sharks have two OXY-like octapeptides, AST and VAT as well as AVT. These two neutral peptides were formerly known as the elasmobranch OXY-like principles (EOP I and II) until they were finally characterized chemically. A different OXY-like neutral octapeptide, GLT, is characteristic, however, of the closely related skates.

Although AVT has been retained in all of the Chondrichthyes, a gene duplication has probably occurred with subsequent mutations resulting through chance in the establishment of at least four different neutral octapeptides. In the case of sharks they would appear to have been an additional variation, but it is not clear from the literature whether one shark can produce both VAT and AST or whether polymorphism exists in these populations.

Neurohypophysial principles from elasmobranchs have been employed in the normal array of mammalian bioassays ever since the first studies were performed in 1908. Extracts prepared from elasmobranch neurointermediate lobes exhibit, milk-ejecting, uterotonic and antidiuretic properties in the ratio of 3:1:0.05. Pressor effects have not been reported. Avian depressor activity is present, however.

Neurointermediate lobe extracts devoid of AVT activity cause contraction in vitro of uteri that were isolated from viviparous sharks. The effects of these extracts presumably containing VAT and/or AST suggest, a primitive role for neutral octapeptides in the induction of oviductal contractions and oviposition or birth. Crude mammalian preparations of neurohypophysial hormones when administered in large quantities to sharks and skates induce a pressor effect, but the physiological importance of this event is questionable.

Holocephalans (*Hydrolagus colliei*) produce OXY-like and AVT-like hormones associated with their neurointermediate lobe. Pharmacological investigations employing mammalian bioassays indicate that the neutral principle may even be OXY, although chemical confirmation is needed. No information is available concerning the physiological roles for neurohypophysial octapeptides in holocephalans.

Class Osteichthyes: Polypteri, Chondrostei, Holostei

The neurointermediate lobe of *Polypterus senegalis* has uterotonic, milk-ejecting, frog water-balance and natriureferic (sodium) activities. The OXY-like activities are presumably due to the presence of IST, and the iono-osmoregulatory actions are due to AVT.

Chondrostean fishes have AVT stored in their neurointermediate lobes and may also possess an OXY-like molecule, most likely IST.

The holostean fishes probably produce both IST and AVT. Extracts from the neurointermediate lobe exhibit uterotonic, pressor, depressor and milk-ejecting activities. Two fractions have been separated from holostean neurointermediate lobes that qualitatively and quantitatively behave like AVT and IST.

In-vivo studies of neurohypophysial octapeptides in polypterid, chondrostean and holostean fishes have not been reported, and such studies are needed to ascribe physiological roles for these hormones.

Class Osteichthyes: Teleostei

Since the first observation of pressor activity in neurointermediate lobes of teleosts, numerous studies have confirmed it and demonstrated antidiuretic, uterotonic and water-balance activities. All of these activities are believed to be due to AVT, however. What was originally believed to be OXY was later discovered to be 4-serine, 8-isoleucine oxytocin or IST. It may not be possible to ascertain whether the uterotonic activity in extracts is due to IST or AVT since purified molecules both possess this activity. Although IST and AVT are the only neurohypophysial octapeptides reported from the teleostean neurointermediate lobe, it should be kept in mind that only a few of the more than 20,000 species of teleosts have been examined. It is likely that in a few cases, at least, a different neutral octapeptide will be discovered.

Arginine vasotocin generally does not produce a water-conserving action in teleosts. Many investigators have demonstrated a strong diuretic action for AVP, LVP, and AVT. This diuretic action appears to be due to an increase in glomerular filtration rate (pressor effect) rather than a renal action. The attachment of biological significance to the large doses of these hormones required to produce diuresis is questioned by studies with the Atlantic eel, *Anguilla anguilla*. High doses of AVT administered to eels cause a pressor effect, intermediate doses cause diuresis and low doses are antidiuretic. It will not be possible to thoroughly evaluate these interesting data, however, until sufficient information concerning circulating levels becomes available for animals maintained under differing environmental conditions. It is possible that all of these activities of AVT are "physiological," depending upon the type of osmotic stress to which the animal is subjected. These data do emphasize the danger of extrapolating from results obtained following injection of arbitrarily chosen doses of a hormone without regard to dose-response relationships or thorough examination of different physiological states.

Numerous reports have implicated AVT in spawning reflexes. Isolated oviducts of several oviparous and viviparous teleosts contract in the presence of small amounts of AVT, IST or OXY. The ovaries of four species of viviparous teleosts (family Cyprinodontidae) also contract in the presence of neurohypophysial octapeptides, and this may be related to expulsion of the young that develop within the ovarian cavity. There is a progressive increase in the sensitivity of these ovaries to AVT during gestation. These observations certainly support a role for AVT in egg laying as well as in the birth process of at least some viviparous species.

Arginine vasotocin is much more potent than IST at inducing the spawning reflex in oviparous killifish *Fundulus heteroclitus*. This does not rule out a physiological role for IST, however, since the important criterion is which hormone is secreted endogenously (if any) during spawning. If only IST is released, the greater potency for AVT would be meaningless physiologically.

Class Osteichthyes: Sarcopterygii

The pars nervosa of the lungfishes is organized anatomically like that of a tetrapod, and like the tetrapods it contains AVT and MST. In addition, some pharmacological studies support the presence of small quantities of OXY. No data have been reported for the living crossopterygian, the coelacanth, but its piscine-like hypothalamo-hypophysial organization suggests AVT and IST or a unique neutral octapeptide might be expected.

Only a few studies have been conducted with respect to the physiological roles for these octapeptides in lungfishes. There appear to be no general water-balance effects or antidiuretic actions on the kidney. Sodium levels may be altered following administration of very large doses of various octapeptides, but the physiological importance of this natriureferic action is questionable. Induction of diuresis by arginine vasotocin presumably is due to a pressor effect and resultant increase in GFR, but further investigation is required to establish a true physiological role. It would be of special interest to examine possible roles of AVT as an antidiuretic principle in estivating African lungfishes while in their mud cocoons.

Class Amphibia

Adult amphibians take up water from the aquatic environment following injection of neurohypophysial octapeptides. This water-balance effect is sometimes termed the Brunn effect after Fritz Brunn, who in 1921 demonstrated this phenomenon in frogs. In *Bufo marinus*, AVT is 50 times more potent than OXY and 200 times more potent than MST in producing a water-balance effect. Furthermore, dehydration of *B. marinus* depletes the pars nervosa of NS material, supporting a physiological role for AVT in water conservation.

The Brunn effect is brought about by the action of AVT at three sites: (1) reduction in urinary losses via effects on the kidney, (2) promotion of water reabsorption from the bladder, and (3) promotion of water absorption across the skin. Of these sites the urinary bladder has received the most attention. The urinary bladder is very sensitive to the actions of neurohypophysial octapeptides, and the toad bladder has become a favourite system in which to examine the mechanism of action of neurohypophysial octapeptide hormones, including the vasopressins. The toad bladder is also one of the standard bioassays for vasopressin-like activities.

The action of octapeptides on the amphibian kidney may be more complicated than previously supposed. Although AVT causes constriction of the afferent renal blood vessels (pressor effect) with resultant increase in GFR (diuresis), MST apparently causes vascular dilation (depressor effect) with a decrease in urine formation (antidiuretic). However, MST increases GFR in hypophysectomized salamander larvae. Arginine vasotocin may prove to be diuretic through a pressor effect, whereas MST could be considered antidiuretic because of its depressor activity on the renal vessels. A direct action of AVT on tubular reabsorption has been demonstrated in bullfrogs but does not occur in *Necturus*. Obviously the simple terms antidiuretic and diuretic for neurohypophysial principles should not be loosely applied since the mechanisms of action and target tissues involved may vary greatly.

Totally aquatic amphibians such as tadpoles of *Rana catesbeiana* or *Bufo bufo* and adult *Xenopus laevis* do not exhibit a Brunn effect. Neotenous and larval urodele amphibians also exhibit little or no Brunn effect, and lower quantities of neurohypophysial octapeptides are present in their neurohypophyses than in more terrestrial individuals.

In addition to employing AVT as a water-conserving hormone, some interesting adaptations have evolved in terrestrial amphibians that are essential for water conservation. Many terrestrial anurans possess a highly vascularized "pelvic patch" that is involved directly in active water absorption. Furthermore, development of this patch in a given species is inversely correlated with the availability of water in their natural habitats.

Arginine vasotocin produces behavioural effects in amphibians. Injections of AVT into male newts causes them to clasp the females (a mating grasp). This action may be mediated by the nervous system. In *Rana pipiens*, AVT stimulates water uptake in females which inhibits the release call. When a female is not ready to spawn, the release call signals the clasping male that he is wasting his time.

Oviductal contractions have been reported in many amphibian species following application of neurohypophysial octapeptides. Arginine vasotocin can induce birth of young in viviparous species such as the salamander *Salamandra maculosa*, and AVT induces contractions in the oviducts of a number of oviparous anurans and urodeles. The presence of oviductal receptors for AVT is dependent on prior exposure to progesterone.

Class Reptilia

The physiological roles for neurohypophysial octapeptides in reptiles are very similar to those observed for birds and amphibians that have the same octapeptide hormones, AVT and MST. These neurohormones have been implicated in reproduction, but associations with water balance and vascular effects are not so well established. Oviposition has been induced with neurohypophysial peptides in several species of snakes and lizards. Birth of young has been induced with AVT in a viviparous lizard. Arginine vasotocin induces oviductal contractions in isolated oviducts of both lizards and turtles. Furthermore, AVT is more effective than either OXY or MST at inducing contractions. No role for endogenous AVT or MST has been reported in relation to oviposition or birth, however.

Some qualitative data have been reported that relate endogenous levels of neurohypophysial octapeptides with dehydration in the ring-necked snake *Diadophis punctatus* and in the garden lizard *Calotes versicolor*. In both cases, dehydration was correlated with a decreasing amount of stainable NS material in the hypothalamus and pars nervosa. Plasma AVT levels are greater in salt-loaded and dehydrated lizards than in water-loaded lizards establishing a possible physiological role for regulating water reabsorption.

Both OXY and MST have been shown to bring about a depressor effect in the alligator, lizards and snakes. A pressor action of OXY, surprisingly, has been reported for turtles as well. These data suggest that crocodilians and squamates are more similar to birds in exhibiting a depressor response to OXY than are the turtles.

Class Aves

Water deprivation or administration of saline (NaCl) solutions to domestic or wild bird species results in depletion of stainable NS material from the pars nervosa. Of those birds examined, only the budgerigar *Melopsittacus undulatus*, a desert-dwelling species, shows no change in NS material following several days of water deprivation. Desert-dwelling mammals exhibit similar insensitivity to dehydration with respect to the quantity of NS material in the pars nervosa. This does not necessarily imply that neurohypophysial hormones are not involved as antidiuretic hormones in these birds. Rather, it illustrates that other adaptations to water deprivation have evolved in desert species, and they are not so sensitive to dehydration when measured by the crude parameter of stainability of the pars nervosa.

Both OXY and AVT are antidiuretic in birds, and a possible role for MST in antidiuretic responses may be inferred. However, it appears that AVT is much more potent than OXY, suggesting that AVT is the endogenous antidiuretic principle. In contrast, AVT is only about 60% as effective a depressor agent in birds as is OXY. The physiological importance of the avian depressor effect is questionable since a much larger dose (10 times) is needed to induce a depressor effect than to cause antidiuresis. Nevertheless, the avian depressor effect has proven to be a useful bioassay for characterizing and elucidating the probable structures of unknown octapeptides.

Arginine vasotocin appears to play an important role in oviposition via its stimulatory effect on oviductal contraction. Oviposition can be accelerated by treating birds with neurohypophysial octapeptides during the appropriate stage of the egg-laying cycled. Arginine vasotocin has about 50 times the potency of OXY in stimulating the avian oviduct. Definitive support for an endogenous role in oviposition is supplied by the observation that blood levels of AVT are elevated in the domestic hen at the time of ovulation.

7

THYROID-PITUITARY INTERRELATION

The embryonic thyroid of higher vertebrates, with few exceptions (rat, hamster, mouse), acquires relatively early the ability to synthesize thyroxine; following this it enters into a period of intrinsic hyperactivity (criteria: cell height, I^{131} uptake), lasting until after hatching or birth. In mammals it attains, and thereafter maintains, a concentration of iodide much higher than blood, and, if the total thyroidal tissue of a litter is considered in sum, it may equal or even exceed the maternal thyroid. For example, the newborn guinea pig has an extraordinarily high BMR for a short time. Observations all point to a very active fetal thyroid and a high titer of thyroxin in the fetal blood at birth. The gland is capable of independent (autonomous) differentiation as shown by transplantation and culture *in vitro*. Once the organ has attained a functional state of differentiation it becomes sensitive to pituitary hormones and antithyroid drugs in a manner similar to that of older or mature thyroids. However, although the thyroid may come under control of the embryo's own pituitary gland, or hypothalamus, this control appears slight at first. Removal of the pituitary produces only a slight retardation of thyroid development, and such hypophysectomized fetuses look like intact animals otherwise, even those that also may have lost their thyroids. Cases of thyroid agenesis in humans appear normally developed in other respects.

There is no known developmental role of the thyroid hormone in warm-blooded embryos, e.g., it does not affect growth of the fetus, deposition of glycogen, or organization of the genital tract. Does the same principle of balance between levels of TSH and thyroid hormone in the blood apply to the fetus as it does in the adult? When does onset of function occur in developing thyroid and pituitary; what is the pattern of morphogenesis; and can morphological and functional differentiation be correlated? If thyroxine crosses the placenta in either direction, as reported in the rat, rabbit, and guinea pig, how is the fetal thyroid able to differentiate, and to maintain, a highly active state? Among placental mammals there is a further probiem: namely, does any interrelationship exist between the maternal and fetal thyroid-pituitary system? Hormonal factors in the development of the chick embryo and mammalian fetus have been discussed.

Functional Development of the Thyroid

There are three distinct steps in differentiation of the vertebrate thyroid: (1) onset of the iodide-concentrating capacity; (2) beginning of organic-binding of iodine and hormone synthesis; and (3) initiation of a period of structural differentiation characterized by follicular and colloid increase, heightened epithelium, and progressive increase in the affinity for radioactive iodide. Young animals concentrate I^{131} more rapidly than old ones, and fetal stages concentrate to a greater degree, proportionately, than mother or nonpregnant adult. The third period parallels the decrease in I^{131} uptake by maternal glands in guppy fish, cow, rabbit, rat, and sheep. The thyroid begins to function at comparatively different times in the gestation periods of various mammals, but there is yet no explanation for tins. Initial chemical differentiation precedes or occurs almost simultaneously with the morphological as shown by all methods employed. Colloid droplets have been detected in the epithelial cells previous to differentiation of even the first primitive follicles, whereas microchemical tests have revealed the presence of iodine at this time. This all indicates that the thyroid enters upon a functional state when structurally it has not advanced beyond the histological state of epithelial-cell aggregates. Intercellular colloid accumulation accompanies follicular development.

Emphasis in this paper will be placed upon the thyroid-pituitary system of the fetal rabbit, with reference made to other forms. Functional differentiation is summarized for the rabbit and certain other warm-blooded vertebrates, including the chick. The first indication of I^{131} concentration was found at 15 days when the minute organ consisted of epithelial cords and plates interspersed with mesenchyme. After 17-18 days of gestation the first radioautographs were obtained, indicating the initial formation of thyroxine from plasma iodide, and the first follicles had appeared. Chromatographs revealed the presence of I^{131}-labelled tyrosine precursors of thyroxine, thyroxine itself, and iodide. At 19 days the concentration of bound iodine attained adult proportions, as well as the relative proportions of MIT, DIT, thyroxine, and iodide, and small follicles were numerous. But only minute quantities of colloid and radioactivity were found until after 21-22 days, when abruptly the thyroid began to accumulate increasing amounts of I^{131}, a tendency which continued to term. This may represent either fetal pituitary stimulation, or attainment of progressively higher functional states by an intrinsically hyperactive gland, and parallels the differentiation of follicles with increasing amounts of colloid. Jost and co-workers favour the view that the fetal pituitary assumes control of thyroid development during the latter third of gestation. The fetal thyroid metabolized very rapidly the radioactive iodide which it fixed, so only a small part was left after 24 hours, and at a faster rate than the mother's thyroid.

Attempts to study experimentally the differentiation of the thyroid have involved a variety of approaches. Jost (1957) found a rapid turnover of I^{131} in the fetal thyroid following intraperitoneal injection. Most of what was accumulated by the gland during the first hour had disappeared 24 hours later, when the fetal blood also contained less radioactivity. In decapitated (hypophysectomized) fetuses the collection of iodine was lower than in the controls, which supported the contention of pituitary control. These fetuses had been decapitated 9 days previously and had grown thereafter in an apparently normal manner. Jost has described morphological changes in the fetal thyroid that are attributable to pituitary deficiency. After birth the maternal thyroid uptake of I^{131} in a 24-hour period quickly returned to the level of the nonpregnant animal, but that of her newborn continued for a time at the high level attained in late fetal life.

A similar temporary period of physiological hyperthyroidism has been found in a number of fetal mammals, including human and cow, by studies of protein-bound serum iodine levels and I^{131} uptake. Suggested reasons for this development have included: (1) increased sensitivity of thyrotrophin, (2) increased production of thyrotrophin, and (3) complex alterations of endocrine balance at birth, none of which is without criticism. No sex differences in embryonic development of the thyroid have been reported. Geloso injected 5 μc of radioactive iodide into the pregnant rat on the twenty-first day of pregnancy and, on examination of the maternal and fetal thyroids 3 hours later, found more thyroxine and triiodothyronine per milligram of gland than in the maternal organ, indicating that the fetus was rapidly synthesizing and utilizing the iodine it received via the placenta. Organic compounds generally become prominent in mammalian fetuses about the time follicles and colloid become conspicuous. The bovine fetal thyroid shows developmental characteristics similar in most respects to those of the rabbit, including progressive hyperactivity, accumulation of more radioactive iodide by the near-term fetal gland than by the mother's thyroid, etc. In many organisms studied it has been observed that follicular arrangement of the thyroid epithelium is not a prerequisite to iodine storage or even for the initial and continued elaboration of thyroxine-like hormone.

The thyroid-pituitary system of the embryonic chick has been studied extensively. At 7 days I^{131} concentration was 15 times that of the blood, almost all nonprecipitable and freely exchangeable with that still in the blood. As in mammals, there was no follicular differentiation at this time, and extracellular colloid was lacking. The greatest iodide accumulation was in the fluid yolk. At nine days droplets of colloid were visible in the cells. Radioiodide appeared in the thyroid within an hour or less after injection. At least half the thyroidal I^{131} was protein-bound in 80 minutes and 90 per cent by 3.5 hours—an important change from the last stage. Concentration of radioiodine averaged over 500 times that in blood, most of this being organically bound. Beginning after about 13 days the thyroid abruptly showed an increased rate of accumulation of radioactivity, which indicated to Maraud et al. (1957) a regulating role by the hypophysis from this time. Functional differentiation of the chick thyroid, as in all mammals tested, is not induced by the embryonic hypophysis, since this occurs also in the absence of the pituitary. It is well known that the chick thyroid will undergo differentiation both morphological and functional in chorioallantoic grafts and culture *in vitro*, which demonstrates an intrinsic quality of development in the gland and its capacity to undergo autonomous development. In fact, the I^{131}-concentrating capacity developing *in vitro* appeared to be of the same order of magnitude and in the same schedule as in the gland developing undisturbed *in vivo*. Control of thyroid morphogenesis by the embryonic hypophysis of the chick has not as yet been clearly demonstrated, even in experiments involving hypophysectomy by decapitation and the injection of appropriate drugs and hormones.

To date, only in the chick has it been shown that development of stages of thyroxine synthesis may occur in a step-wise pattern. In the rabbit the earliest chromatograms showed the presence of thyroxine and several of its precursors, but Trunnell and collaborators have reported in the chick embryo the presence of distinct successive tyrosine and thyroxine stages of biochemical differentiation: monoiodotyrosine at 8.5 days, diiodotyrosine at 9.25 days, and thyroxine at 9.75 days. Gonzales found mono- and diiodotyrosine and thyroxine at the end of 2 days' cultivation of thyroids explanted from 7-day embryos. Very probably a similar condition exists in the mammal, but the study of closer stages in an animal with a relatively long gestation period may be necessary to demonstrate it (e.g., the pig or sheep). Employing 8.5-day chick embryos, anterior pituitary explants from 1-year-old cocks and thyrotrophic hormone, Gaillard (1955) came to the conclusion that colloid formation and follicle differentiation principally are dissociable processes.

Role of the Embryonic Pituitary Gland

Attempts to correlate fetal thyrotrophic activity, and histogenesis of the developing anterior pituitary, with thyroid activity have led to the general conclusion that some control of the thyroid comes about towards the end of incubation or gestation. While the anterior pituitary undergoes some degree of cytological differentiation anticipating ultimate postnatal function, the cells show a very weak chromophilia at first, fine granulation (if any), and small size; consequently they are far from distinct. In the rabbit fetus PAS-positive cells and acidophils were present in large numbers after Susa fixation, at least during the latter third of gestation. The former were of larger size than the more numerous acidophils, and by 28-30 days they were numerous and scattered throughout the gland. Since no distinction could be made between thyrotrophs and gonadotrophs, these cells probably represented an early stage in differentiation. Then, during the first two days postpartum large, lightly staining, aldehyde-fuchsin-positive cells (thyrotrophs) appeared, subsequently increasing in number and staining intensity and acquiring their typical polyangular shape. They occurred singly or in small groups of a few cells to many, being more abundant in the periphery and occasionally scattered in the central region. They were less numerous than the neighbouring acidophils. Oval to rounded gonadotrophs (delta cells) were also found, even at earlier stages. At 5 weeks postpartum the numerous thyrotrophs stained more deeply and hence could be distinguished with greater clarity, as was also the case with the other cell types. They were visibly larger, with increased amount of cytoplasm and abundant coarse granules. By 8 weeks they had attained practically a fully differentiated state in respect to size, granulation, and staining intensity.

Jost and co-workers have reported the presence of weakly staining PAS-positive cells in the fetal rabbit anterior pituitary and their abrupt diminution in number between 24 and 26 days. This, and the fact that following decapitation at 19 days the thyroid differentiated in a typical pattern until 21 days, seemed to indicate the time when TSH became liberated in effective quantity and hypophyseal control of the thyroid was established. However, the thyroid continued to exhibit a continuous and progressive increase in differentiation thereafter with only slight evidence of retardation when relatively few PAS-positive cells were present.

Whether there is a similar active participation of the fetal rat anterior pituitary in thyroid development is not clear. Supporting evidence has been given by Jost (1957). Its cytology parallels that of the rabbit. Although the fetal rabbit thyroid shows functional differentiation early in development, that of the rat does not until around the eighteenth day. Possibly the scarcity and weakly staining properties of the thyrotrophs of the rat during the first week of life may be related to the late development of endocrine activity in this animal. Siperstein and co-workers have concluded that, if thyrotroph secretion is necessary for thyroid activity, the hypophyseal elaboration apparently takes place before birth, but not in amounts sufficient for storage of histologically demonstrable amounts of the hormone. Hwang and Wells (1958) believe that the hypophyseal-thyroid system begins to function before birth. Congenital absence of the pituitary gland in the human fetus provoked hypoplasia of thyroid and suprarenal glands, and PAS-positive cells have been found at 19 weeks. Aron (1931) has described the fetal pituitary cytology in sheep, pig, guinea pig, calf, and man. TSH has been demonstrated (tadpole test) in effective quantity in the fetal pig hypophysis toward the end of gestation (260-280 mm) when all cell types were present; and Nelson (1933) found acidophils and basophils in 70-100-mm stages, followed by a marked rise in number of basophils after 160-170-mm stages, when the bioassay test with hypophysectomized tadpoles was negative.

Contrary to earlier views it is now believed that TSH does not cross the placenta. Greer (1951) has postulated the existence of two thyroid-stimulating anterior pituitary hormones: one controlling I-concentration, the other the growth of the thyroid. When TSH was injected into intact and subtotally decapitated chick and mammalian fetuses, the thyroid was stimulated to hyperactivity. Injection of TSH into the gravid female does not cause any changes in the fetal thyroid. Hypophysectomy of the gestating rat does not interfere with pregnancy, nor does it provoke any developmental modification of the fetal thyroid. Pregnant animals do quite well when deprived of their pituitaries in comparison with the poor behaviour of hypophysectomized nonpregnant animals; Greer failed to find TSH in the latter. Effect upon fetal weight has led to different conclusions. The placenta does not secrete a thyrotrophic hormone.

Evidence points to the liberation of TSH from the chick anterior pituitary as early as 8-9 days of incubation, or around 11 days or later. Removal of the pituitary primordium from 33-38-hour stages resulted in a retardation of thyroid development. When thyroid glands from hypophysectomized chick embryos of 12 days were transplanted to the chorioallantoic membrane of intact and hypophysectomized 8-day hosts, the grafts showed typical development of the intact hosts whereas on the hypophysectomized hosts they were retarded, as was also the thyroid of the host. Thyroid and pituitary explants cultured in contact showed stimulation of the thyroid; certain other tissues explanted with them either favoured the response of the thyroid (muscle, liver) or apparently totally inhibited the response (spleen, hypothalamus of late embryo chick).

To summarize briefly: histogenesis of the embryonic pituitary lends little support for the possibility of early functional hypophyseal control of the developing thyroid in chick and mammal: much of the confirmatory evidence comes from studies with decapitated embryos. Efforts to test the late fetal rabbit hypophysis for the presence of TSH by the cockerel test of Smelser have been more exploratory than definite. Thus far little clear-cut information has been obtained even when as many as 6 late fetal glands were transplanted Subcutaneously at one time. This method of bioassay may not be sufficiently sensitive for the low level of TSH involved. The developing thyroid is endocrinologically labile and reacts somewhat like the adult gland, and it would appear that the gland is not wholly autonomous in its development, as shown by decapitation and hormone injection studies. On the other hand, it might be pointed out that such stressful factors as decapitation itself, anesthesia, and manipulation of fetus and mother could provoke developmental retardation of so sensitive an organ as the thyroid, as it has been shown that the degree of retardation is slight. There is a growing opinion that various traumatic agents, including anesthesia and laparotomy, may affect thyroid function in the adult animal.

Goitrogens

The main effects of goitrogenic agents (propylthiouracil, thiouracil, thiourea) in warm-blooded vertebrates (chick, goat, guinea pig, man, mouse, rat) during incubation or gestation are: (1) retardation of hatching and yolk sac retraction, (2) decreased growth, and (3) enlargement of the thyroid with associated hypertrophy and hyperplasia. These effects are similar to those provoked in mature thyroids. This is evidence for: the placental passage of antithyroid drugs, sensitivity of the thyroid to TSH, and the secretion of TSH by the developing anterior pituitary. Effects of propylthiouracil administered during pregnancy were counteracted by the simultaneous injection of thyroxine. When goitrous offspring of propylthiouracil-treated guinea pigs were followed in postnatal development, the thyroids underwent involution but were still enlarged after eight months. There was no effect on growth and sexual differentiation.

Experiments employing antithyroid drugs are numerous. Methylthiouracil crossed the placenta of the rat. Thiouracil and methimazole depressed the avidity of maternal and fetal mouse thyroids for I^{131} and protected the fetal glands against radioiodine necrosis by causing greater excretion of the isotope. The use of thiouracil has shown that the guinea pig thyroid is unresponsive until around 30 days of gestation, and that of the goat, prior to midterm. Potassium thiocyanate inhibited the iodide-concentrating mechanism in the fetal mouse thyroid, as it does in the adult. KIO_4 produced large goiters in guinea pig fetuses, but did not cause maternal goiter. Kauffman et al. (1948) on the basis of such experiments suggested the fetal mouse pituitary may begin secretion of thyrotrophin on or before the 16th day. Jost (1957) found that decapitated rabbit fetuses from mothers given propylthiouracil had small, poorly developed thyroids rich in colloid, while intact fetuses of mothers treated with the drug possessed hypertrophied glands with little or no colloid. He concluded from this that the goitrogen hypertrophy could only result from hyperactivity of the fetal hypophysis. Sethre and Wells (1950) reported no results from thiouracil injected into the fetal rat. Delay in the differentiation of the nervous system, and, in the postpartum rat, a delay in myelinization, has been reported by Barrnett; perhaps thiouracil may have effects in early development other than those mediated by the pituitary and thyroid. Thiouracil decreased, and thyroxine increased, the metabolic rate of the hen's egg. Even in chicks hatched from eggs of hens fed thiouracil the thyroid glands showed increased weight. Hypertrophy of the chick thyroid evoked by thiourea caused an increase in thyroid weight amounting to as much as 300 per cent during the embryonic period, which dropped to 72 per cent during the postembryonic period. Occasional results have been reported of the use of antithyroid agents in the treatment of the pregnant human female: enlarged fetal thyroid, no effects, normal thyroid in an anencephalic monster.

To summarize: Goitrogens affect developing and older vertebrates similarly in most ways, and their use has confirmed the conclusion that the fetal anterior pituitary liberates effective quantities of TSH and hence is in a position to regulate hormonally the function and even development of the fetus' own thyroid.

Maternal Thyroid During Gestation

Decreased activity of the maternal thyroid late in gestation has been reported for a number of animals: rabbit, rat, sheep, mouse, cow, and the viviparous cyprinodont fish, *Lebistes reticulatus*. This points to a possible influence of fetal thyroid secretion on the mother's thyroid and parallels the period in gestation when embryonic thyroids are increasingly active both structurally and functionally Maternal total thyroidal I^{131} uptake may be about half that of the litter (cow, rabbit). In early pregnancy the histologic picture is one of relative inactivity; uptake of I^{131} decreases, and thyroxine synthesis occurs at a slower rate than in the fetus. Radioactivity of maternal and fetal plasma never attains equilibrium, but the fetal plasma is always more radioactive, indicating the possibility of an iodide-concentrating mechanism in the placenta. Decreased maternal thyroidal activity may be caused by the high titer of diffusible fetal thyroxine in festal blood, reported in several mammals, which escaped into the maternal blood stream to induce functional quiescence.

Several other theories advanced to account for this decrease in maternal thyroid activity include: (1) difference in thyrotrophic secretion of maternal and fetal pituitaries, (2) competition for available I^{131} by fetal and maternal thyroids, (3) loss of iodine by the fetus through the placenta which favours the fetus, and (4) robbery of the maternal system by the fetal system. Gorbman and co-workers considered it remarkable that when thyroid glands of the bovine mother and fetus have accessible to them at the same time a common pool of radioiodide, the smaller fetal gland accumulated more than twice as much as the large maternal gland after 24 hours of equilibration. Actually there may not be a common pool because of the presence of an iodide-concentrating mechanism in the placenta. The latter condition has also been found in the late fetus of the dogfish, *Squalus acanthias*. Disappearance rate of thyroxine in pregnant rats' thyroids can not be ascribed to an accumulation of thyroxine in the fetuses.

Hyperactivity of the maternal thyroid has been reported during pregnancy in a number of animals, extending in some cases even into the period of lactation: dog, guinea pig, cat, thirteen-lined ground squirrel, and human. Laboratory ground squirrels showed lower activity than field animals. In the human the condition appeared in the first trimester, but was not accompanied by symptoms of hyperthyroidism in spite of the elevated BMR in late pregnancy. At present there is no adequate explanation for human pregnancy goiter or for elevated serum PBI.

Cachectic symptoms in thyroidectomized pregnant animals are postponed until delivery of full-term fetuses, indicating an action of the fetal thyroxine, and its passage across the placenta. Development of mammary glands and disappearance of the myxedemic condition both occurred in pregnant heifers shortly after onset of fetal thyroid function. In the thyroidectomized pregnant bitch the puppies showed enlarged and precociously functioning thyroids, and preserved the mother's health until birth. As stated earlier in this paper, there is abundant evidence that exogenous thyroxine crosses the placenta. To examine this problem further, pregnant rabbits were thyroidectomized at 7-9 days of gestation and the state of the thyroids of their newborn young was studied histologically and by means of I^{131}. Thyroids of first-generation litters showed no acceleration of development nor activity; rather, the uptake of radioiodide by the thyroids was depressed. Thyroids from subsequent litters from these thyroid-deficient females continued to show neither stimulation nor depression. The animals continued to mate well after each delivery. When examined 6-7 months following thyroidectomy, nine of the twenty-one cases showed no signs of thyroid tissue in the customary location of the gland. Other locations of thyroid tissue have not been described for the rabbit. Thus, partial or complete absence of thyroid tissue in the pregnant doe provoked no stimulatory effect upon the fetal thyroids; probably the amount of fetal thyroxine is typically more than enough to compensate for any deficiency or lack of activity by the mother's thyroid during the latter part of gestation.

If it might be assumed that the fetal hypophysis exerts no effective control over the autonomously-differentiating, hyperactive thyroid in the same animal until very late in gestation or after birth, possibly due to the incompletely differentiated state of the feedback mechanism or some comparably responsible factor (hypothalamus?), a possible explanation could be visualized for the decreased activity of the maternal gland during the later part of gestation as well as for the progressively increasing development of structure and function of the fetal thyroid within an environment where only thyroid-stimulating hormone is unable to cross the placenta. The differentiating thyroid, though under no or partial hypophyseal control as yet, very gradually increases in secretory activity; lack of response of the fetal pituitary or hypothalamus to the resulting increase in thyroxine level of the fetal and maternal bloods permits continued fetal TSH secretion. However, the thyroxine acts upon the maternal anterior pituitary to reduce the output of its TSH. In this sense the fetal thyroids could be looked upon as accessory sources of thyroxine in the maternal system. This reasoning is not free from criticism, and also fails to explain the cases of hyperactivity during pregnancy in some animals. It might be mentioned that most reports of hyperactivity have been based on cytological (cellular) evidence, whereas those of decreased maternal thyroid activity have come from functional analyses with radioactive iodide. Severe chronic hyperthyroidism of the pregnant rat produced cretinoid progeny and the retarded growth of the cretinoids was overcome by the administration of thyroxine.

8

THYROID GLAND

The thyroid gland is unique among vertebrate endocrine glands in that it stores its secretory products (thyroid hormones) extracellularly. Two separate hormones are synthesized by thyroid cells from the amino acid tyrosine: *triiodothyronine* (T_3) and *tetraiodothyronine* or *thyroxine* (T_4). These hormones contain iodide ions (I^-) bound to the phenolic rings of the tyrosines.

Thyroid hormones influence reproduction, growth, differentiation and metabolism. These actions often occur cooperatively with other hormones, and the thyroid hormones enhance the effectiveness of these other hormones. This cooperative role for thyroid hormones is referred to as a *permissive action* whereby thyroid hormones produce changes in target tissues that "allow" these tissues to be more responsive to another hormone, to neural stimulation or possibly to certain environmental stimuli such as light. The major role for thyroid hormones in adult organisms may be to maintain this state of well-being in many types of tissues so that maximal sensitivity to other regulating agents is retained. Thyroid hormones also are essential for normal development.

SOME HISTORICAL ASPECTS

Either deficient or excessive production of thyroid hormones may lead to serious pathological states. The first description of thyroid disease was of abnormal enlargement of the thyroid in man recognized by Chinese physicians about 3000 B.C. As a remedy they recommended ingestion of seaweed and burned sponge or desiccated deer thyroids. The first two substances contained therapeutic quantities of iodide and the last sufficient thyroid hormones to alleviate the pathological symptoms in most cases. Hypothyroid deficiencies of this sort were recognized in Western culture as clinical disorders many centuries later. In 1526 the *cretinism syndrome* was described clinically in Europe. Cretinism is manifest very early in life as a consequence of severe thyroid deficiency. This syndrome is characterized by dwarfism and a number of other physical abnormalities in addition to severe mental retardation, slow mental and physical activity, bradycardia (slowing of heart beat) and hypothermia. In 1880-1890 another classic clinical disorder in adults, *myxedema*, was linked to hypothyroid function. Myxedematous symptoms in adults are related to abnormal accumulation of water and protein throughout the body as well as to other disturbances in general metabolism. These accumulations of protein and fluid alter facial features, causing the patient to appear expressionless. In later stages of the disorder the sufferer becomes less interested in both self and environment, and if untreated would eventually enter a coma and die. *Juvenile myxedema* is similar to cretinism except that early growth and development are normal but become severely retarded in later childhood. All of these different clinical syndromes have the same basic cause: hypofunction of the thyroid gland.

Bauman discovered in 1896 that an organic iodide-containing compound could be extracted from thyroid glands. Subsequently it was demonstrated that this "thyroidin" substance could reverse the adverse effects of iodide deficiency. In the early 1900s the thyroid gland and its hormones were implicated in elevating basal metabolic rate, primarily through effects on certain tissues, for example, liver, kidney and muscle. This observation has strongly influenced the direction of thyroid research in mammals as well as in many nonmammalian vertebrates. The action of thyroid hormones on metabolism is reflected in clinical thyroid states.

The iodide-containing hormone, *thyroxine*, was isolated and crystallized by Edward C. Kendall in 1915. This event marked a significant point not only in thyroid research but in endocrinology as a whole for thyroxine was the first hormone to be isolated in pure form. It was not until 1952, however, that the second thyroid hormone, T_3, was identified by J. Gross and R. Pitt-Rivers. The importance of this discovery will become evident as the mechanisms of synthesis and action for thyroid hormones are discussed. It was the discovery of the so-called antithyroid drugs in the early 1940s as well as the ready availability of radioactive isotopes of iodide (radioiodide) following the nuclear fission of uranium that made it possible to elucidate the details of thyroid hormone synthesis, metabolism and mechanisms of action.

Embryonic Development and Organization of the Mammalian Thyroid Gland

The entire mammalian thyroid gland consists of many follicles encapsulated with a connective tissue sheath. The thyroid gland is highly vascularized with a dense capillary network surrounding each follicle. Sympathetic innervation of the thyroid has been described, and it appears that both the vasculature and the follicle cells may be innervated.

Development of the thyroid gland begins by formation of a ventral bud in the floor of the embryonic pharynx (endoderm) between the first and second pharyngeal pouches. The gland initially differentiates as cellular cords that later separate into clusters of cells destined to become thyroid follicles. The cells of a cluster secrete a proteinaceous fluid termed *colloid* that accumulates extracellularly in the center of the cluster. This secretory activity eventually leads to a colloid-filled space, the *lumen* of the follicle, surrounded by a single layer of epithelial cells, the epithelium of the follicle. The portion of the follicular cell that borders on the lumen of the follicle is known as

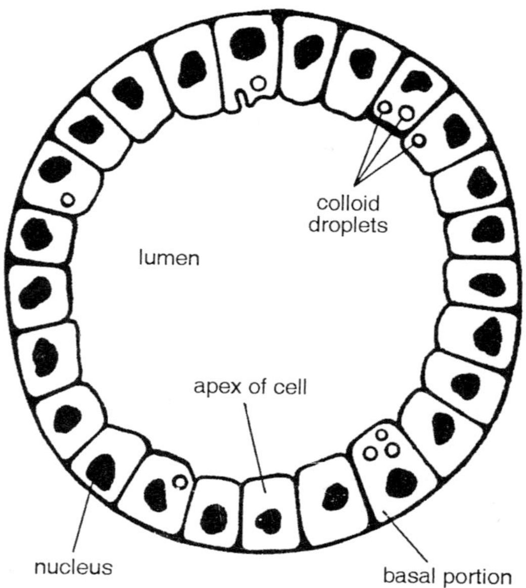

Fig. 8.1. Structure of a thyroid follicle.

the *apical part*. The nucleus is generally found in the *basal* portion of the cell that is farthest from the lumen and closest to the capillaries.

In addition to capillaries and follicles, *parafollicular* or *C cells* occur in the regions between or adjacent to the follicles. Parafollicular cells may occur within follicles or may even form follicular structures in some species. These cells are derived from another pharyngeal derivative, the *ultimobranchial body*, and secrete a hormone, *calcitonin*, that influences calcium metabolism. A comparison of parafollicular cells and follicular cells emphasizes their different structural and functional features. In some mammals the parathyroid glands may be embedded within the mass of the thyroid. The parathyroids, like the parafollicular cells, have their origin nearby from the embryonic pharynx and in some species get incorporated into the mass of thyroid follicles during development.

Biochemistry of Thyroid Hormones

The events related to the ability of thyroid follicles to synthesize and release thyroid hormones are discussed separately for simplicity, but it is important to keep in mind that many of these events may be occurring simultaneously. The processes discussed in this section include:

1. Accumulation of inorganic iodide by follicular cells.
2. Synthesis of thyroglobulin, a glycoprotein containing tyrosine residues for hormone synthesis.
3. Binding of inorganic iodide to tyrosine residues in thyroglobulin.
4. Synthesis of T_3 and T_4 from iodinated tyrosines.
5. Storage of thyroglobulin containing the thyroid hormones in the lumen of the follicle.
6. Engulfing of the colloid by follicular cells and hydrolysis of thyroglobulin to release thyroid hormones.
7. Diffusion of T_3 and T_4 into the general circulation and their transport to target tissues.
8. Mechanisms of action for thyroid hormones.
9. Metabolism and excretion of thyroid hormones.

Dietary Iodide and Iodide Uptake

The principal source for inorganic iodide is dietary. In certain portions of the world, environmental iodide is in short supply, for example, the Great Lakes and Rocky Mountain regions of the United States and northeastern Europe. Consequently unsupplemented human diets are low in naturally occurring iodide, and hypothyroid states commonly are encountered unless an iodide supplement is used. At one time hypothyroid goiters or enlarged thyroids were common in people who inhabited these low-iodide regions or "goiter belts," but the use of iodized salt has almost eliminated this condition. (The term goiter or goitre originally meant any tumor or abnormal glandular enlargement in the neck but has come to mean an enlarged thyroid.)

Inorganic iodide is readily absorbed from the intestine into the blood, from which it is selectively accumulated by thyroid follicular cells. There is an energy-dependent, active transport mechanism or "iodide pump" in the follicular cell basal membrane that is specific for iodide. This uptake of iodide is enhanced by rapid conversion of inorganic iodide within the follicular cell to organically bound forms (that is, iodinated tyrosines).

The events of iodide uptake, accumulation and binding to tyrosine have been examined with the aid of radioiodide. The radioisotope of iodide employed most frequently for iodide uptake studies is ^{131}I, a strong gamma-emitting isotope with a short radiation half-life (8.3 days). Radioiodide is generally administered in the form of a sodium salt ($Na^{131}I$). This particular isotope of iodide can be detected in blood or tissues with unsophisticated detection equipment because of the high-energy gamma radiation it emits. Another isotope of iodide, ^{125}I, is often employed in thyroid studies that make use of the lower energy radiation produced by this isotope and its longer radiation half-life (57.4 days). This isotope emits beta and gamma radiation and is suitable for high-resolution autoradiography at both the level of the light and electron microscopes. Because of its lower energy emission and longer radiation half-life, ^{125}I is more suited for metabolic studies, for it is much less destructive to cells than ^{131}I. Uptake of radioiodide, like that of the normal isotope (^{127}I), is stimulated by thyrotropin (TSH) from the adenohypophysis, and there is no discrimination among the various isotopes in the formation of organically bound iodide associated with thyroid hormone synthesis.

Calculation of a rate for radioiodide accumulation following administration of a given dose provides a quantitative estimate of the degree of TSH stimulation and a reflection of pituitary TSH release. Hence, measurement of radioiodide uptake and accumulation provides a simple and rapid method for estimating endogenous activities of the hypothalamo-thyroid axis as well as a means to assess responsiveness of thyroid follicular cells to exogenous TSH. Usually radioiodide uptake is expressed as percent uptake of the injected dose at some predetermined time, such as 24 hours following administration of radioiodide.

The ability to bind iodide is not a feature unique to thyroid follicular cells, and most cells will accumulate some iodide. Cells such as melanophores (melanocytes), pigmented retinal cells and the epithelial cells of sweat glands, salivary glands, lactating mammary glands and kidney tubules readily accumulate radioiodide following injection of either radioactive isotope. Furthermore, oocytes of many oviparous (egg-laying) vertebrates accumulate large amounts of injected radioiodide. This ovarian accumulation is associated with the normal process of ensuring a source of iodide in the egg that can be used by the young animal for early synthesis of thyroid hormones until an adequate dietary source becomes available. Marsupial and placental mammals and other live-bearing vertebrates probably transfer sufficient iodide to the developing young from the maternal blood or via the milk to suckling newborns.

Thyroid hormones may not be released into the circulation in proportion to the uptake and binding of radioiodide, however. Uptake, binding and release of thyroid hormones are separate events independently influenced by a variety of factors, as evidenced in the following discussions. Nevertheless, measurement of radioiodide uptake is a rapid and convenient method to assay for TSH activity with respect to endogenous levels or exogenous treatments, and it is widely employed in thyroid research.

Biosynthesis of Thyroid Hormones

The actual synthesis of thyroid hormones in the follicular cells involves, first, the synthesis of thyroglobulin containing tyrosine residues and, second, the binding of accumulated inorganic iodide to the tyrosines. The final step in synthesis of thyroid hormones is the linking together (coupling) of two iodinated tyrosines of thyroglobulin to form the iodinated hormones T_3 and T_4. Most of these events that are described below were elucidated with the aid of ^{125}I.

Synthesis of thyroglobulin

Thyroglobulin is a large, globular glycoprotein that is not especially rich in tyrosine residues. Actually the term as used here may embrace more than one molecular form of thyroglobulin. Thyroglobulins extractable from follicles occur in several sizes, ranging from 12S to 27S. (S stands for Svedberg sedimentation coefficient, which is related to size, shape and, to a lesser degree, electrostatic charge of a molecule. These parameters determine the migration and final position of molecules in a molecular gradient following high-speed centrifugation.) Comparative analysis of thyroglobulins from different vertebrates indicates different proportions in the various size classes of thyroglobulins. Most mammalian thyroglobulin preparations exhibit a predominant 19S component (88-100% of the total iodinated protein in thyroid preparations) with a small proportion of larger 27S and occasionally a small quantity of 12S thyroglobulin (rabbit, rat and especially the guinea pig). The 19S form consists of two 12S-subunits, and the 27S form is composed of three subunits. Regardless, thyroglobulin will be treated in the following discussions as though it were a single molecular species. An emphasis on the true form of thyroglobulin is irrelevant to its essential role in thyroid hormone synthesis.

Thyroglobulin synthesis occurs at the rough endoplasmic reticulum and is packaged into membrane-bound secretion granules in the Golgi apparatus. It appears that noniodinated tyrosines are incorporated into thyroglobulins first, since no transfer RNAs for iodinated tyrosines have been demonstrated in follicular cells. Iodination of tyrosine residues in the completed thyroglobulin molecule occurs later at the cell-colloid interface.

Iodination of tyrosine residues in thyroglobulin

Organic binding of iodide begins with conversion of inorganic iodide to "active iodide," a form of inorganic iodide that readily binds to the phenolic ring of tyrosine. Although the exact chemical nature of active iodide has never been determined with certainty, it is apparently formed in the follicular cell by an enzymatic *peroxidase system* that involves glucose oxidation and reduction of pyridine nucleotides to form hydrogen peroxide, H_2O_2. Inorganic iodide reacts with H_2O_2 to form active iodide, which in turn binds immediately to tyrosine residues in thyroglobulin. Treatment of thyroid hormone-synthesizing systems with the enzyme catalase specifically hydrolyzes H_2O_2 to water and oxygen. This blocks iodination, supporting the role of peroxides in formation of organically bound iodide.

The binding of one active iodide to tyrosine at position 3 on the phenolic ring yields *3-monoiodotyrosine* or MIT. A second active iodide may attach at position 5 of the same tyrosine residue, resulting in conversion of MIT to *3,5-diiodotyrosine* or DIT. The 3 position is always iodinated more readily and 3-monoiodotyrosine but never a 5-monoiodotyrosine is formed. The proportion of MIT to DIT formed

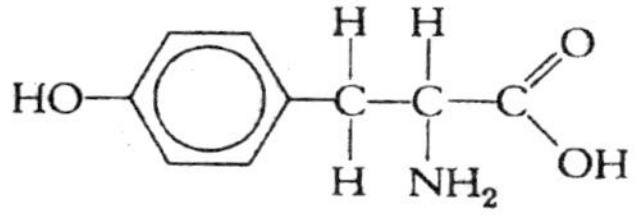

Tyrosine

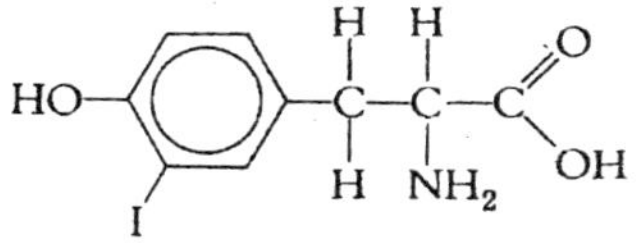

3-Monoiodotyrosine (MIT)

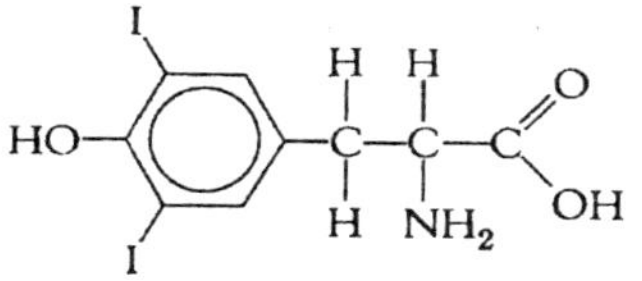

3,5-Diiodotyrosine (DIT)

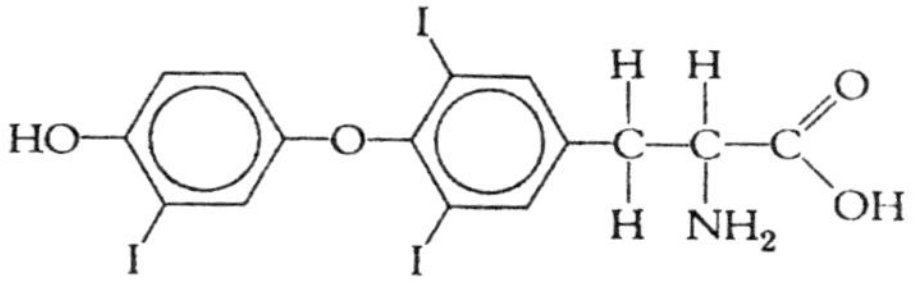

3,5,3′-Triiodothyronine (T_3)

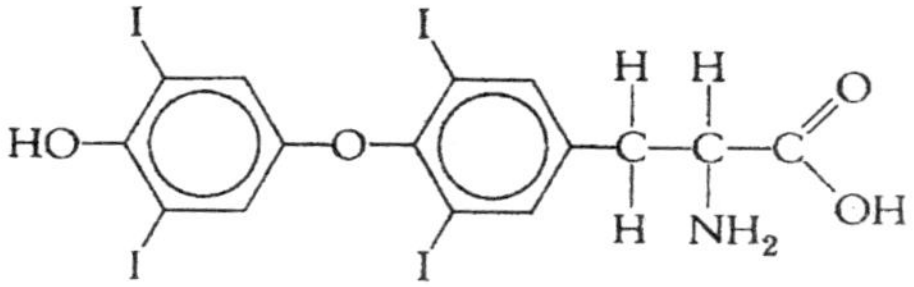

3,5,3′,5′-Tetraiodothyronine (T_4, Thyroxine)

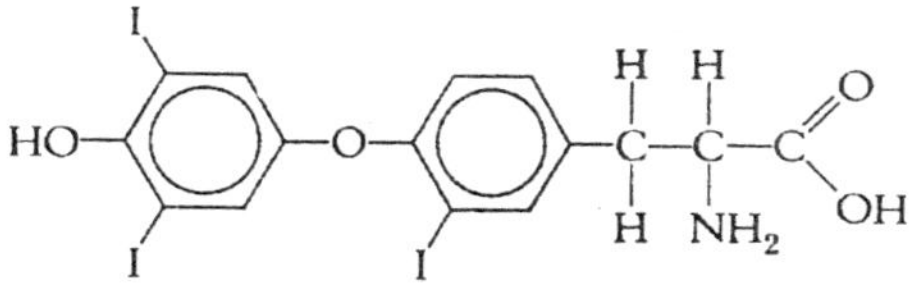

3,3′,5′-Triiodothyronine (rT_3, Reverse T_3)

Fig. 8.2. Chemical structures of thyroid hormones (T_3, T_4), precursor (tyrosine, MIT, DIT) and reverse T_3 (rT_3).

will be determined by the amount of active iodide available, which in turn depends upon the size of the inorganic iodide pool.

Most of the available evidence supports the interpretation that iodination of tyrosine residues in thyroglobulin occurs at the surface of microvilli present on the apical surface of the follicular cell. The peroxidase enzymatic system believed to be responsible for the formation of active iodide is present in the membranes of the apical surface. Presumably iodination occurs as thyroglobulin is being secreted into the colloid for storage.

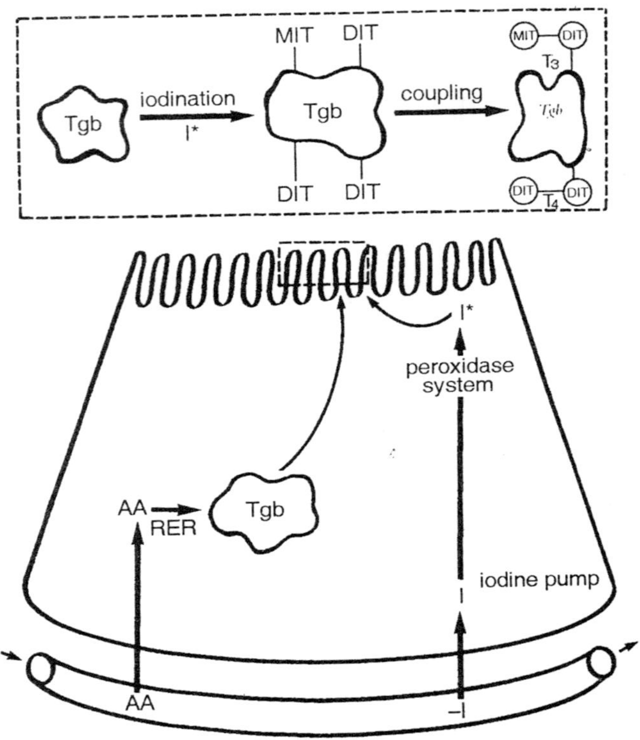

Fig. 8.3. Cytological locations for major events in the biosynthesis of thyroid hormones, MIT and DIT.

Coupling of iodinated tyrosines

The exact way in which thyroid hormones are formed from iodinated tyrosines is not known. Coupling appears to be an enzymatically controlled process that involves two iodinated tyrosines, either two DIT molecules or one DIT plus one MIT. The alanine side chain of one of the iodinated tyrosines is cleaved off, and the remaining iodinated phenolic ring is joined to the other iodinated tyrosine through formation of an ether (—O—) linkage. The resultant structure is known as an iodinated *thyronine*. Appropriate coupling of MIT and DIT yields the resultant 3,5,3´-triiodothyronine or T_3. Similarly, coupling of two DIT molecules results in formation of 3,5,3´,5´- tetraiodothyronine or T_4. The proportion of MIT and DIT available for coupling will influence the proportions of T_4 and T_3 formed. Normally much more T_4 than T_3 is synthesized, but the relative proportion of T_3 may increase markedly if iodide is in short supply.

Coupling of iodinated tyrosines presumably occurs between adjacent residues in the folded, globular thyroglobulin molecule or possibly between iodinated tyrosines in adjacent thyroglobulin molecules. Only a fraction of the iodinated tyrosines are actually coupled, however.

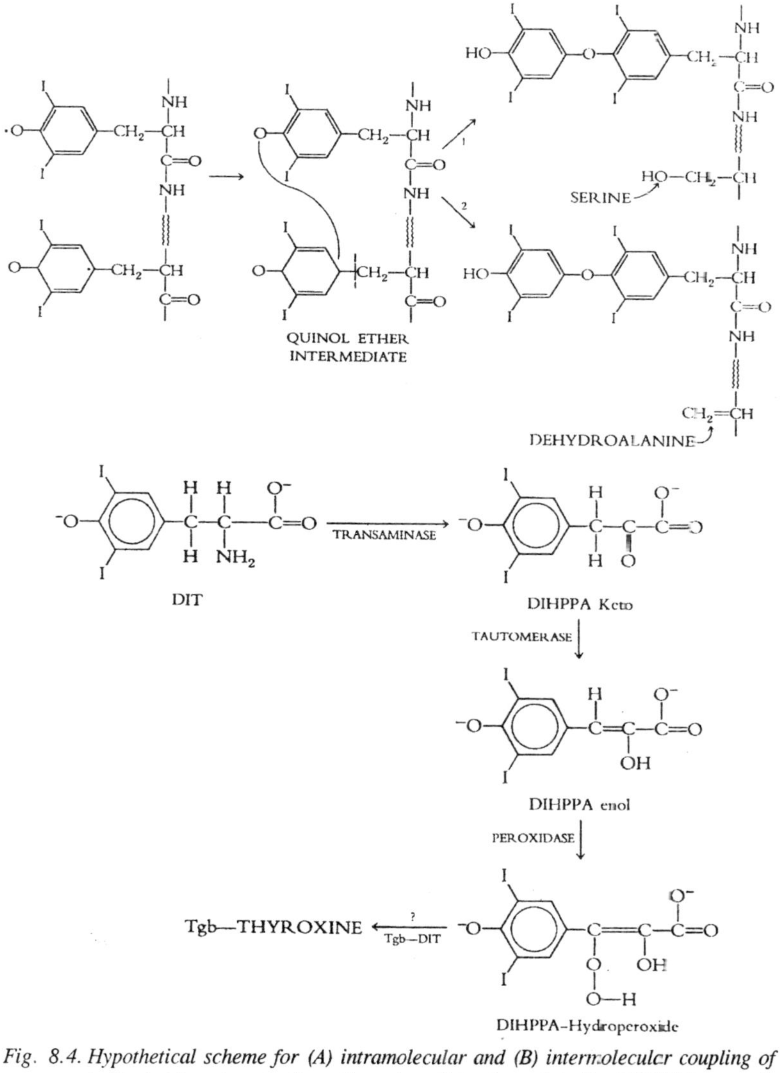

QUINOL ETHER INTERMEDIATE

SERINE

DEHYDROALANINE

DIT

TRANSAMINASE

DIHPPA Keto

TAUTOMERASE

DIHPPA enol

PEROXIDASE

Tgb—THYROXINE

Tgb—DIT

DIHPPA-Hydroperoxide

Fig. 8.4. Hypothetical scheme for (A) intramolecular and (B) intermolecular coupling of iodinated tyrosines to form thyroxine. Tgb, thyroglobulin, DIHPPA, pyruvic acid analog of DIT.

Formation of iodinated thyronines occurs for only about 10% of the iodinated tyrosine residues present in thyroglobulin, and about 90% of the extractable organic iodide is still in the form of MIT and DIT.

The specificity of the coupling reaction implies an enzymatic conversion of iodinated tyrosines to thyronines. Only one triiodothyronine is ever formed in the thyroid, although it would be possible to form 3,3′,5′-triiodothyronine, depending upon the relative positions of MIT and DIT and from which molecule the alanine side chain was cleaved. This isomer of T_3 can be formed by deiodination of T_4 in the liver. The other possible combinations are ruled out because formation of 5-MIT does not occur. There is a requirement that DIT be positioned "on the right," whereas either MIT or DIT may occur to the "left" for the coupling reaction to proceed. The stereospecificity for DIT is further supported by the failure of thyroid systems to synthesize, even under conditions of severe iodide deficiency, any diiodothyronines (T_2) utilizing two MIT molecules. Two enzymatic schemes have been proposed to explain observations on the coupling mechanism. One of these schemes employs the same peroxidase system involved in the formation of active iodide, making it the more likely endogenous mechanism.

Hormone Release: Hydrolysis of Thyroglobulin

Release of thyroid hormones following administration of TSH is not linked to thyroid hormone synthesis. Thyrotropin stimulates engulfment of colloid by the follicular cell and its intracellular hydrolysis to amino acids, MIT, DIT, T_3 and T_4. Autoradiographic studies indicate that the first event observed following TSH administration is the engulfment of colloid through a process termed *endocytosis* (essentially like phagocytosis or pinocytosis). These colloid droplets migrate from the apical portion of the cells toward the basal portion where they become associated with electron-dense organelles that appear to be lysosomes. These lysosomes contain a number of hydrolytic enzymes, including acid phosphatase. Fusion of the colloid droplets with lysosomes results in formation of "fusion droplets" or *phagolysosomes*. As the phagolysosomes migrate toward the basal portion of the cell they become progressively degranulated, presumably because of hydrolysis of thyroglobulin and diffusion of the hydrolysis products into the cytosol.

Thyroglobulin hydrolysis within the phagolysosome releases MIT, DIT, T_3, T_4 and amino acids, which diffuse into the cytosol. Essentially it is only T_3 and T_4 that are "allowed" to diffuse from the cell into the vast capillary network surrounding the follicles. A cytoplasmic

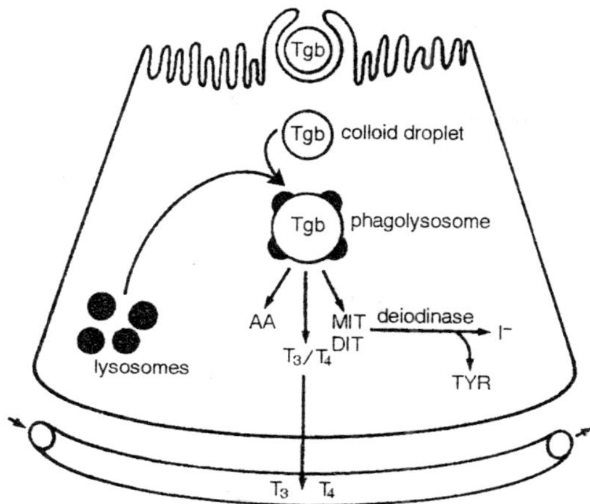

Fig. 8.5. Cytological events in the endocytosis of colloid, hydrolysis of thyroglobulin (Tgb) and release of thyroid hormones into the circulation. AA, amino acid; TYR, tyrosine.

enzyme, deiodinase, hydrolyzes MIT and DIT to tyrosine and inorganic iodide, which are lacking in any thyroidal hormone activity. Deiodinase may also deiodinate a very small proportion of T_3 and T_4 as well. Deiodination is apparently a conservation mechanism to reuse inorganic iodide for later iodination. The iodinated tyrosines (MIT and DIT) cannot be used in thyroglobulin synthesis, as stated previously, and they must be either deiodinated or allowed to diffuse from the cell. Approximately 85 to 90% of the iodide released through deiodination of MIT and DIT enters a "second iodide pool" within the follicular cell, which is then available for iodination of newly synthesized thyroglobulin. The remainder of this inorganic iodide diffuses out of the cell.

It has been proposed that hydrolysis of thyroglobulin could occur extracellularly in the colloid as well as intracellularly. Hydrolytic enzymes, catheptases, capable of releasing iodinated compounds from thyroglobulin have been localized in the colloid. These catheptases, however, require very low pH (pH 3-4) for optimal activity, and they may not be active in the colloid where the pH conditions are approximately neutral (pH 7.0). Nevertheless, one could argue that localized pH changes sufficient to activate these enzymes might occur in the colloid that would not be detected by standard pH determinations. Although extracellular hydrolysis might contribute to overall thyroglobulin destruction following TSH stimulation, the majority of hydrolysis occurs in the phagolysosomes.

Transport of Thyroid Hormones in the Blood

Most of the circulating thyroid hormones (about 99%) are bound reversibly to serum proteins. Serum binding and transporting of thyroid hormones are essential, because T_3 and T_4, like the steroid hormones, are hydrophobic and are not highly soluble in blood. Being hydrophobic, free thyroid hormones readily cross cell membranes and are rapidly removed from the blood and metabolized, especially by the liver and kidneys. Several different serum proteins are capable of binding and transporting thyroid hormones. In man, for example, about 75% of the bound hormones are linked to the α_2-globulins and 15 and 10% respectively are bound to prealbumin and albumin. Only a very small fraction (<1%) are transported free in the blood.

Prior to development of sensitive radioimmunoassays and competitive protein-binding assays for T_3 and T_4, the standard method for assessing thyroid hormone levels was to determine the total *protein-bound iodide* (PBI) in the plasma. In many species, especially anamniotes, inorganic iodide is also carried by serum proteins. Determination of PBI may provide an overestimate of thyroid hormone levels, which also vary with changes in circulating iodide levels. Extraction of whole serum with butyl alcohol (butanol) removes the iodinated thyronines with the butanol phase and leaves the inorganic iodide behind in the aqueous phase. Thus; the *butanol-extractable iodide* (BEI) is used as a better estimate of circulating thyroid hormone levels, especially in nonmammals. Techniques for radioimmunoassay of T_3 and T_4 in very small volumes of plasma or serum have been developed, and the use of PBI and BEI is becoming of historical interest only.

Under normal conditions, circulating T_4 levels are much greater than T_3 levels. Only about one-seventh to one-half of the circulating T_3, however, is of thyroid origin, and the remainder is produced through peripheral deiodination of T_4. This deiodination is accomplished primarily through the efforts of the liver, kidneys, and skeletal muscle. Free T_3 or T_4 in the serum is in equilibrium with bound hormones so that as free hormones are metabolized there is a proportionate dissociation of bound hormone to replace the free hormone lost through metabolism.

Thyroxine is more tightly bound than T_3 to the serum proteins, and consequently T_3 is more readily eliminated from the blood. The biological half-life for T_3 in humans is about 24 hours, whereas T_4 has a much longer biological half-life (about 7 days), attributable to the greater affinity of T_4 for the serum-binding proteins. The reduced serum protein binding and more rapid clearance of T_3 from the circulation is believed in part to provide the explanation for the observed greater effectiveness of T_3 over T_4 when administered to humans or rats. Thus T_4 may prove to be only a storage reservoir, bound as it is to the serum proteins. Following conversion from T_4 through deiodination, T_3 can readily enter target cells.

Peripheral Metabolism of Thyroid Hormones

Thyroxine has several metabolic fates after being released from the thyroid gland. Approximately 33 to 40% is converted to T_3, and this deiodination may be important in the mechanism of action for thyroid hormones. Peripheral deiodination of T_4 is the major source for circulating T_3. About 15 to 20% of the circulating T_4 is converted to tetraiodothyroacetic acid, which has no significant physiological activity and is excreted in urine or bile. Approximately one half of the circulating T_4 is eventually converted by deiodination to a unique form of T_3 with the structure of 3,3′,5′-triiodothyronine. It has no biological activity. This form of T_3 is known as *reverse* T_3 (rT_3), and it is more rapidly degraded than normal T_3. As a result of this rapid clearance, rT_3 levels in the blood are rather low. Increases or decreases in circulating T_3 levels are always accompanied by reciprocal changes in rT_3 levels.

Mechanism of Action of Thyroid Hormones

The molecular mechanism of action for thyroid hormones appears to be similar to that described for steroids. Thyroid hormones, because of their hydrophobic nature, readily enter target cells where they bind to receptor proteins. Following binding to receptors they may influence the synthesis of new proteins via an effect on nuclear gene transcription or on mitochondrial protein synthesis. They may also produce effects on oxidative phosphorylation in the mitochondria.

Nuclear receptors for thyroid hormones have been isolated and characterized. They have greater affinity for T_3 than for T_4, supporting the hypothesis that conversion of T_4 to T_3 is a necessary requisite for thyroid hormone action. Mitochondrial receptor proteins have also been demonstrated, and these mitochondrial receptors may be associated with observed effects on mitochondrial protein synthesis and oxidative metabolism. Unoccupied receptors for thyroid hormones have not been demonstrated in the cytosol.

The permissive actions of thyroid hormones related to that "sense of well-being" may be a consequence of effects of thyroid hormones on the nuclear-directed synthesis of adenyl cyclase or on availability

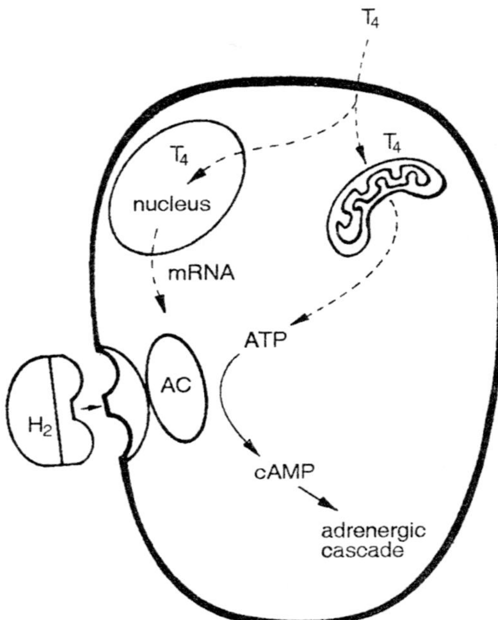

Fig. 8.6. Postulated permissive action for thyroid hormones by influencing availability to ATP, the enzyme adenyl cyclase or both, thus influencing the ability for another hormone (hormone 2 [H_2]) to produce its characteristic action.

of ATP through their actions on mitochondria or on both. The levels of adenyl cyclase and ATP would certainly influence the effects of any of the hormones that normally produce their actions through some cyclic adenosine 3´,5´-monophosphate (cAMP)-dependent mechanism.

Factors That Influence Thyroid Function in Mammals

A great variety of factors may influence thyroid state in mammals and thus influence many processes controlled by other hormones as well. Dietary iodide and the general level of hypothalamo-hypophysial-thyroid functioning are two very important factors. The latter is influenced by a host of different environmental factors as well as by other endocrine systems. A number of naturally occurring chemicals and synthetic drugs have been discovered that have profound effects on thyroid function. Nonendocrine factors are discussed first, including dietary factors, physical factors and internal rhythms. Under the endocrine factors affecting thyroid function are included the hypothalamic and hypophysial hormones as well as hormones produced by other endocrine glands.

Nonendocrine Factors

Dietary iodide

Low iodide availability reduces the synthesis of thyroid hormones and leads to development of hypothyroid states. Most seriously affected by the reduction in iodide is the synthesis of DIT, which in turn reduces the proportion of T_4 that can be synthesized. Conversely, an excessive level of blood iodide inhibits uptake and accumulation of iodide by the follicular cells, presumably by poisoning the iodide-pumping mechanism. There may be little importance to this observation outside of the laboratory since in nature it would be most unusual for a mammal to be subjected to an excess of iodide.

Inhibitors of iodide uptake

Certain anions are effective in blocking accumulation of iodide by the follicular cells through competitive inhibition of iodide transport. Thiocyanate (SCN^-) and perchlorate ions ($HClO_3^-$) are particularly effective at blocking iodide uptake. These agents are used commonly by investigators to block thyroid function and particularly to block iodide uptake mechanisms.

Inhibitors of formation of "active iodide"

Compounds that interfere with thyroid hormone synthesis by inhibiting iodination of tyrosines are often termed *goitrogens*. The resultant reduction in circulating hormones causes increased TSH secretion as a consequence of reduced negative feedback. Continuous stimulation of the thyroid gland by TSH results in enlargement of the thyroid and production of a goiter. Goitrogens, in addition to blocking formation of active iodide, also induce goiter formation. Many of these compounds secondarily inhibit iodide uptake by increasing the size of the intracellular inorganic iodide pool. Because agents that selectively inhibit iodide uptake also block synthesis and can lead to goiter formation, the anions, such as SCN^- and $HClO_3^-$, are often termed goitrogens.

Several synthetic drugs are capable of blocking formation of active iodide, including thiourea (TU), propylthiouracil (PTU) and other thiocarbamide derivatives. This list is continuously being expanded as additional drugs are tested. These drugs interfere with the peroxidase system responsible for generation of H_2O_2. Treatment with such drugs can be used to "chemically thyroidectomize" an animal reversibly. Certain reduced compounds such as ascorbic acid, reduced glutathione and reduced pyrimidines remove H_2O_2 from the system and also block formation of active iodide.

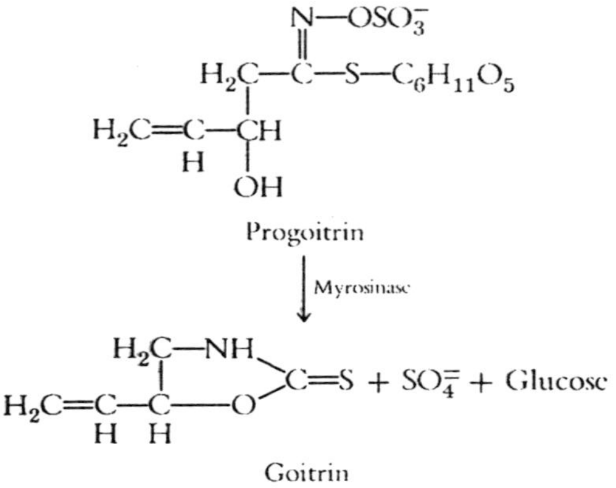

$$N{-}OSO_3^-$$

$$H_2C{-}C{-}S{-}C_6H_{11}O_5$$

$$H_2C{=}C{-}CH$$

H OH

Progoitrin

↓ Myrosinase

$$H_2C{-}NH$$

$$H_2C{=}C{-}C{-}O \; C{=}S + SO_4^= + \text{Glucose}$$

H H

Goitrin

Fig. 8.7. Structure of progoitrin and the product of its enzymatic hydrolysis.

Many flowering vascular plants of the family Brassicae (cabbage, brussel sprouts, rutabaga and turnips) naturally contain a compound known as *progoitrin* that can be enzymatically converted to a goitrogenic compound called *goitrin*. If sufficient quantities of goitrin are absorbed from the intestine into the general circulation the synthesis of thyroid hormones is impaired and a hypothyroid state ensues. People or animals that consume large quantities of these plants tend to exhibit hypothyroidism, which may be accentuated if coupled with an iodide-poor diet. Cooking the plants normally destroys the enzyme that converts progoitrin to goitrin, but progoitrin is not affected by the quantity of heat applied in cooking the vegetables. Bacteria in the human intestinal flora are capable of converting all ingested progoitrin into goitrin, however.

Some vascular plants such as cauliflower contain a glycoside of thiocyanate that can be converted to free thiocyanate in the body. One would have to ingest about 10 kg of cauliflower per day to produce any serious effects on thyroid function unless dietary iodide was extremely low.

Environmental Factors

Environmental factors such as photoperiod and temperature may influence thyroid hormone secretion rates through nervous or endocrine agents. Such factors may influence synthesis and release of hypothalamic and hypophysial hormones or may alter thyroid function directly through sympathetic innervation of the gland.

Internal biological clocks may be related to the actions of environmental factors in regulating thyroid cycles. Cyclical variations have been reported for thyroid hormones on both a diurnal and seasonal basis. Internal secretory rhythms of hypothalamic factors might be influenced by environmental factors or might regulate the sensitivity of other effectors to the external factors.

Prolactin and Its Interactions with the Thyroid Axis

A general antagonism has been reported between prolactin (PRL) and the thyroid axis of amphibians, reptiles and birds. A number of studies have demonstrated what appears to be a goitrogenic action of PRL directly on thyroids of teleosts, amphibians, lizards and birds, and the peripheral antagonisms of thyroxine and PRL in certain tissues of larval amphibians are well known. Reports of enhancement of thyroid secretion following injections of large doses of mammalian PRL to a teleostean fish and a urodele amphibian are probably related to contamination of these preparations with TSH. The reported enhancement by mammalian PRL of thyroxine-induced molting in a lizard requires further investigation. The significance of an antagonistic interaction between thyroid hormones and PRL in fish, reptiles, birds and mammals is not clear. In amphibians the administration of anti-PRL agents (PRL antibodies or ergot derivatives) enhances responses of larval amphibians to endogenous and exogenous thyroid hormones. These observations support an endogenous role for PRL in preventing premature metamorphosis.

It was discussed already that mammalian synthetic thyrotropin-releasing hormone (TRH) causes PRL release from pituitaries of bullfrogs, turtles, birds and mammals but not from the pituitaries of red-spotted newts. These observations further complicate the picture and raise some important questions concerning the biological significance, if any, of demonstrations that PRL produces antithyroid effects.

Surgical and Chemical Thyroidectomy and Radiothyroidectomy

A basic approach employed in thyroid studies involves hypophysectomy or thyroidectomy or a combination of the two followed by classical replacement therapy. Sometimes, however, it is desirable to make an animal only slightly hypothyroid or reversibly hypothyroid or both. Chemical thyroidectomy involves administration of a chemical goitrogen (for example, PTU or TU) at a predetermined dose for a given period. Withdrawal of the goitrogen may then allow the animal to return to a euthyroid condition for comparisons. With this approach, changes in thyroid function before, during and after treatment may be examined in each individual. Caution must be exercised in that some goitrogens have been shown to produce effects on other tissues (especially the liver) that do not appear following surgical thyroidectomy. Such "nonspecific" actions of chemical inhibitors are well known to investigators, and when possible other approaches should be used.

Large doses of ^{131}I are often employed as therapeutic agents to destroy excessive amounts of thyroid tissue in certain hyperthyroid conditions. Accumulated ^{131}I destroys cells because of the destructive effects of gamma radiation. Very large doses can be used to completely destroy thyroid tissue, especially where it is difficult to remove it surgically. Radiothyroidectomy must be interpreted with caution since radioiodide accumulation may result in destructive changes in other tissues that may not be thyroid related.

Endocrine Factors Affecting Thyroid Gland Function

TRH Function and TSH

The hypothalamus exerts regulatory control over release of TSH from the pars distalis via secretion of TRH. Thyrotropin release is stimulated by TRH, and in turn produces an increase in circulating thyroid hormones. Synthesis and release of TSH are regulated through a cAMP-dependent protein kinase mechanism. Adenyl cyclase activity in TSH-secreting cells of the adenohypophysis is stimulated by TRH, and this activation of adenyl cyclase is related to TSH release. This cAMP-dependent mechanism is also dependent upon an influx of calcium ions following binding of TRH to the thyrotropic cell.

Thyrotropin enhances uptake of radioiodide and the synthesis of thyroglobulin and thyroid hormones. In addition, TSH induces engulfment and hydrolysis of thyroglobulin, causing thyroid hormones to be released into the blood. Continued stimulation by TSH causes structural changes in the follicular cells that are related to thyroid hormone synthesis and release. The follicular cells in an inactive or unstimulated follicle are usually flat or squamous cells. Thyrotropin can cause such flat cells to assume a cuboidal or even columnar shape, resulting in visible thickening of the follicular epithelium. Much of this enlargement of the follicular cells is due to an increase in rough endoplasmic reticulum and Golgi apparatus for thyroglobulin synthesis. This increase in follicular or cellular size due to increased cellular growth is referred to as *hypertrophy*. Chronically stimulated thyroid glands may exhibit

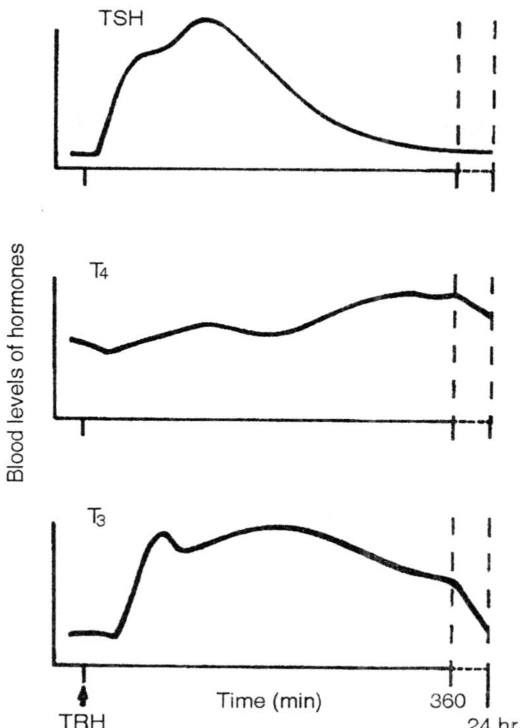

Fig. 8.8. Relative responses in serum TSH, T_4 and T_3 in euthyroid human volunteers following rapid intravenous infusion of synthetic TRH.

hyperplasia as well, which is an increase in cellular numbers due to mitotic divisions by the stimulated cells. Mitosis is a rarely observed event in healthy thyroid glands, but highly stimulated glands may exhibit mitotic figures. Hypertrophy or hyperplasia or both can lead to formation of a goiter.

Follicles of a stimulated gland often exhibit a reduced proportion of colloid and hence a smaller lumen because of increased endocytosis and hydrolysis of stored thyroglobulin to meet the demands for thyroid hormones. Vacuolated colloid is often observed in the follicular lumina of stimulated glands next to the epithelium, and this condition may be related to increased endocytotic activity. Colloid droplets are often seen in the apical portions of cells in the epithelium of a stimulated follicle. These conditions are often used as evidence for stimulation.

The increase in the cellular portion of the follicle due to hypertrophy and the reduction in colloid are reflected in a change in the diameter of the follicle with respect to the thickness of the epithelium or to the volume of the lumen. The ratio of follicle diameter to thickness of the epithelium or diameter of the follicular lumen changes predictably with TSH levels and is frequently used as a measure of the degree of stimulation by TSH. Generally a "stimulated" histology is indicative of thyroid hormone deficiencies and enhanced TSH secretion to compensate for these deficiencies. Other factors, such as cold stress, may be operating at the hypothalamus, however, to elevate TSH secretion above that normally maintained through negative feedback by the thyroid hormones.

One of the first cellular events that occurs in follicular cells following administration of TSH is activation of adenyl cyclase and resultant increase in the intracellular levels of cAMP. It is not clear which of the succeeding cellular events are mediated by cAMP (that is, iodide uptake, thyroglobulin synthesis, formation of organic iodide or engulfment and hydrolysis of colloid). Coincident with increased iodide uptake, formation of organic iodide and endocytosis of colloid is an increase in glucose oxidation that may be caused by cAMP. Glucose oxidation is thought to be the "driving force" for both endocytosis and iodination, the latter involving oxidation of pyridine nucleotides and formation of H_2O_2. By controlling reactions such as glucose oxidation, cAMP could mediate several different cellular events associated with the action of TSH on the follicular cells.

T_3 and T_4 Feedback Effects

The release of TSH is regulated by negative feedback produced by thyroid hormones, and administration of exogenous thyroid hormones decreases circulating TSH and associated thyroid gland activities. The major site for negative feedback is on the thyrotropic cells directly and not the hypothalamic thyrotropic center responsible for TRH production. Thyrotropes contain receptors for both T_3 and T_4, and thyroid hormones are believed to interfere with the cAMP-dependent releasing mechanism by stimulating synthesis of an inhibitory protein or peptide. Thyrotropin levels seem to be maintained by direct negative feedback, and the role of TRH may be to override the system during times of increased demand for thyroid hormones. In other words, thyroid hormones determine the "set point" in the adenohypophysial TSH-secreting cells for daily regulation of thyroid gland activities.

Evidence has been reported of a stimulatory role for thyroid hormones on hypothalamic TRH release. These experimental observations have not established this as a major regulatory pathway in mammals, but they do provide the basis for further investigation into the mechanism whereby hypothalamic control can override the adenohypophysial set point under conditions of increased demand for thyroid hormones.

Prostaglandins

The prostaglandins are small lipids that appear to be involved in the mechanisms of action of many hormones. Prostaglandins directly stimulate synthesis of thyroid hormones without altering circulating TSH, indicating they may be involved in the mechanism of action of TSH. Prostaglandin E_1, cAMP and TSH stimulate increased glucose oxidation, and it is possible that prostaglandins mediate the action of TSH on cAMP formation.

Epiphysial Complex

The pineal gland of the epiphysial complex in mammals has been implicated as a factor regulating thyroid function. This action on thyroid function is probably mediated through effects on hypothalamic TRH release. Numerous studies have demonstrated an inhibitory action for melatonin, one of the principles synthesized and released from the pineal gland. Photoperiodic influences on thyroid activity may also be mediated through the pineal.

Biological Actions of Thyroid Hormones in Mammals

Thyroid hormones affect many diverse tissues and influence major processes such as metabolism, growth, differentiation and reproduction. They are responsible for maintaining a general state of well-being for many cells so that they are capable of maximal responses to other hormones. Although other effects of thyroid hormones will be discussed here, this "permissive action" of thyroid hormones may be their single most important role.

Metabolic Actions

The effects produced by thyroid hormones on mammalian metabolism include a so-called *calorigenic* or *thermogenic action* and specific effects related to carbohydrate, lipid and protein metabolism. These actions of thyroid hormones become more meaningful when considered together with the actions of other hormones on metabolism, but an overview of these effects will be discussed here. Many of these metabolic actions are possibly permissive actions occurring in cooperation with other hormones such as epinephrine and growth hormone.

Thyroid hormones cause calorigenic or heat-generating actions in certain tissues and may be involved in certain physiological responses to cold stress. They can accelerate the rate at which glucose is oxidized and thus increase the amount of metabolic heat produced in a given time. This excess heat production could be used to warm the body. Increased glucose oxidation is reflected in an increased BMR as measured by the change in rate of oxygen consumption. In addition, thyroid hormones have been claimed to "uncouple" oxidative phosphorylation, which decreases the efficiency of ATP synthesis in the mitochondria and increases the quantity of heat released per mole of glucose oxidized. Although such uncoupling has been observed in a hyperthyroid pathological state known as thyroid storm, its role in cold stress is probably not so important as the ability to increase the total rate of glucose oxidation.

Thyroid hormones may not be important in acute cold responses that are probably mediated by epinephrine from the adrenal medulla. Thyroid hormones do induce increased synthesis of mitochondrial respiratory proteins, especially cytochrome C, cytochrome oxidase and succinoxidase. This mitochondrial action would be advantageous in adapting to chronic cold stress. In general, thyroid activity in mammals is greater during prolonged periods of cold stress (winter) than during warmer periods.

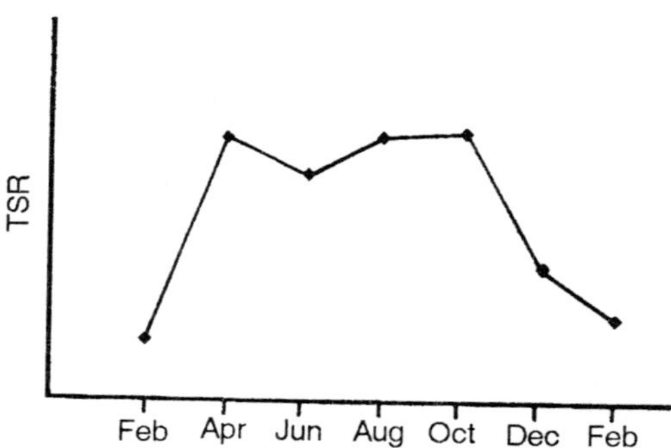

Fig. 8.9. Seasonal variations in thyroxine secretion rate (TSR) in the tammar wallaby Macropus eugenii.

In many nonhibernating mammals such as beaver and muskrat, thyroid activity is depressed during the winter months. Hypothyroidism has been described for hibernating ground squirrels and badgers, but there does not appear to be a causal relationship between reduced thyroid function and the onset of hibernation. Additional field studies employing sophisticated methods for assessing thyroid functions are needed before the endocrine factors related to either onset or termination of hibernation will be firmly established.

In addition to increasing glucose oxidation, thyroid hormones cause hyperglycemia and may secondarily stimulate lipid oxidation (hydrolysis of fats, or lipolysis). These actions may in part be associated with potentiation of the hyperglycemic and lipolytic actions of epinephrine. Thyroid hormones alter nitrogen balance and are either protein anabolic or catabolic, depending on the tissue being examined and under what experimental conditions it is examined. These actions are probably related to enhancement of effects normally produced by other hormones.

Growth and Differentiation

Thyroid hormones are essential for normal growth and differentiation in mammals as evidenced in cretinism and juvenile myxedema. These growth-promoting actions of thyroid hormones are closely related to the role of pituitary growth hormone (GH), and they probably represent a permissive action on GH-sensitive target cells. Thyroid hormones may also stimulate somatomedin production and hence augment the action of GH.

The major tissue affected by the lack of thyroid hormones during differentiation is the nervous system. Normal development of the nervous system as well as attainment of normal mental capacities is strongly influenced by thyroid hormones. Hypothyroidism during early development seriously impairs differentiation and functioning of the nervous system. A reduction in mental capacity can occur in hypothyroid adults.

Replacement of hair in adult mammals is stimulated by thyroid hormones. The postnuptial molt cycle in harbor seals *Phoca vitulina* involves thyroid hormones and cortisol from the adrenal cortex. Hair loss is correlated with low thyroid function and high cortisol levels, whereas resumption of hair growth is correlated with increased T_4 and return of cortisol to basal levels. Thyroid activity also is related to molting of hair in other mammals including red fox and mink.

Reproduction

Another cooperative role for thyroid hormones occurs with respect to gonadal development; in general, sexual maturation is delayed in hypothyroid mammals. In hypothyroid males, spermatogenesis may occur, but androgen synthesis is low. Ovarian weight is reduced and ovarian cycles irregular in hypothyroid females. These correlations to hypothyroidism have been attributed to reduced gonadotropin levels and can be alleviated by treatment with thyroid hormones. Experimental studies support the notion that thyroid hormones influence gonadotropin release through an effect at the level of the hypothalamus. Reduction of gonadotropin, however, was not observed in hypothyroid female rats.

Clinical Aspects of Thyroid Function

Thyrotoxicosis and Hyperthyroidism

Thyrotoxicosis is a general term referring to an excess of thyroid hormone. If this condition results from thyroid hypersecretion, it is known as hyperthyroidism. Primary hyperthyroidism may be due to *toxic multinodular goiters* consisting of multiple aggregates of small, hyperactive follicles (Marine-Lenhart Syndrome) or several large TSH-dependent hyperactive follicles (Plummer's disease). Follicular adenomas are sometimes autonomously hyperactive as well. Circulating TSH levels typically are low when autonomously hyperactive nodular goiter or adenomas are present.

Hyperthyroidism may be of a secondary nature caused by a rare pituitary adenoma of TSH-secreting cells. Certain cancerous tumors such as choriocarcinomas elaborate TRH-like or TSH-like molecules that stimulate thyroid activity. Such tumors typically are insensitive to any feedback by thyroid hormones.

Graves' disease is a secondary hyperthyroid state that may be mediated by an immunoglobulin known as LATS: *long-acting thyroid stimulator*. The basis for production of LATS is not understood.

Rarely, hyperthyroidism is due to a nonthyroid source of thyroid hormones. For example, ovarian dermoid tumors can synthesize sufficient thyroid hormones to bring about hyperthyroidism.

Juvenile thyrotoxicosis is characterized by nervousness, tremor, accelerated heart rate and thyroid enlargement (goiter). This syndrome is manifest in children beyond age 10 (80% of the cases). Exophthalmus (protrusion of the eyeballs) occurs in about half of these children. Weight gain is usually retarded.

A somewhat rare but dramatic condition is thyrotoxic crisis or *thyroid storm*. This disorder involves a sudden increase in thyroid secretion, severe hypermetabolism, fever and some other more variable symptoms. It may be precipitated in hyperthyroid patients following incomplete thyroidectomy, interruption of antithyroid therapy or even as a reaction to an infection or tooth extraction. It may also be induced by periods of excessive summer heat. The actual cause of thyroid storm is not certain and the condition could encompass a variety of different disorders.

Myxedema and Hypothyroidism

Myxedema is the condition where no thyroid hormones are secreted. In these patients there is swelling of the skin and subcutaneous tissues caused by the extracellular accumulation of a high-protein fluid. *Hypothyroidism* refers to conditions of insufficient thyroid hormones of primary (at the thyroid) or secondary (hypothalamus or pituitary) origins. It is especially serious in children because of marked effects on development. The term *juvenile hypothyroidism* refers to cases of hypothyroidism in children that do not lead to severe retardation in somatic and intellectual development. When development is markedly retarded, it is called *cretinism*. This syndrome is rather common and 1 in every 8500 births exhibits cretinism. However, if recognized early, it can be alleviated with thyroid hormone therapy so that growth and development are normal.

There are a number of symptoms characteristic of hypothyroidism including rough and dry skin, yellow pallor, coarse scalp hair, hoarse voice, slow thought and action. However, sometimes the hypothyroid person exhibits none or only a few of these symptoms. Obesity is often listed as a characteristic but it does not always accompany hypothyroidism. Persons suffering from secondary hypothyroidism are often thin. Exophthalmus which is usually correlated with hyperthyroid states may occur in primary myxedema.

Goiters

Any enlarged thyroid is referred to as a goiter regardless of the cause or nature of the enlargement. Actually there are four kinds of clinical goiters. The first kind is an hypothyroid goiter caused by failing hormone production resulting in a shortage of T_3 and T_4. Circulating TSH levels are elevated because of reduced negative feedback, and increased TSH causes enlargement of the thyroid gland and formation of a goiter.

The second and third types are hyperthyroid goiters: the hyperfunctioning goiter and the hyperfunctioning goiter of pregnancy. They are not as common as hypothyroid goiters. In the first case circulating thyroid hormones are high and TSH levels are low. Diffuse thyrotoxic goiter is often termed Graves' disease whereas toxic nodular goiters are associated with Plummer's or Marine-Lenhart syndromes. The second case of hyperthyroid goiter occurs during normal pregnancy. There is an increase in thyroxine-binding globulins and a consequent decrease in free T_4 and T_3 in maternal plasma. This results in elevated TSH and a slight thyroid enlargement in order to maintain normal functional levels of thyroid hormones. This condition usually returns to normal after pregnancy.

The last kind of goiter develops in people with otherwise normal thyroid function. Such enlargements have many different causes including infiltration of the gland with tuberculosis or syphilitic bacteria or parasites and the presence of adenomas or carcinomas. Inflammation due to autoimmune disease (thyroiditis) can also cause enlargement of the thyroid.

Thyroiditis

Thyroiditis is a general term applied to a collection of autoimmune disorders involving production of antibodies that attack thyroid proteins, especially thyroglobulins. The most common types are *Hashimoto's thyroiditis* (*struma lymphomatosa*) and *Reidel's thyroiditis* (*struma fibrosa*). Hashimoto's disease is characterized by the presence of numerous lymphoid follicles, diffuse infiltration of normal cells, extensive increase in the connective tissue components of the gland and some changes in follicular structure. One of the most common causes for goiter among adolescents is Hashiomoto's thyroiditis. In Reidel's thyroiditis the gland is progressively replaced by fibrous connective tissue. In either type of thyroiditis there is a gradual reduction in the thyroid's levels of thyroid hormones. Elevation of TSH secretion contributes to goiter formation. Antibodies produced against thyroid iodoproteins and microsomes usually are demonstrable in the blood.

Inherited Disorders

More than twenty heritable defects in thyroid gland metabolism are known and may be classified according to the type of metabolic defect. For example, there are several defects related to unresponsiveness of the thyroid cell to TSH including insufficient numbers of receptors or abnormal receptors and defects at or in the adrenergic cascade. Defects also occur in iodide transport mechanisms, iodination and coupling of iodinated tyrosines. Several defects have been identified in relation to synthesis of thyroglobulin. Deficiencies in thyroid deiodinase and total body deiodinase are known. Abnormal plasma transport proteins and tissue unresponsiveness to thyroid hormones sometimes occur as heritable errors.

Euthyroid Sick Syndrome

Many nonthyroid conditions can alter thyroid function. Even though the person is actually euthyroid, he/she will appear to be hypothyroid.

For example, fasting, anorexia nervosa, protein-calorie malnutrition and untreated diabetes mellitus are all metabolic disturbances that bring about marked decreases in circulating T_3 and increases in rT_3. Thyroid hormones also are altered in a variety of liver and renal diseases, as a consequence of numerous infections and following myocardial infarctions. Some conditions may alter levels of T_4, TSH, T_3, and/or rT_3. These alterations in thyroid parameters by peripheral disorders constitute the so-called *euthyroid sick syndrome*, reflecting normal responses of the euthyroid person to the generalized disease state.

Evolution of the Thyroid Gland

The thyroid gland structurally is a conservative endocrine gland in vertebrates. It is generally found as one or two masses of highly vascularized follicles surrounded by a connective tissue capsule or as scattered follicles throughout the pharyngeal region (most fishes). Regardless of any gross morphological differences, follicular structure and function is mammalian-like in all the gnathostomes with respect to iodide metabolism, hormone synthesis, storage of thyroglobulin and hormone release. Biochemical differences are quantitative and not qualitative, and the same thyroid hormones, T_3 and T_4, are synthesized and released in all vertebrates.

The most primitive condition is considered to be that found in the cyclostomes in which only a few scattered follicles may be present. Extracellular storage of thyroid hormones does not occur in the cyclostome follicle, and iodoproteins are retained in the follicular cells of these primitive fishes.

Origin of the Thyroid Gland

Although many biochemical aspects of the synthesis of thyroid hormones are well know in vertebrates, the evolutionary origin of the vertebrate thyroid and acquisition of functional significance for these iodinated compounds is uncertain. Acceptance of the simple notion that the thyroid cells evolved from specialized cells of the endostyle of protochordates, leaves some basic questions unanswered. Why were these iodinated compounds first synthesized? What was their primitive role? What selective force were responsible for adoption of these iodinated compounds as metabolic regulators? Certainly the ability of iodide to be bound to tyrosine residues in a protein and even formation of MIT, DIT and iodinated thyronines occur repeatedly among the invertebrate phyla with no suggestion of a distinct regulatory role for

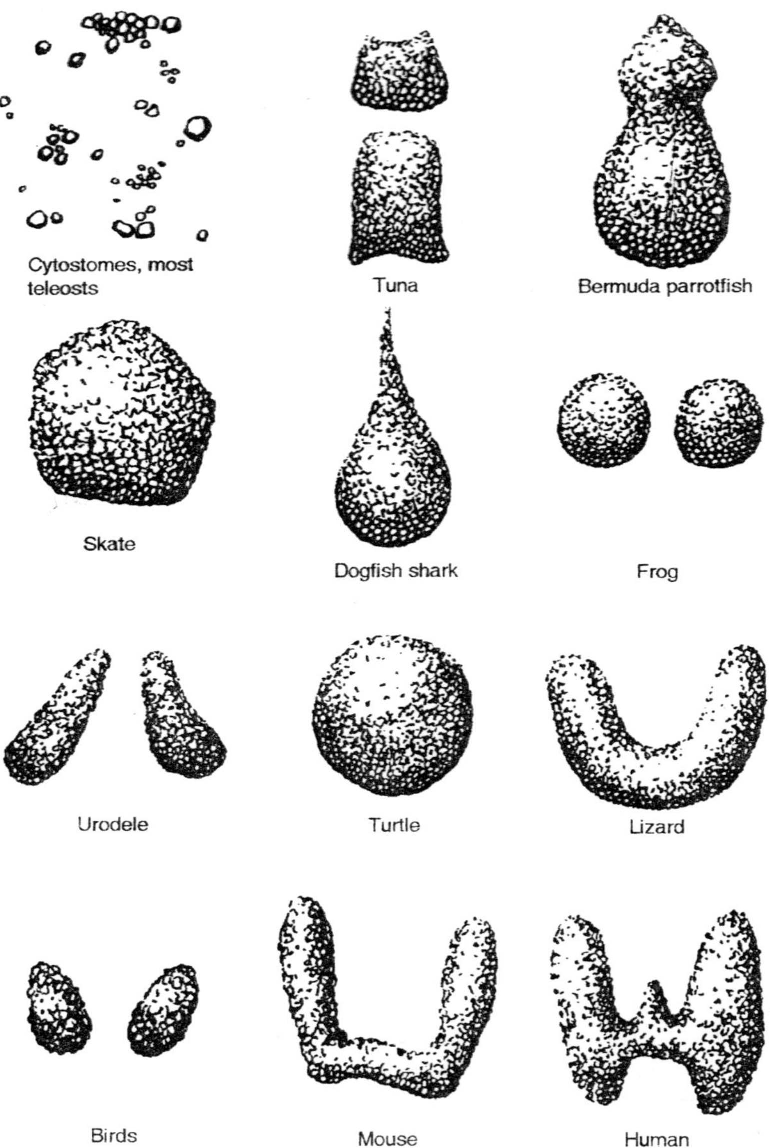

Fig. 8.10. General outline for thyroid glands in diverse vertebrate groups.

these iodinated molecules. One exception may be the participation of iodinated compounds in metamorphic changes of certain cnidarians.

It has been hypothesized that these iodide-containing compounds were obtained originally from feeding activities of protochordates. These dietary compounds supposedly were first utilized opportunistically as

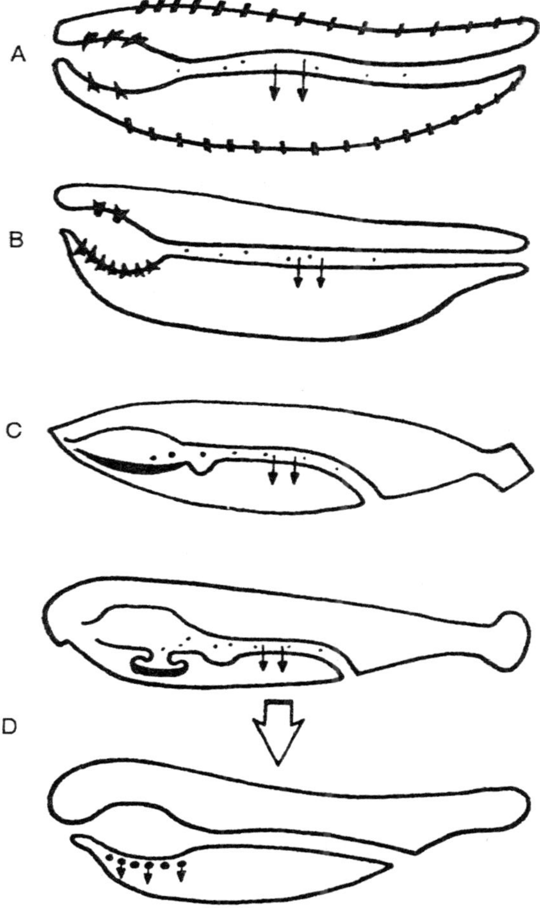

Fig. 8.11. Generalized distribution of iodinated proteins among invertebrate and vertebrate animal suggesting an evolutionary pattern from a general distribution of iodinated proteins (A) to a more localized relationship to the digestive tract (B). Amphioxus (C) illustrates an even greater tendency for localization by concentrating iodoprotein production to the endostyle and with digestion of swallowed iodoproteins occurring in the intestine where enzymatically release iodinated tyrosines and thyronines could be absorbed. Finally the alteration of the endostyle of the ammocetes larva to the thyroid gland of the adult lamprey during metamorphosis is illustrated (D).

regulatory substances of some sort. Later the ability to synthesize these specific compounds was acquired, culminating in the concentration of such a mechanism in certain cells of the endostyle. These specialized cells later became the follicular cells of the thyroid gland as evidenced from studies in lampreys. Iodide-concentrating cells of the endostyle in the larval ammocete are incorporated into the thyroid gland of the lamprey during metamorphosis. The production of extracellular colloid and its resultant accumulation in the follicular lumen do not occur in modern cyclostomes and presumably were acquired at a later time. This hypothesis, although interesting, has been criticized for its failure to provide an adequate explanation as to how a food substance whose production is not controlled in any way by the organism could assume a regulatory function. A possible alternative is that primitive iodoproteins first served an enzymatic or structural function and that the basic mechanism became modified into a hormonal synthetic pathway. An additional problem lies with the origin of the vertebrate thyroid and the iodide-concentrating endostyle of invertebrate chordates. If the ammocoetes endostyle is not homologous to the protochordate endostyles, then accumulation of iodide and formation of iodoproteins by protochordates cannot be used to explain evolution of the vertebrate thyroid.

There is a basic similarity among thyroid iodoproteins in all vertebrates. Thyroglobulins from birds are similar to those of rats with about 94% as 19S, about 5% as 27S and a trace of 12S thyroglobulin. Turtles, teleostean fishes and selachians exhibit lower amounts of 19S (73%-84%), but a significant quantity of 12S thyroglobulin is present (6%-16%). Cyclostomes are the most primitive living vertebrates, and they exhibit what may be considered the most primitive condition. Lampreys produce largely a 12S component (68%), and the remainder of iodinated protein is in the form of a much smaller component (5.4S). Iodoproteins of about 7.6S have been isolated from urochordates and yield monoiodotyrosine, diiodotyrosine, T_3 and T_4 upon hydrolysis. These data suggest that thyroglobulin exists as 12S monomers (or an even smaller unit) that aggregate to form 19S and 27S complexes under certain conditions. These data may also be interpreted as reflecting degrees of disruption from different vertebrates of the intact thyroglobulin during the extraction procedures.

Origin of Thyroid Function

There appears to be no consistency in the actions of thyroid hormones when the entire subphylum Vertebrata is surveyed, although in part the inconsistency may be due to lack of thorough investigation in lower vertebrates, especially the fishes and amphibians. It has been proposed that thyroid function evolved hand in hand with endocrine control of reproduction and that the basic function for thyroid hormones is associated primitively with gonadal maturation. This hypothesis is supported by a variety of observations, including the close parallel of thyroid activity and reproductive cycles in elasmobranch and bony fishes. Pituitary thyrotropes and TSH are apparently absent in agnathans, and these cells may have evolved later from gonadotropes. This proposed origin for thyrotropes is supported by their cytological similarity to gonadotropes, their location in the adenohypophysis of elasmobranchs (pars ventralis) and teleosts (proximal pars distalis) and by the biochemical similarity of TSH to the gonadotropins. Exogenous thyroid hormones and gonadal steroids both have inhibitory actions on thyrotropes as well as on gonadotropes in teleosts. Similar influences of thyroid hormones on gonadal function and, reciprocally, gonadal steroids on thyroid function have been reported in amphibians, reptiles and birds. Furthermore, mammalian gonadotropins stimulate thyroid function in fishes, although these same gonadotropins are ineffective when tested on mammalian thyroids. The effects of thyroid hormones that modify behaviour through actions on the central nervous system presumably evolved from the general feedback effects on the hypothalamus originally associated with the gonadal axis. The actions of thyroid hormones related to growth, metabolism, development and the integument would be later evolutionary events. Location of the major feedback of thyroid hormones on TSH release in the adenohypophysis rather than at the hypothalamus would be a consequence of this evolutionary sequence.

Comparative Aspects of Thyroid Function in Nonmammalian Vertebrates

The following account stress the biological actions of thyroid hormones on target tissues. The secretion of TSH and the effects of hypothalamic hormones on TSH secretion were discussed previously, and only the functions of T_3 and T_4 will be considered here.

Class Agnatha: Cyclostomata

No function for thyroid hormones has been verified in cyclostome fishes, although in lampreys the binding of iodide by the larval endostyle increases following administration of T_4, and thyroid hormones may participate in metamorphosis. Circulating levels of T_4 decrease markedly as ammocetes of sea lampreys *Petromyzon marinus* undergo metamorphosis.

As mentioned previously, hormone synthesis in agnathans differs from events in other vertebrates in that organic binding of iodide and storage of the iodinated proteins occurs intracellularly. Hypophysectomy does not alter thyroid function in adult lampreys or in hagfishes, and it has been concluded that TSH is absent in agnathans. These observations suggest that evolution of thyrotropes and TSH occurred first in jawed fishes, possibly in the placoderms. Thyroxine is present in serum from mature female sea lampreys comparable to levels reported for ammocetes of this species. Much lower levels of T_4 are reported for mature male sea lampreys. Triiodothyronine occurs in the serum of both immature and mature sea lampreys, but the levels are higher in the mature individuals. These data suggest a relationship to reproductive maturation, but definitive data are needed before a firm relationship can be accepted.

Class Chondrichthyes

Only limited data are available with respect to thyroid function in sharks, in which the pituitary-thyroid axis appears to be well established. Most studies on sharks are related to reproduction or oxygen consumption.

Reproduction

Thyroid cycles are positively correlated with reproductive cycles in sharks. However, increased thyroid activity observed in at least one species is correlated with migratory behaviour related to reproduction rather than with reproduction or gonadal maturation per se.

Oxygen consumption

Late embryos of *Squalus suckleyi* exhibit a transient increase in oxygen consumption following treatment with T_3 or T_4, but this response cannot be maintained by continued treatment with thyroid hormones. Treatment with a goitrogen, propylthiouracil, has no effect on oxygen consumption in this species. These observations do not provide strong support of a role for thyroid hormones in oxidative metabolism.

Differentiation

Differentiation of hypothalamic neurosecretory centers is accelerated in the embryos of the oviparous shark, *S. suckleyi*, following treatment with T_3 or T_4. This effect of thyroid hormones is manifest in both the preoptico-hypophysial fiber tracts and in the neurohypophysis. These data are suggestive of a role for thyroid hormones in nervous tissue differentiation and maturation of the hypothalamo-hypophysial system.

Class Osteichthyes: Chondrostei

Some limited data for sturgeons indicate peak thyroid activity coincides with spawning behaviour. Furthermore, thyroid treatments can reverse the degenerative changes that occur in gonads of captive sturgeon, suggesting a direct relationship between thyroid hormones and reproduction. Additional studies are needed in chondrostean fishes to examine other possible roles of thyroid hormones and to confirm this suggestive relationship with reproduction.

Class Osteichthyes: Teleostei

Thyroid function has been studied intensively in teleostean fishes and considerable information is available with respect to iodide uptake, hormone synthesis, secretion rates, clearance rates and metabolism of thyroid hormones. Thyroid hormones are implicated in reproduction and behaviour of teleostean fishes. Their role with respect to oxygen consumption is not so clear, however. Thyroid hormones are essential for normal development and growth. They are also implicated in metamorphosis of young fishes and in the parr-smolt transformation of salmonid fishes. Although thyroid hormones have been shown to influence osmoregulation, carbohydrate metabolism, nitrogen metabolism and growth, other hormones are known to play more direct roles. In fact, many of the reported effects of thyroid hormones associated with these processes may be "permissive" in nature.

Heterotopic thyroid tissue in teleosts

The teleostean thyroid consists of scattered thyroid follicles throughout the pharyngeal region. The diffuse nature of the teleostean thyroid gland, with only a few exceptions (for example, the tuna and the Bermuda parrot fish), makes assessment of thyroid function difficult and renders surgical thyroidectomy impossible. Most of the thyroid tissue is located in the pharyngeal area where follicles are found between the second and fourth aortic arches. Because of the absence of a covering connective tissue sheath that holds the follicles into one or more masses, thyroid follicles are frequently found outside this pharyngeal region. These extrapharyngeal thyroid follicles are termed accessory or *heterotopic* thyroid because of their location outside the normal site. Heterotopic thyroid occurs in species of some of the more recent teleostean families. Relatively large numbers of thyroid follicles may be found embedded within the head kidney of some species and occasionally in other locations such as the ovary. In such species it may be important to examine the activity of heterotopic thyroid as well as of pharyngeal thyroid when attempting to assess thyroid function.

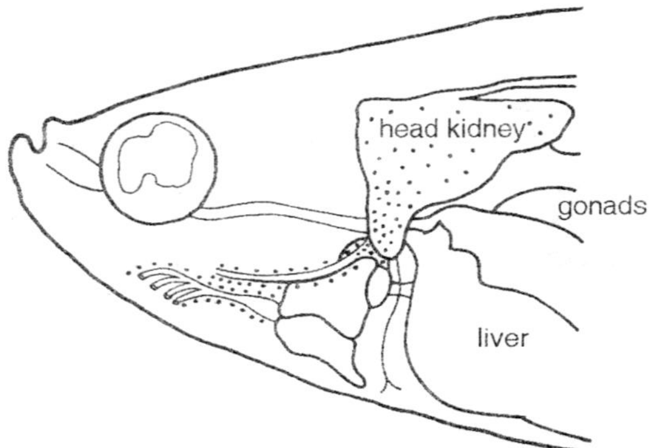

Fig. 8.12. Distribution of thyroid follicles in the pharyngeal region of a teleostean fish, Xiphophorus maculatus.

Reproduction

There is a strong positive correlation in many teleostean species between thyroid state and reproductive cycles. Thyroxine consistently stimulates precocial gonadal maturation, whereas radiothyroidectomy or treatment with goitrogens inhibits gonadal development. In several species thyroid activity is greatest at spawning. In spawning Atlantic salmon, however, iodide uptake, PBI levels and T_4 secretion rates are low, an observation that bears repeated investigation in lieu of more recent data and the development of newer techniques for assessing thyroid function. These observations may relate to other migratory species such as codfish in which increased thyroid activity is more closely correlated with the spawning migration rather than reproductive changes.

Oxygen consumption

Observations have yielded opposing results with respect: to the importance of thyroid hormones in oxygen consumption by teleostean fishes. Most studies suggest that neither thyroid hormones nor goitrogens have any effects on oxygen consumption. A positive correlation has been reported for cyclic seasonal variations in thyroid activity and in oxygen consumption by *Heteropneustes fossilis*. Studies of oxygen consumption are difficult to interpret because temperature, changes in behaviour, and potential side effects of treatments used to render the fish hypothyroid (goitrogens, radiothyroidectomy) may have direct effects on metabolism.

Osmoregulation

Thyroid hormones may enhance seawater adaptation and may influence migratory behaviour in species that migrate between salt and fresh water. These actions may be related to a permissive type of action rather than to a causative role for thyroid hormones.

Behavioural actions of thyroid hormones

Thyroid hormones may influence many behaviours through demonstrated actions on the nervous system. Such behaviours might include spawning, premigratory and migratory behaviours. Increased thyroid function has been reported for migrating fishes, and both a preference for water of higher salinity and premigratory restless behaviour have been correlated with increased thyroid function. Again the action of thyroid hormones may only enhance the sensitivities of neural components to environmental stimuli.

Class Osteichthyes: Sarcopterygii

The function of thyroid hormones in lungfishes has received only a cursory examination. Thyroid hormones have been linked to a particularly fascinating aspect of the life history of African lungfishes: the ability to survive periods of drought while encased in a "cocoon." Awakening of estivating lungfishes *Protopterus annectens* from the cocoon of dried mud when moistened may involve a neuroendocrine mechanism associated with the thyroid axis. One hypothesis, based upon limited experimental data, suggests that increasing humidity activates the hypothalamic apparatus controlling TSH release and stimulates thyroid hormone secretion. Thyroid hormones in turn would increase the sensitivity of olfactory centers that evoke normal feeding behaviour as well as other behaviours associated with wakening. Additional studies of thyroid function in these fishes are needed to substantiate this interesting hypothesis.

Class Amphibia

Thyroid hormones influence many processes in both anurans and urodeles, including reproduction, metamorphosis, metabolism, growth and molting. The most dramatic and best studied event among amphibians is the endocrine induction of metamorphosis from an aquatic larva to a terrestrial or semiterrestrial form.

Reproduction

Experimental reduction of thyroid function, such as thyroidectomy or administration of goitrogens, accelerates gonadal development in several anurans. In one urodele, *Ambystoma tigrinum*, circulating levels of thyroxine are inversely correlated with gonadal development, although such correlations do not substantiate cause-effect relationships. Some conflicting results with respect to thyroid hormones and seasonal maturation have been reported for some adult anurans, but the general relationship in amphibians appears to be an antagonistic one. All of the data reported to date are circumstantial and an absolute antagonism by endogenous thyroid hormones in natural populations is yet to be demonstrated.

Oxygen consumption

The relationship between thyroid hormones and metabolism is not clear. Some studies have shown a positive correlation between thyroid state and oxygen consumption, whereas most studies show no effects. Oxygen consumption is not elevated during either spontaneous metamorphosis caused by elevated endogenous thyroid hormones or during induced metamorphosis caused by exogenous thyroid hormones.

Liver slices prepared from T_4-treated adult frogs, *Rana pipiens*, exhibit significantly greater oxygen consumption in vitro than appropriate controls at 25°C. Oxygen consumption of treated slices is the same as controls when observed at 15°C. These data suggest that the respiratory response of amphibian tissues to thyroid hormones may be temperature dependent and that the response occurs only at higher temperatures. Most studies reporting no effect for thyroid hormones on oxygen consumption were performed below 20°C. What the biological importance is for enhanced oxygen consumption at higher temperatures is open to speculation.

Metamorphosis

Thyroid hormones induce metamorphosis, a marked biochemical, physiological, morphological and behavioural transformation from an aquatic larva to a terrestrial or semiterrestrial form. These events occur in the life history of every amphibian, although the developmental sequence and duration of the various stages may be modified extensively in those species that incubate their eggs on land or are live bearing.

Amphibian metamorphosis has been studied at all levels of organization and provides an excellent illustration of hormonal interactions of thyroid hormones, several additional hormones and a wide variety of tissues. Furthermore, this process is closely linked to the interaction of complex environmental factors that together with endocrine-related events constitute the metamorphosis phenomenon.

Thyroid hormones are also involved in a second metamorphic event, the so-called water drive associated with reproduction in newts. Water-drive behaviour is a well-known bioassay for PRL activity. Small quantities of thyroid hormones facilitate the PRL-induced water drive and associated morphological changes in the integument. Larger amounts of thyroid hormones inhibit water drive, and thyroid hormones appear to be responsible for land-drive behaviour of recently metamorphosed amphibians.

Growth

The involvement of thyroid hormones in growth of amphibians, like the relationship to reproduction, deviates from the general vertebrate pattern. The surge of thyroid activity that initiates metamorphosis in larval amphibians arrests growth. Treatment of anuran tadpoles with goitrogens or mammalian PRL accelerates larval growth and blocks metamorphosis. Growth in natural populations is arrested during metamorphosis. The complicated metamorphic process itself involves many drastic physiological changes and tissue rearrangements that may be responsible for arrested growth. These observations associated with larval growth do not eliminate participation of low levels of thyroid hormones in growth of the subadult or postmetamorphic juvenile forms. Such participation for thyroid hormones in the growth of metamorphosed amphibians has not been reported.

Molting and other integumentary effects

Molting or shedding of skin (ecdysis) in larval and adult urodeles is under direct stimulatory control by thyroid hormones, and frequent molting accompanies and follows metamorphosis in these animals. Molting in adult anurans, however, does not appear to be influenced by thyroid hormones. Here molting is stimulated by corticosteroids or indirectly by ACTH operating through its action of increasing corticosteroid secretion. The reasons for this marked difference in hormonal control of molting between urodeles and anurans are not known.

Thyroid hormones cause a number of skin changes related to invasion of the terrestrial environment. Both a thickening of the epidermis and keratinization are induced by thyroid hormones in urodeles and anurans. In the newts, thyroid hormones antagonize the effects of PRL on the skin of the eft prior to its return to water. Prolactin reduces the keratinization and helps return the skin to a smooth, moist condition characteristic of aquatic newts.

Class Reptilia

Changes in thyroid state in reptiles are correlated with reproduction, environmental temperature and activity, although cause-effect relationships have not been clearly established. Exogenous thyroid hormones also stimulate oxygen consumption, growth and molting under certain conditions.

The presence of iodinated tyrosines in the blood is a situation unique to reptiles. Blood of the striped racer ***Elaphe taeniura*** and the cobra *Naja naja* contains MIT and DIT. Thyrotropin stimulates in vitro release of MIT, DIT and T_4 from chunks of thyroid tissue from a turtle (*Geochemys reevsii*), a lizard (*Gekko gecko*) and a snake (*E. rachiata*). These studies suggest that reptilian thyroid glands may lack any deiodinase to convert MIT and DIT to inorganic iodide and tyrosine. These data, however, are open to other interpretations, and additional research is needed.

Reproduction

Seasonal changes in thyroid function have been correlated positively with a number of reproductive events. An active thyroid, usually assessed histologically, is associated with spermatogenesis, ovulation and mating in a number of lizards. Similar observation have been reported in snakes and in the least one turtle.

Among live-bearing reptiles there is no evidence for increased thyroid function during gestation. However, thyroidectomy of pregnant lizards *Lacerta vivipara* 6 weeks prior to term caused premature discharge of some eggs, and retained eggs failed to hatch.

Environmental temperature

In general, thyroid function in reptiles varies proportionally with changes in temperature, although the cause-effect relationships are still obscure. Behaviours such as basking in lizards make it difficult to estimate body temperature when assessing thyroid state. Thyroid activity in temperate lizards is highest during the warmer seasons. Lowering the environmental temperature artificially during the summer months causes reduction in thyroid activity, and lizards maintained in the laboratory at high temperatures (35°C), exhibit greater thyroid function than when maintained at 15°C. Thyroid function in snakes is greater during warm periods and lowest during hibernation. The reverse is reported for lizards. The greatest level of thyroid activity correlates with reproductive events in some lizards and snakes and not with temperature. Similar data have been reported for turtles.

An exception to the relationship of thyroid activity and temperature described above is the situation found in several species of lizard inhabiting warmer climates. These lizards do not exhibit depressed thyroid function at lower temperatures. Thyroids of such lizards tend to be more active during cool periods. There appears to be a positive correlation also between thyroid state and activity in lizards and snakes. Increased humidity and activity are correlated with increased thyroid function in the lizard *Agama agama savattieri*, suggesting that a more complex relationship exists between thyroid state and environmental factors.

Oxygen consumption

The relationship between thyroid state and oxygen consumption is temperature dependent. Thyroid hormones, TSH and thyroidectomy have little effect on lizards maintained at 20°C, but a positive relationship appears between oxygen consumption and thyroid state at 30°C. Heart, brain, liver, muscle and lung tissues from lizards incubated in vitro at 30°C respond to T_4 with increased consumption of oxygen. Homogenates of both liver and skeletal muscles from T_4-treated snakes *Natrix piscator* exhibit higher oxygen consumption at 30°C than do tissues from untreated snakes. Thyroidectomy causes a decrease in oxygen consumption in these animals that is restored to normal by treatment with T_4.

Molting

Shedding of the skin is stimulated in lizards by thyroid hormones and is retarded by thyroidectomy. Implants of thyroid tissue into muscles of thyroidectomized *Lacerta* spp. restore molting that has been interrupted by thyroidectomy. Apparently either mammalian TSH or PRL is capable of restoring molting in some hypophysectomized lizards, and PRL enhances the effect of T_4 on intact animals. Such a relationship between PRL and the thyroid axis in lizards is unlike the role for PRL in the amphibians.

In snakes thyroidectomy increases molting frequency, and cessation of molting follows administration of thyroid hormones. No explanation for this marked difference between snakes and lizards has been offered. This situation is different from that found in amphibians in which thyroid hormones and corticosteroids, seem to be involved in urodeles and anurans respectively.

Growth

Although detailed studies of the relationship of thyroid state to growth are lacking for reptiles, a variety of studies employing embryonic, juvenile and adult reptiles suggest a relationship between thyroid hormones and growth that is similar to the mammalian pattern. Careful studies are still needed to substantiate this relationship.

Class Aves

The structure and function of avian thyroid glands are similar to those of mammals. Thyroid hormones have been reported to affect reproduction, growth, metabolism, temperature regulation, molting and various behaviours. Most studies on avian species have concentrated heavily on certain domestic birds (chicken, duck, pigeon), and few data are available for wild bird species. Nevertheless, with respect to most of these processes, the relationships are similar for wild and domestic species.

Reproduction

Domestic birds require thyroid hormones for normal gonadal development. For example, T_4, stimulates testicular growth, whereas thyroidectomy or goitrogen administration impairs testicular Junction or induces gonadal regression. Thyroxine can stimulate testicular growth out of season in some wild birds as well. Exogenous T_4 under some circumstances may exert suppressive actions on gonads.

In several wild species, thyroidectomy prior to the time for normal gonadal growth results in precocious gonadal growth and maintenance at maximum condition. Treatment with T_4 results in gonadal regression in thyroidectomized birds at any time of year. In other wild species thyroidectomy simply prolongs the active gonadal phase and shortens the time during which the gonads are regressed. Obviously one cannot generalize about the relationships between thyroid hormones and reproductive events in birds since each species may prove to be a special case.

Thermogenesis and oxygen consumption

It appears that thyroid hormones are directly involved in cold adaptation. Thermogenesis (heat production) in birds as in mammals is closely linked to oxygen consumption. Wild birds in temperate regions exhibit heightened thyroid activity in late autumn and early winter as evidenced by histological examination and changes in thyroid gland Weight. Thyroidectomy of adult birds depresses their ability to produce heat, and treatment with thyroid hormones increases oxygen consumption.

Carbohydrate metabolism and growth

Thyroid hormones reduce glycogen stores in liver, increase free fatty acid levels and induce mild hyperglycemia. Thyroidectomy causes a decrease in blood glucose and liver fatty acid levels but an increase in blood cholesterol levels. These effects may not be of primary importance since a number of other hormones, such as glucagon and epinephrine, are known to be more important than thyroid hormones with respect to controlling carbohydrate metabolism in birds. Thyroid hormones may produce a permissive effect with respect to the actions of these other "glucohormones," possibly through an effect on adenyl cyclase levels in the target tissues. These effects on carbohydrate metabolism may be related to the thermogenic action of thyroid hormones.

There is evidence that thyroid hormones act cooperatively with GH. Thyroidectomy causes depression or retardation of growth in birds and seasonal changes in circulating GH and T_4 are correlated.

Molting

Thyroid hormones produce stimulatory effects on the skin and feathers that are usually associated with the molting process. It is generally accepted that the gonadal axis provides the factors that directly regulate the molting process in birds. Unlike the condition for urodele amphibians and lizards, the role for thyroid hormones in the molting process of birds may be permissive rather than causative.

Migratory behaviour

Migratory birds have been shown to possess active thyroid glands as compared to nonmigrating individuals of the same species, suggesting some interaction in the migratory process. Although some of the older literature indicates that thyroid hormones may directly influence migratory behaviour, the nature of this influence has not been confirmed. Thyroid hormones may alter metabolic patterns associated with energy requirements during migration. Another suggestion is that thyroid hormones "tune" the nervous system so that it is more sensitive to environmental or other endocrine cues or both.

9

THYROID IN INVERTEBRATES

The significant findings in this topic which relate to fishes have been covered very recently by excellent reviews. Analyses of the role of the thyroid gland are admittedly fragmentary, and sometimes controversial. Morphogenetic effects, involving differential growth of fins and body wall, can be elicited by thyroid hormone treatment. In contrast, antithyroid drugs depress growth and morphogenetic processes. Generally the effects of hypophysectomy and of antithyroid drug treatment require a long time to be made evident. In *Lebistes*, inhibition of the thyroid gland results in prolonged sexual immaturity; contrariwise, thyroxine treatment stimulates early sexual maturation.

The metamorphoses of certain fishes are accompanied by increased levels of thyroid gland activity, e.g., the parr to smolt transformation of the Atlantic salmon. Thyroxine treatment accelerates some of the aspects of this transformation, such as changes in relative growth rates and increased rate of guanine deposition, but it fails to enhance the increased tolerance to salinity which is a normal part of the transformation. Antithyroid drugs inhibit the same kinds of changes which thyroxine accelerates. Metamorphosis in the herring is also accompanied by increased thyroid gland activity. Analogous observations have been reported in the Gobiiformes in their change to forms with semiterrestrial adjustments. Harms also reported that thyroid hormone induces an early transition in these forms.

The transformation in eels from the leptocephalus to the elver stage was thought to depend on increased thyroid hormone output, just as is metamorphosis in amphibians, but it has been reported that the increased thyroid gland activity, rather than preceding the changes, is to be found only at the end of the transformation, in company with the migration from salt water to fresh water, and with the increased deposition of pigment. Since the original change of leptocephalus to elver cannot be induced by thyroid hormone treatments, the acceleration of thyroid gland activity at this time cannot be considered a sufficient cause for the transformation. Thyroid activity of the elver declines, to rise again years later, at sexual maturity, at the time of the downstream migration of the eels. A change to a silvery condition, thickening of the skin, transformation of the pectoral fins, and a return to salt water accompany this hyperplastic phase of thyroid gland activity.

The study of amphibian metamorphosis has attracted the attention of many investigators. The more recent reviews have covered all but the most recent literature quite adequately, from the pioneer feeding experiments and the implication of the pituitary gland to the series of studies by many investigators devoted to elucidating the responsive capacities of the various tissues which undergo transformation.

In anurans, generally, responses to thyroxine treatment have not been made evident until one day or so after the completion of spiracle formation and the onset of feeding. In the period from 24 to 58 hours after spiracular formation, at 18°C, a series of structures begins to change, with immersion in strong thyroxine solutions. However, the particular sequence of events which is observed is quite different from the sequence which is obtained when *readiness to react* is determined, i.e., the rate of response influences the early or late detection of the response, but gives no reliable clue as to the period when the structure is first ready to respond to strong solutions of the hormone. Recently, in our laboratory, we were able to demonstrate late onset of reactivity in a previously uninvestigated reactive system, the mesencephalic nucleus of the trigeminal nerve. In tadpoles, *Rana pipiens*, up to 18 mm long, the cells of this nucleus show no capacity to respond to thyroxine treatment; the capacity to respond is weakly developed at a length of 30 mm (stages III and IV), and at a length of 40 mm (stage V) the responsive capacities are not yet fully developed. These findings emphasize that not all responding cells become sensitive together, or nearly so, as indicated by Etkin, and tend to support the assertion of progressive increase in sensitivity (or reactivity) which was indicated by Allen (1932) and Geigy (1941), and later expanded, particularly with respect to the regressive changes of the mouth apparatus and the tail.

Very substantial interest has centered about the problems of control and intimate correlation or integration of the metamorphic processes.

Etkin (1939) has shown that thyrotropic activity sufficient to bring about substantial stimulation of an *adjacent* thyroid gland is present in the vicinity of the hypophysis a few days after feeding begins in *Rana pipiens*, but only undetectable amounts of this hormone get into the blood. Not until 5 to 7 weeks later is there indication, evidenced by slightly accelerated leg growth, of activation of the normal thyroid gland. The growth of the thyroid gland had been studied in detail in anurans, and its increased activity during metamorphosis reported upon several times. It is clear from work reported by Etkin (1932, 1935) that thyroid hormone concentration in tadpole blood is probably very low in the early phases of metamorphosis, that it increases gradually throughout the process and drops quite rapidly at or near the end of the transformation. The particular sequence of events observed in unmodified metamorphosis is then to be interpreted as the resultant of the particular hormone concentrations present in the blood and their rate of increase, coupled with the threshold values for response of the various responding tissues; and the individually different rates of increase of these are dependent upon the gradually increasing thyroid hormone concentration. Nonetheless, Etkin (1955) writes that he "found any effective concentration of thyroxine was capable of inducing both early and late metamorphic changes if allowed to operate for a long enough period." His statement reflects a similar expression of Alien (1938). This view is no longer tenable, since it has been shown in our laboratory, that definite thresholds for particular events can be established, and failure to exceed these thresholds results in failure to produce further metamorphic progress, even though treatment continues for many months. In particular instances individual tadpoles, at constant temperature and at a constant daily dosage, would achieve a particular metamorphic stage and remain at this particular stage of development for months, responding by further change to even very small increases in hormone dosage or to increases in temperature. This further progress, in turn, came to a permanent halt, subject again to increase in dosage. These earlier results with DL-thyroxine have been largely duplicated by more recent work with 3, 5, 3´-triiodothyropropionic acid and 3, 5, 3´, 5´-tetraiodothyroformic acid, showing for each reacting system a particular maximum level of expression for each of the lowest levels of concentration which were used.

One aspect of metamorphic integration but little explored as yet is the role of temperature. Temperatures of 5°C or lower are known to inhibit metamorphosis in anurans and urodeles. Etkin (1955) indicates that tadpoles in cold water grow to larger body sizes than when maintained in warm water. Our own experience with *Rana pipiens* confirms this. That these responses to temperature may be wholly or largely independent of the reactions of the thyroid gland is demonstrated by the responses of hypophysectomized tadpoles subjected to identical thyroxine concentrations at two different temperatures. Not only do the cold animals take longer to transform, but (at the thyroxine concentrations used—inadequate to complete metamorphosis) they never progress as far through metamorphosis as do the animals kept at the higher temperature.

Since the report of Cotronei (1914) on metamorphic disharmony induced by thyroid gland feeding, it has been obvious that the various tissues of the tadpole which are thyroid-sensitive are more or less dissociable from one another in their responses to the hormone. Direct evidence of this was first provided by Hartwig (1940) on skin and tail fin reduction in urodeles, following local implantation of thyroxine-containing pellets. Further work by Kollros (1943) on certain brain cells in anurans was extended by others. With improvement of the technique to permit longer release of the hormone from the pellet, local implants were used repeatedly in analyzing certain of the responding systems in anurans. The special value of the technique of local action was especially well displayed in studies on the thinning and perforation of the skin window of the opercular integument. At least five theories concerning the cause of opercular perforation have been advanced, but the work of Newth on *Xenopus* and of Kaltenbach on *Rana* have established that perforation occurs as a result of local specification of the integument to atrophy in response to a thyroid hormone stimulus. These results, as Newth points out, throw doubt on the process of opercular perforation as a case of double or triple assurance. In contrast, perhaps, one should point out that no objection has yet been taken to Helff's findings that tympanic membrane formation is only indirectly dependent upon thyroid hormone, while being directly dependent upon influences from the annular tympanic cartilage.

As an adjunct notion to that of dissociability of responses in metamorphosis is that of the discreteness of the response of the various parts. Thus Champy (1922) showed growth and regression responses of leg bud and opercular chamber lining at a cell-sharp boundary, without a transition zone. Similarly, degeneration of Mauthner's cell, during metamorphosis, is accompanied by growth of the adjacent hindbrain neurons. It is quite clear that some tissues do not respond to thyroid hormone by obvious morphological change; others grow; and still others degenerate. Some tissues respond by sharp increases in the rate of mitotic activity. This was reported in detail by Champy (1922). Weiss and Rossetti (1951) reported increased mitotic activity in the hindbrain of tadpoles in response to a local hormone source. Baffoni (1957; 1958) has demonstrated a similar rate change in *Bufo* and *Triturus* larvae occurring normally during phases of metamorphic climax. In *Bufo* both the alar and basal plates participate in this increase, whereas in *Triturus* only the alar plate shows such an increase. These changes in mitotic rate could be duplicated to some degree by generalized stimulation in strong thyroxine solutions.

It has long been established that the metamorphic action of thyroxine and related compounds need not be correlated with the capacity of these compounds to stimulate metabolism in the mammal. This has been especially sharply pointed out recently by the finding that triiodothyropropionic acid is, on an equal weight basis, only as effective as thyroxine in mammals, but is 290 times as effective as thyroxine in bringing about shortening of the tadpole in metamorphosis. They also report that tetraiodothyroformic acid has less than one-fifth the metamorphosis-stimulating activity of DL-thyroxine.

Recent studies in our laboratory, using these two last named analogues in very dilute solutions to stimulate metamorphosis in hypophysectomized tadpoles, yield two extremely interesting results. First, the relative activity of these compounds, tested over an effective period of months, rather than days, is much less different than the factor of some 1500 times suggested by Roche and his co-workers. Perhaps 300 or 400 times would be a much more likely figure. Second, these compounds show some degree of selectivity in their action, so that certain sequences of events proceed somewhat differently for one compound than for the other. The propionic acid derivative shows greater precocity, relative to other events, of tail fin reduction and skin window formation and perforation. The formic acid derivative shows unusually early stimulation of skin gland and plical growth. It also tends to produce an unusual twisting of the reducing tail tip, seen only slightly in the tadpoles treated with the triiodothyropropionic acid and never seen after thyroxine treatment. Both derivatives produce gross enlargement of the skin windows, compared to tadpoles treated with thyroxine. Under the circumstances of the treatment, apparently, the capacity for degeneration of the opercular skin is not limited to the area destined to form skin window alone, but is widely distributed over essentially all of the ventral and lateral skin coverings of the branchial chamber, resulting in the massive enlargement of the skin windows and their fusion in the midline. Throughout this period the gills remain functional. They become clearly visible from above and are somewhat reminiscent of gills in urodele larvae, except that they are deficient in pigment cells, thus appearing intense red in colour,

In addition to the kinds of studies on thyroxine analogues carried out by Roche and his co-workers, and the ones just reported above, are those by Kaltenbach (1957). She utilizes a large series of thyroxine analogues, compressed into pellets with cholesterol, and places these into the dorsal fin of frog tadpoles. Effective compounds, released from the pellets, produce localized fin reduction. Tests of this sort permit her to relate effectiveness of action to specific chemical structure. Recently she has demonstrated that cortical steroids, mixed with the thyroid hormones in the pellet, enhance the activity of the thyroid hormone in bringing about fin regression.

A summary listing of factors involved in the control of amphibian metamorphosis must include:

1. The concentration of the hormone in the circulation. The amount is negligible at first, but detectable about one-third of the way through larval life. The amount increases slowly, then more rapidly at metamorphic climax. These details are presumably under control of the hypophysis.
2. Readiness to respond. Presumably sensitivity to the hormone is well established before appreciable quantities of the thyroid hormone are released.
3. Thresholds. These differ from tissue to tissue, and with increasing hormone concentration are partly responsible for the sequence of events.
4. Specific changes in rate of response of the different tissues as hormone concentration increases.
5. Some temperature thresholds. Below 5 or 7°C metamorphic changes apparently cannot occur. It would be instructive to see if this temperature differed for different tissues. Above this critical level, temperature affects specific thresholds for particular events.
6. Role of the different active compounds released by the thyroid gland. Each of these may be thought of as partially responsible for each of the events of metamorphosis. To the extent that production and release of these various entities may vary during the course of metamorphosis, a slightly different role, from time to time, must be assigned to each entity.

The transforming tadpole must be viewed as an extremely complex mosaic of parts, many of which are sensitive to thyroid hormones, and respond to them in specific ways. The manner, nature, and rate of the response varies with hormone concentration, and perhaps to some degree with different proportions of the active molecules released. Ordinarily these various reactions are so marvelously integrated as to produce the normal juvenile amphibian.

10

THYROID IN VERTEBRATES

In recent years the attention of biochemists has been focused on three separate aspects of thyroid hormonal physiology: first, the nature of the hormones produced in and secreted by the gland; second, the nature of the hormones in the circulating blood plasma; and third, the nature of the hormone which acts in the peripheral tissues. Within the thyroid the attention of investigators has shifted from monoiodohistidine, mono- and diiodotyrosine, and thyroxine to the other thyronines which are produced in much smaller proportion. To date, thyroxine, 3:5:3′ triiodothyronine, 3:3′:5′ triiodothyronine, and 3:3′ diiodothyronine have been demonstrated within the thyroid, and all of these, as well as several unknowns, in the plasma. Thyroxine and 3:5:3′ triiodothyronine are the only two of the secreted thyronines definitely known to have biological activity. The others occur in minute proportions, if at all, and are of questionable activity.

The complex of hormones deriving from the gland has proved easier to identify than the form which is active in the peripheral tissues. Because of its surprisingly high biological activity, 3:5:3′ triiodothyronine has been suggested as the final form of the hormone but some recent data contradict this possibility. It is, of course, conceivable that the peripheral hormone is not one—but a group of substances including thyroxine, triiodothyronine, and their deaminated metabolites. So far, however, the deaminated analogues, such as triiodothyroacetic acid have been demonstrated only in peripheral tissues and only after the administration of radioactive thyroxine: they were never found after injection of radioiodide. Thus, the picture of thyroidal hormone synthesis which seemed so clear a few years ago has been obscured, as much as clarified, by the discovery of the numerous hormonal variants which, though small in proportion, are of high physiological activity and unevaluated significance.

As might be expected, most of the information that has been gathered concerning the nature of the thyroid hormones was derived from work with adult mammals, in particular the rat, although the high potency of 3:5:3´ triiodothyronine was well demonstrated in the tadpole metamorphosis assay. It now becomes important to learn the phyletic distribution of thyroid hormones and to recognize possible ontogenetic and phylogenetic progressions of change in hormone synthesis and hormone function in embryos and in lower vertebrates. The first question to present itself is this: is thyroxine a characteristic hormone of all vertebrate thyroids, or do different vertebrate groups utilize different ones of the iodinated amino acids? Until recently, before most of the circulating thyroid hormones were known, there seemed to be no reason to suspect that some thyroids might make a different hormone from others. Now, however, this possibility has been advanced by several kinds of experimental evidence. Smith and Matthews (1948) have shown a metabolic response in the fish *Bathystoma* after treatment with thyroids of the parrot fish. Thyroxine treatment has failed to elicit a metabolic response in any one of the many species of fish in which it has been tried. This is at least suggestive that the thyroid in fishes manufactures a hormone other than thyroxine. A similar conclusion might appear suggested by the relatively extreme sensitivity of the frog tadpole to triiodothyropropionic acid. This possible metabolite of thyroxine is 300 times as potent as thyroxine itself in inducing metamorphosis in tadpoles, *Rana pipiens*.

In order to explore the possibility of the existence of specific evolutionary patterns of thyroidal hormone synthesis, a large number of species must be examined. Fortunately, sufficient analyses of thyroxinogenesis have now been made to permit some generalizations about the most abundant iodinated compounds (iodides, monoiodotyrosine, diiodotyrosine, 3:5:3´ triiodothyronine, thyroxine). Most of the work is recent, and none older than eleven years, but progress in thyroid research has been so rapid that chromatographic analyses made in 1953 did not in most cases carry either diiodothyronine or triiodothyronine as standards. For this reason slightly older information provides little or no enlightenment with regard to these two important thyroid compounds.

IODINE METABOLISM IN NONVERTEBRATES

It appears a safe generalization that of the many organisms that can be shown to metabolize iodine, all make at least monoiodotyrosine and diiodotyrosine with this element. This is even true of marine algae, although here the mechanism of iodide accumulation and oxidation of iodide to iodine may perhaps differ from that in the thyroid gland. However, despite the possible difference in mechanisms, the first iodinated amino acid formed is monoiodotyrosine.

Mono- and diiodotyrosine, as well as iodinated histidines, have been reported as part of iodoproteins throughout the invertebrate world. A high percentage of the protein-bound iodine is in the horny or fibrous structures, but some is in the soft organs—most often in epithelial tissues. Granting the iodination of tyrosines, one might expect thyroxine to be occasionally formed, but it was a surprise to find that both the bivalve *Musculium partumeium* and a number of insects have as much of the total tracer iodine in the form of thyroxine as a mammal. A small part of the high proportion of thyroxine in insects could have been due to unseparated triiodothyronine. The cockroach, the only insect examined organ by organ, concentrated thyroxine mainly in the nerve cord, fat body, and muscle. Nerve cord is of special interest since, in the mammal, thyroxine is both accumulated and metabolized there. Mammalian skeletal muscle, too, has been found to have a protein that binds thyroxine. It is not yet known whether thyroxine is produced in the nerve cord and muscle of the insect, or is produced elsewhere—nerve cord and muscle being sites of concentration. Two other thyronines have been identified chromatographically in invertebrates, 3:5:3′ triiodothyronine in a gorgonian, *Eunicella verrucosa stricta*, and both triiodothyronine and diiodothyronine in the snail, *Planorbis corneus*, an unusual animal since it made two relatively rare iodinated thyronines, but no thyroxine.

These facts, demonstrating the widespread occurrence of thyroid hormones and of hormonal precursors in thyroidless invertebrates, must be taken into account in any theory of evolution of thyroid function. The mere occurrence of iodoproteins, even in high concentration, cannot be considered as evidence that their thyroxine, or thyroxine-like constituents are being used in any particular invertebrate organism. In fact, no clear evidence has yet been presented to show that treatment of an invertebrate with thyroid hormone results in any kind of response. Thus, the most basic properties of thyroid glands, iodide transport, and protein binding of iodine, occur for no known purpose in many plant and animal organisms. The question of the evolution of a special vertebrate organ containing these properties is, therefore, really one of development of a metabolic use for this common substance, thyroxine.

Radioiodine Metabolism in Fish

Only two aspects of thyroidal metabolism of fish will be considered: radioiodine accumulation and radiohormone production. In general, thyroids of salt-water fishes living in an iodine-rich environment accumulate less radioiodine and thyroids of freshwater fish more, but there is a considerable range of overlap. The least efficient freshwater thyroids, i.e., those with the lowest avidity for iodine, were found in the goldfish, *Carassius*, and the sunfish, *Lepomis*. Chavin (1956) has shown the goldfish to have heterotopic thyroid tissue in the head kidney, and when both thyroids were taken into account the goldfish accumulated almost as much of an injected dose of radioiodine as the iodine-starved freshwater fish of the Great Lakes area. A more detailed quantitative study in different seasons of *Fundulus heteroclitus*, a fish that can live in fresh as well as salt water, has resolved some questions. Thyroidal

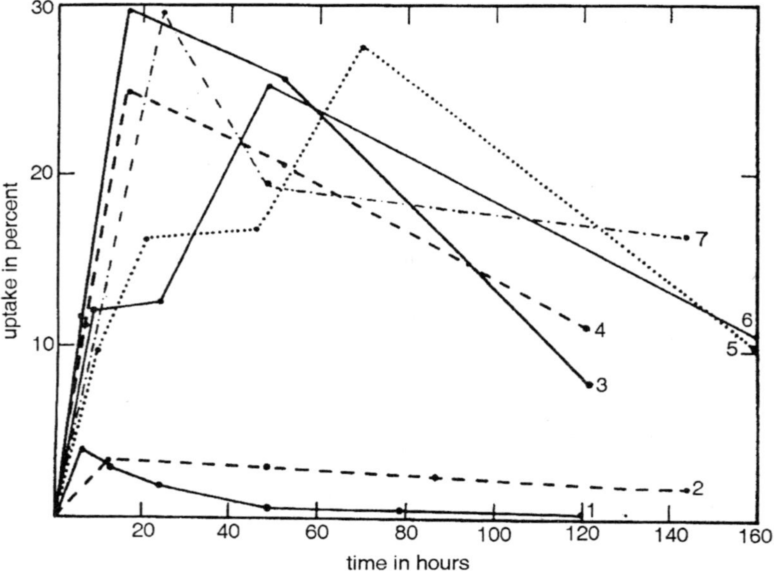

Fig. 10.1. Radioiodine uptake and loss over a period of time in seven species of fresh water teleosts maintained at 21°C. 1. Carassius; 2. Lepomis gibbosus; 3. Percina caprodes; 4. Notropis deliciosus; 5. Umbra limi; 6. Umbra pygmaeus; 7. Xiphophorus maculatus.

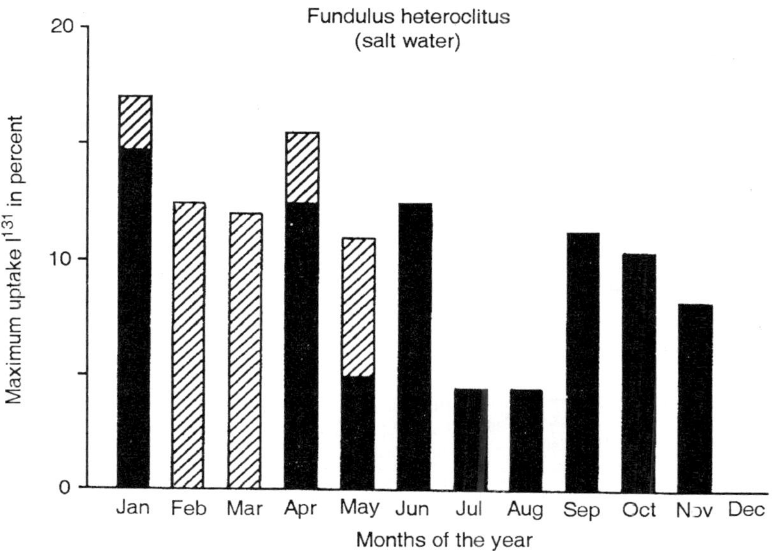

Fig. 10.2. Maximum accumulation of radioiodine by the thyroid of the killifish during eleven months of the year.

I^{131} uptake varied with season, being four times as great in January as in July. This quantitative variation was not well correlated with the seasonal variation in thyroxine production, nor was it correlated with thyroidal epithelial cell height or with photoperiod. It was not well correlated with sexual activity. The seasonal rise and fall was found over several seasons; it occurred whether the fish were kept in sea water, in freshwater, or in diluted sea water whose salinity was restored with iodine-free salts. The latter was specially prepared to check whether radioiodide uptake by the fish thyroid responded to the general salinity of the sea water or to its high iodine content. Under our experimental conditions no thyroidal response to salinity *per se* was seen, possibly because the experiment was terminated after ten days. The response of the mammalian thyroid to sodium chloride does not become evident for several weeks. From this we can only conclude that radioiodine accumulation by the thyroid gland is not a good exclusive index of thyroidal activity in fishes because it is influenced by the iodine content of the water and by the season, and does not seem to be correlated with the amount of thyroxine produced.

Thyroxine has been found in almost every fish tested, from the ammocoetes larva to the lungfish *Protopterus*. Our own negative results with several species may be accounted for, at least in part, by the

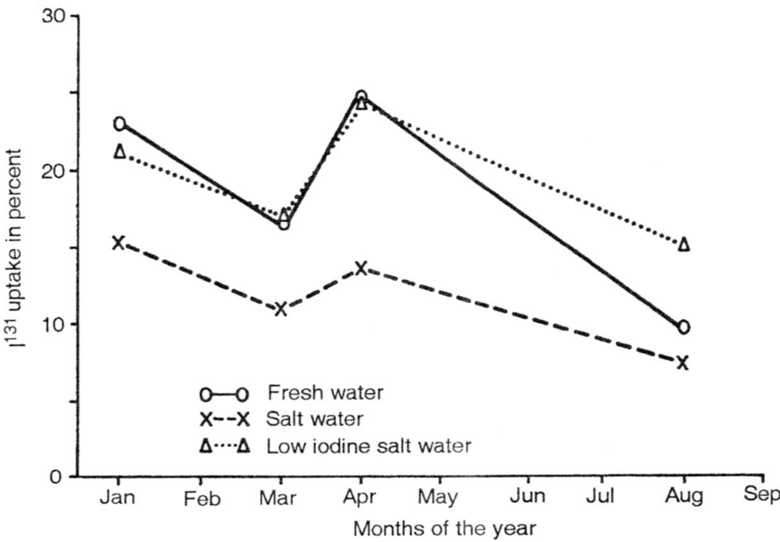

Fig. 10.3. Radioiodine uptake and loss over a period of time in Fundulus heteroclitus.

fact that thyroxine is often slow to appear and is present only in trace amounts or only in certain seasons. Looking for a seasonal pattern in thyroxinogenesis we analyzed iodine utilization in *Fundulus heteroclitus* each month for a year. *Fundulus* made no newly labeled thyroxine within test periods of more than five days in August through November; its thyroxine production started slowly in January, was steady until April, reached a peak in June, and tapered off until it reached zero again in midsummer. The above data were from *Fundulus* kept in salt water, but *Fundulus* simultaneously captured and kept in fresh water in March made about as much radiothyroxine. In this laboratory experiment, temperature was not a factor for thyroxine production, since under our experimental conditions all fish were acclimatized and maintained at the same constant temperature. The *Fundulus* with the most active thyroids (in respect to thyroxine production) were actively breeding, and those with quiescent thyroids had spent gonads, so sexual cycle correlates well with thyroxine production. Photoperiod is another possibility, but was not tested separately.

Triiodothyronine has proven difficult to demonstrate in fish. It was present in *Fundulus* only when the gland was most active, and then only in trace amounts, and has been demonstrated in *Protopterus* in relatively small proportion. We were, therefore, surprised to find one group of mud minnows, *Umbra limi*, which made a considerable

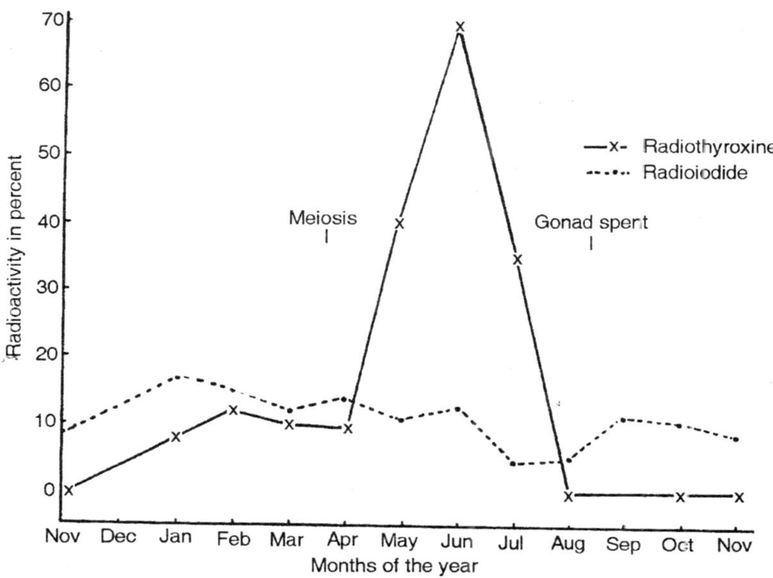

Fig. 10.4. Maximal radioiodine uptake and maximal radiothyroxine production in Fundulus heteroclitus during eleven months of the year.

amount of 3:5:3′ triiodothyronine, and no thyroxine. Three series of experiments were conducted, one in Michigan in August, using Michigan *Umbra limi*, and two in New York in March and December, using *Umbra* shipped from Minnesota and Wisconsin. In the Michigan series, 54 hours after injection of fish kept at 15°C, about 7 per cent of the thyroidal I^{131} was converted to radiotriiodothyronine, but no radiothyroxine was produced. At room temperature, at approximately the same time interval (48 hours) another series of fish from Minnesota had converted about 10 per cent of the I^{131} to radiothyroxine in addition to 4.5 per cent in the form of radiotriiodothyronine. Two groups of *Umbra* from Wisconsin kept at different temperatures made no appreciable amounts of radiotriiodothyronine, but at the lower temperature labeled thyroxine production was very high. Leloup and Lachiver (1955) suggest that a paucity of environmental iodine may determine triiodothyronine production. This condition certainly prevailed for the mud minnows in Michigan, where iodine was low, while the requirement may have been quite high, as the fish were used shortly after breeding.

Temperature was not a variable in these experiments, but there are indications that it has an effect on hormonogenesis in nature—

although not necessarily the same effect in different species. The *Umbra* mentioned above, which formed radiotriiodothyronine, formed more of it at a lower temperature and less at a higher temperature. Similarly, the December-March *Umbra limi* formed more new thyroxine at lower temperatures and less at higher temperatures. This may be an example of temperature acclimatization in poikilothermic organisms and suggests a possible mechanism for such acclimatization.

Only Leloup has examined the blood of fishes for iodine. He has demonstrated iodide in the red blood corpuscles, and iodide, thyroxine, and triiodothyronine in the plasma. There is no uniformity in the blood iodine distribution in various fishes—some have erythrocytes which are very permeable to iodide, in others the erythrocytes are almost free of iodine. In some species thyroxine appeared rapidly in the plasma, in others only after many hours. Once again, the tests defined a possibly transitory functional state of the thyroid and will remain difficult to interpret until patterns are available for all seasons.

With our present knowledge there is no indication that the thyroid hormone in fishes is qualitatively different from that of mammals. The peripheral target of the hormone is unknown, and so is its physiological effect.

Iodine Metabolism in Amphibia

Frogs are remarkably sensitive to thyroxine analogues—as judged by their metamorphic rate. Triiodothyropropionic acid is the most potent of all the analogues, being 300 times more efficient than thyroxine in *Rana pipiens* and 130 times more effective in *Rana catesbeiana*. In general the 3:5:3′ triiodinated analogues are more effective than their tetraiodinated counterparts, just as triiodothyronine is usually more effective than thyroxine. In addition to the analogues, injections of iodinated proteins speed up metamorphosis.

In species that respond so sensitively to the analogues the question of the nature of the "true" hormone is raised. It is necessary to remember that the exceptional potency of an analogue may be an indication that it is not a natural metabolite. A receptor cell must have both a means of binding a hormone and a mechanism for removing it or its waste product. If the analogue is bound in the same manner as the hormone but is not removed by the usual cellular mechanism, then a small quantity may give continued stimulation creating the illusion of superior specificity. Therefore, no analogue should be considered as a possible thyroid hormone until it can be shown that it is made by the animal.

As part of a larger study on iodine metabolism in neotenous urodeles, we implanted thyroxine pellets into *Amphiuma means* adults. Two weeks later, they were given a tracer dose of 1^{131}, and 2 to 4 days after injection they were sacrificed. On the basis of these experiments, the thyroid of *Amphiuma* is very inactive. Although the salamanders were kept at 20°C, iodine accumulation was very slow, and the steps of conversion to thyroxine even slower. Ninety-six hours after injection there was more mono- than diiodotyrosine, and no triiodothyronine or thyroxine. This may be directly referable to the season (the experiment was done in winter). In any event, thyroxine is not without effect in *Amphiuma*, since it decreased radioiodine uptake quite sharply even below its normal low value of less than 1 per cent.

A similar study of *Necturus maculosus*, a neotenous salamander that retains its gills, showed an even more sluggish thyroid than that of *Amphiuma*. A tracer dose of 20 μc of I^{131} was given one week after implantation of thyroxine pellets into some of the *Necturus*, and the animals were sacrificed 24 to 192 hours later. Uptake of radioiodine was very low (maximum, 1.5 per cent of the injected dose), and no thyroxine at all was produced within eight days after injection. It would be tempting to conclude that neoteny in *Amphiuma* and *Necturus* is due to thyroidal inactivity. However, as is well known, neither of these species responds to treatment with large doses of thyroxine. These tests were made from April to June; the mating season is in the fall. There is some indication, on the basis of radioactive tracer studies, that in a salamander, as in mammals, thyroxine is excreted through the liver. Schmidt (1956) has shown some organic iodine in the bile of *Ambystoma gracile*; the chemical form of the biliary iodine is not yet known.

Radioiodine Metabolism in Reptiles and Birds

The thyroid gland of turtles kept without water, although very slow, eventually accumulates the most iodine of an injected dose of any vertebrate species tested to date. According to Shellabarger et al. (1956) the unusual maximum uptake of 80 per cent of the injected dose of I^{131} may be made possible by reabsorption of radioiodine from the urinary bladder, which is seldom voided when the turtles are kept in a dry environment. There is a correlation between season and thyroidal iodine binding, but the high iodine accumulation is not linked with a high thyroxine production. A complete seasonal study has not been made, but in turtles, as in fish, there are indications that thyroxine is produced in some seasons and not in others.

Both species of turtles tested by Shellabarger et al., one terrestrial the other aquatic, behaved similarly in terms of thyroidal accumulation of radioiodine during periods of dessication. To learn whether this is a feature common to other reptiles, we have completed a series of tracer-iodine studies, under wet and dry conditions, with two lizards: *Anolis carolinensis* and *Sceleporus occidentalis*. Peak uptakes under dry conditions of 12 per cent and 25 per cent, respectively, are well in line with nonreptilian species. The absence of water does not seem to have the effect on lizards that it has on turtles. Both species of lizards produced some thyroxine and a trace of triiodothyronine, differing again from the turtles. In addition, thyroids of both lizards produced traces of two unknown substances. In the lizards both unknowns were present in minute quantities; they were of possible interest because they were common to both species. Furthermore, these substances characterized

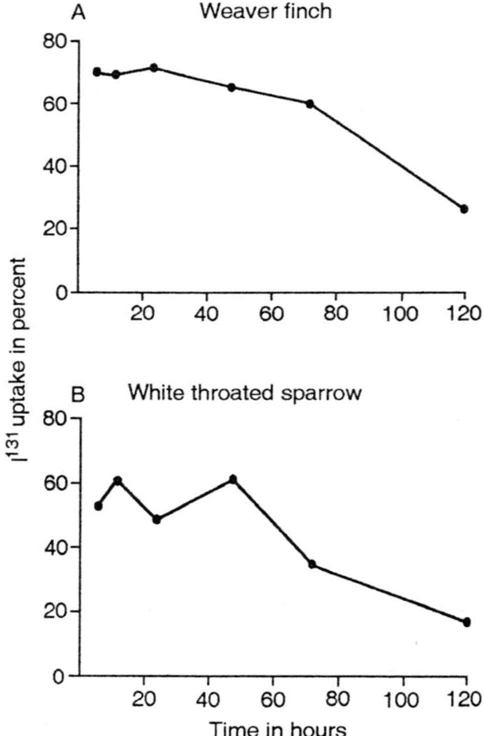

Fig. 10.5. Radioiodine uptake and loss over a period of time in the weaver finch and the white throated sparrow.

clearly by chromatographic analysis, were found not only in lizards, but also in even greater proportions in two species of birds tested in our laboratory.

There have been several reports on utilization of radioiodine by chicks, the most extensive being a study of cockerels by Vlijm (1958). Iodine utilization by the thyroid of the adult cockerel is not remarkable, except that the cockerels produced 3:5 diiodothyronine as well as mono- and diiodotyrosine, 3:5:3′ triiodothyronine, and thyroxine. Both the triiodothyronine and the diiodothyronine deserve comment. The former has also been identified recently in the chick by Shellabarger and Pitt-Rivers (1958). Its presence is not unexpected, but: the fowl is the only form tested in which the triiodinated compound is not more potent than its tetraiodinated homologue, thyroxine.

The domesticated chick may not be a very good index of avian thyroidal function in general. In the white throated sparrow (*Zonotrichia albicollis*) and the weaver finch (*Euplectes pyromelana*) tested by us shortly after capture, thyroidal iodine accumulation is extremely high and rapid (maximum of 60-70 per cent of injected dose less than six hours after I^{131} injection). Their diet in our experiments (canary seed, 2 micrograms of iodine per 100 grams of seed) was low in iodine, but this may not have been the only influential factor. Uptake was rapid, and was followed by slow discharge, and most of the iodine was gone from the gland by 120 hours after injection.

At one point the weaver finch had 20 per cent of its thyroidal radioiodine in the form of thyroxine.

11

Cortical and Chromaffin Cells

The responses an animal makes to any stressful stimulus include the release of adrenal hormones. These hormones induce changes in metabolism that work to combat stress. Knowledge of this system contributes to an understanding of how animals adapt physiologically to physical and psychological traumas.

Mammals typically possess two adrenal glands, one located superior to each kidney (adrenal or supra renal). An adrenal gland consists of an outer portion composed largely of lipid-containing steroidogenic *adrenocortical* cells, the *adrenal cortex*. The cortex surrounds an inner mass of *chromaffin* cells, the *adrenal medulla*. The adrenocortical cells are derived from the coelomic epithelium in the pronephric region of the embryo adjacent to the genital ridge that gives rise to the gonads. These cells produce the corticosteroid hormones and are under the influence of corticotropin (ACTH) from the pituitary gland. In contrast, the chromaffin cells are of neural crest origin, and the medulla functions essentially like a sympathetic ganglion. The adrenal medulla is under direct neural control (cholinergic) and releases into the blood the normal neurotransmitter of most postganglionic sympathetic neurons, *norepinephrine*. In addition, the chromaffin cells secrete an important derivative of norepinephrine into the blood, *epinephrine*. (It should be noted that these substances can be considered neurotransmitters or neurohormones according to the site of release.)

There is no functional significance for the close anatomical relationship of these two distinctly different tissues (adrenocortical and

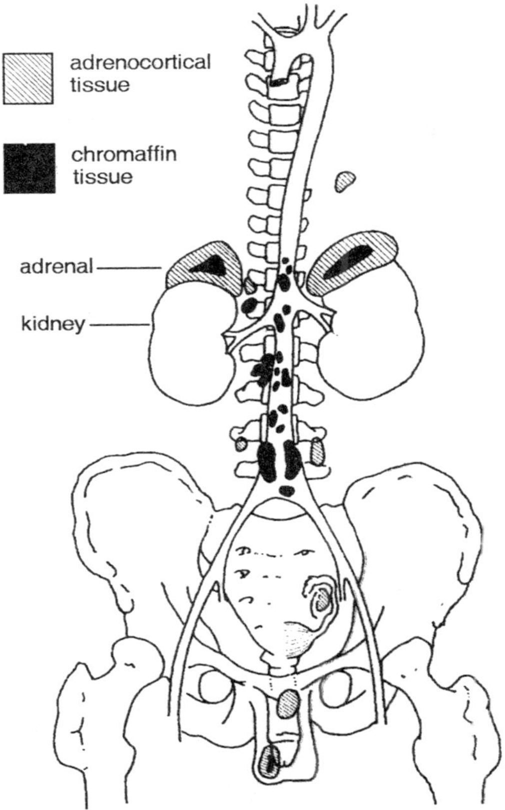

Fig. 11.1. The adrenal glands and distribution of adrenocortical and chromaffin tissues in humans.

chromaffin). Although there is participation of both systems with respect to adaptations to stressful stimuli, the factors controlling release of their secretions are not obviously related nor do their biological actions overlap. The anatomical closeness of the cortex and medulla in mammals as well as relationships of their homologous tissues in nonmammals may simply be a function of the physical closeness of their embryological sites of origin. There are some actions of ACTH and glucocorticoids on catecholamine synthesis by chromaffin cells, however.

The cortex and medulla are treated here as entirely separate endocrine glands, which is the case for many vertebrates. The adrenal cortex of mammals is discussed first and is followed by a comparative account of adrenocortical homologues in nonmammalian vertebrates. An account of the mammalian adrenal medulla and chromaffin tissues of nonmammals completes the chapter.

Mammalian Adrenocortical Cells

The adrenal cortex of adult mammals may be subdivided by means of histological criteria into three well-defined regions: *zona glomerulosa*, *zona fasciculata* and *zona reticularis*. In addition, a functional *fetal zone* is present in some fetal mammals and is related to maintenance of gestation.

The cells of the outermost region of the adrenal cortex, the zona glomerulosa, are smaller, more rounded and contain less lipid than those of the more central zona fasciculata. The zona glomerulosa is responsible for synthesis of the corticoid *aldosterone* as well as some other corticosteroids. There are few cytological changes in the zona glomerulosa following hypophysectomy or administration of ACTH, suggesting that the secretion of aldosterone is independent of pituitary control. Although ACTH is not necessary for steroidogenesis and release

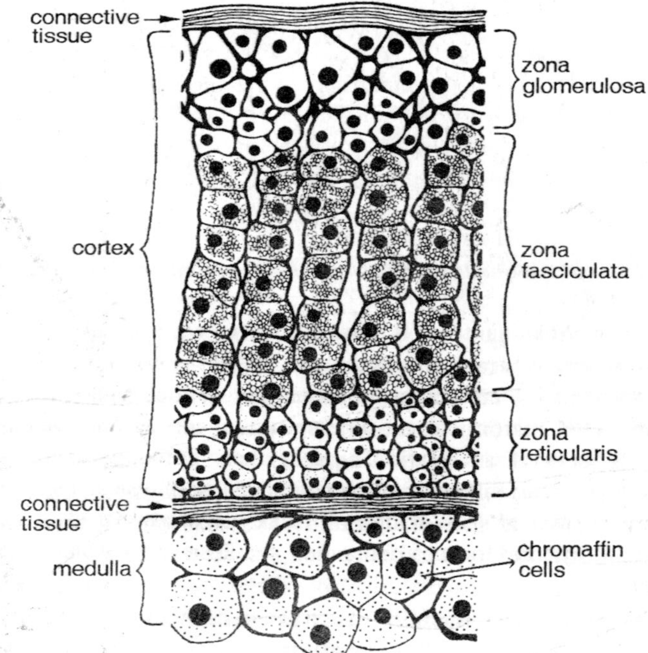

Fig. 11.2. Zonation in the mammalian adrenal.

of aldosterone, the responsiveness of the glomerulosal cells to agents that normally elicit these events is reduced in hypophysectomized mammals and is enhanced with ACTH treatment.

The zona fasciculata consists of polyhedral (many-sided) cells that are sources of the so-called glucocorticoids such as *cortisol* and *corticosterone*. This region is characterized by many blood sinusoids that allow the cells to be bathed with blood. The proportion of cortisol and corticosterone secreted differs markedly, from secretion of primarily cortisol (man), through mixtures of both (cat), to primarily corticosterone (rat). The zona fasciculata is located between the zona glomerulosa and the zona reticularis and is histologically distinct from both.

The zona reticularis typically borders the adrenal medulla, and it contains numerous reticular fibers (hence its name). Some glucocorticoids and androgens are synthesized in this region, but the majority of the glucocorticoids are secreted by the zona fasciculata. Unlike the zona glomerulosa both the zona reticularis and zona fasciculata atrophy following hypophysectomy, and treatment with ACTH restores them histologically and functionally to normal.

It should be noted that this "typical" anatomical pattern within the adrenal cortex and the relationship of cortex to medulla varies considerably within mammals as a group. Furthermore, ectopic nodules of functional cortical tissue are not uncommon, and this accessory adrenocortical tissue may become an important source for corticosteroids following adrenalectomy.

The cortex of the mouse adrenal contains a unique *X-zone* between the zona reticularis and the medulla. The X-zone degenerates in males at puberty and in females during the first pregnancy. The function of this region is not known.

Fetal Zone

In many mammals, especially the human, a very conspicuous zone occupies the bulk of the adrenal gland prior to birth. This region has been named the *fetal zone* and is responsible for the large size of the adrenal at birth. In humans the adrenal of the neonate may be as large as the total adrenal gland of a 10- to 12-year-old child. During the gestation period the fetal zone synthesizes and releases relatively large quantities of the steroid *dehydroepiandrosterone sulfate* (DHEA), which serves as a precursor for synthesis of estrogens by the placenta. Failure of the fetal zone to produce adequate amounts of DHEA results in premature termination of gestation. Following birth the fetal zone ceases to function and degenerates rapidly. The zona reticularis

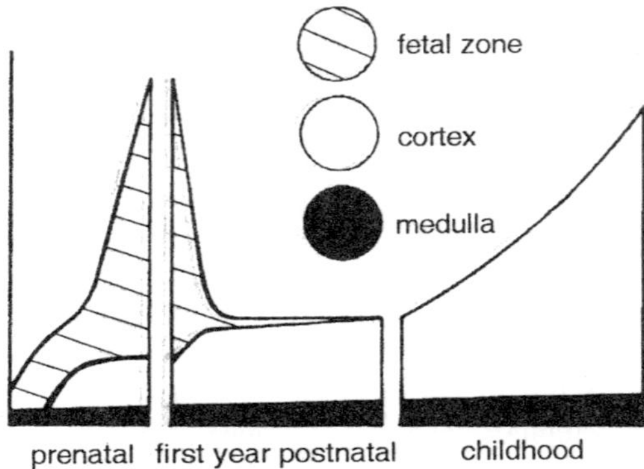

Fig. 11.3. Relative proportions of the adrenal gland represented by the fetal zone during gestation and childhood.

continues to synthesize DHEA until about age 20 when its production declines. Although it has been suggested that DHEA may be an antitumor substance and/or a precursor for synthesis of other androgens or estrogens, the function of DHEA in adults is unknown.

Biological Actions of "Glucocorticoids"

Secretion of glucocorticoids from the zona fasciculata and zona reticularis is under direct control of the hypothalamo-hypophysial axis involving corticotropin-releasing hormone and ACTH respectively. Circulating ACTH levels are depressed by elevated levels of glucocorticoids and are increased following adrenalectomy, establishing the existence of direct negative feedback of corticosteroids on ACTH release. Corticotropin is released by direct neural input to the hypothalamus as a result of such stressful events as physical trauma (injury, surgery), and this response may occur even in the presence of elevated glucocorticoid levels. Thus, stressful stimuli can bring about and maintain a severalfold increase in glucocorticoid levels. The ability of traumatic stimuli to override the normal negative feedback mechanism verifies the importance of neural control over ACTH release.

Glucocorticoids produce marked effects on energy metabolism at physiological doses as a result of changes in transport of materials into cells and induction of new enzyme syntheses. Their major effect is to supplant and conserve the energy normally derived from circulating glucose by (1) inhibiting glucose utilization by peripheral tissues (for example, muscle), (2) stimulating entry of amino acids into cells and their conversion to glucose and storage as glycogen and (3) stimulating mobilization of fat stores. It has been suggested that the normal physiological role for glucocorticoids is that of "permissive agents" that through changes in membrane permeabilities to important metabolites and the synthesis of new enzymes provide the appropriate cellular environment in which other hormones such as growth hormone and glucagon may operate.

Secretion of glucocorticoids follows a circadian pattern that is thought to be entrained by the light-dark cycle. Studies in rats and humans suggest these cycles are regulated by food intake rather than by light-dark or wake-sleep cycles. Elevated corticoids may be related to their metabolic actions.

Glucocorticoids and stress

The responses characteristic of adaptations to chronic stress are mediated by glucocorticoids. These responses have been incorporated by Selye into a general theory of adaptation to stress termed the *general adaptation syndrome*. According to Selye, there are three stages of adaptation by an organism to stressful stimuli. Stress is used as an all-encompassing term to include all stimuli that are harmful or potentially harmful to the organism. The first phase of the general adaptation syndrome is the *alarm reaction*, which includes a generalized increase in sympathetic stimulation involving the adrenal medulla as well as increased secretion of glucocorticoids. This phase is frequently marked by enlargement of the adrenal glands, primarily due to hypertrophy of the zona fasciculata and to some extent the zona reticularis of the cortex.

Under the influence of glucocorticoids the organism adapts to the continued presence of the stressful stimuli. This phase of adaptation is termed the stage of *resistance* and is characterized by prolonged increased secretion of glucocorticoids. (It can be argued whether this is truly adaptation or the lack of adaptation.)

Finally, under continuation of extremely stressful conditions, the ability of the organism to function normally is impaired. The continuous presence of the stressful stimuli causes the organism to enter the final stage of *exhaustion* that leads to death.

Stress of experimental animals is important when attempting to interpret data on corticosteroid levels or nutrient levels (amino acids, glucose, etc.). Laboratory conditions alone can influence the adrenal axis. Stress can also influence the levels of other hormones. For example, the order in which groups of rats were removed from a common holding facility and killed at a remote site influenced the mean levels of PRL that were measured for each group. Since it is not possible to undertake experimental work on animals free from stress, detailed knowledge of how animals were maintained and all procedures involved with an experiment is essential when interpreting results.

Pharmacological actions of glucocorticoids

The glucocorticoids are better known for their pharmacological actions and therapeutic side effects than for their biological actions. The tremendous potential for glucocorticoids in the treatment of the rare hypoadrenocorticism described by Thomas Addison in 1885 (Addison's disease) was not recognized for many years. However, the discovery of the anti-inflammatory effects of cortisone, a synthetic glucocorticoid, and its use for treatment of rheumatoid arthritis spurred a tremendous explosion in therapeutic applications of glucocorticoids. The debilitating symptoms of rheumatoid arthritis relieved by glucocorticoid therapy are the consequence of inflammation associated with an autoimmune response in which the patient produces antibody against his own connective tissue. Glucocorticoid therapy alleviates painful inflammation occurring as a result of the immune reaction but does nothing to correct the causative factors.

One of the most recent therapeutic applications of glucocorticoids is suppression of the entire immune response following tissue and organ transplantations. Although normal therapeutic doses (anti-inflammatory) of glucocorticoids do not interfere with normal antigen-antibody interactions, very high doses depress new antibody synthesis.

Glucocorticoids interfere with the elaboration of histamine or with its actions in mediating the inflammatory response, which includes local hyperemia and resultant edema, or with both. One postulated mechanism for this interference is the inhibition of the *kallikreins*, which catalyze formation of *kinins* from a plasma precursor protein. Kinins also induce inflammation. The release of histamine normally observed following the combination of antigen and antibody is caused by kinins. Another suggestion for glucocorticoid anti-inflammatory activity stems from observations of their effects on lysosomes. Glucocorticoids stabilize lysosomal membranes, thereby reducing release of hydrolytic enzymes following cell injury and hence reducing the spread of the inflammatory reaction. Glucocorticoids may inhibit the synthesis of leukotrienes and reduce inflammation.

Mechanism of glucocorticoid action

The initial requirement for glucocorticoid action on liver cells is binding to intracellular receptors and eventual stimulation of nuclear RNA synthesis (both messenger RNA and ribosomal synthesis).

Approximately 2 to 4 hours following application of glucocorticoids there is an increase in new enzymes that bring about the changes in cellular metabolism characteristic of glucocorticoid action. In liver cells these new enzymes include those associated with the conversion of amino acids into glucose and the polymerization of glucose to form glycogen. Amino acid transport into the liver cell is also stimulated. In contrast, glucocorticoids inhibit the uptake of amino acids and the metabolism of glucose in peripheral tissues such as skin and adipose cells.

Excessive doses of glucocorticoids inhibit protein synthesis in certain tissues (for example, muscle, bone and lymphoid tissue), which has been related to some of the adverse side effects of glucocorticoid therapy. These protein catabolic effects are not manifest in liver cells even at very high doses. The basis for this difference is not known.

Aldosterone: The Principal Mammalian "Mineralocorticoid"

The zona glomerulosa secretes aldosterone independently of direct pituitary control, although, as mentioned earlier, ACTH appears to play a permissive role in maintaining the responsiveness of these cells to other controlling factors. The major action of aldosterone is maintenance of the normal sodium-potassium balance in body fluids, and its secondary action is to regulate extracellular fluid volume. Aldosterone stimulates sodium reabsorption by the nephrons in the kidney. The mechanism controlling secretion of aldosterone involves a most complex and seemingly round-about series of events involving both liver and kidney, the *renin-angiotensin system*.

Control of aldosterone secretion

Renin-angiotensin system

Renin is an enzyme produced in the kidney by the *juxtaglomerular body*, a modified group of cells in the afferent arteriole carrying blood to the glomerulus. This enzyme is a glycoprotein (about 40,000 daltons) possibly secreted as a larger, inactive form (mol wt 63,000) or prorenin. Conversion of prorenin to renin may be accomplished by the activity of kidney kallikrein. Development of renin activity from prorenin also occurs following mild acidification of the plasma. The juxtaglomerular body is intimately associated with a modified region of the distal convoluted portion of the nephron known as the *macula densa*. Together these two structures comprise the *juxtaglomerular apparatus*.

Blood volume (pressure in the renal arterioles) and sodium concentration in the glomerular filtrate as it enters the proximal

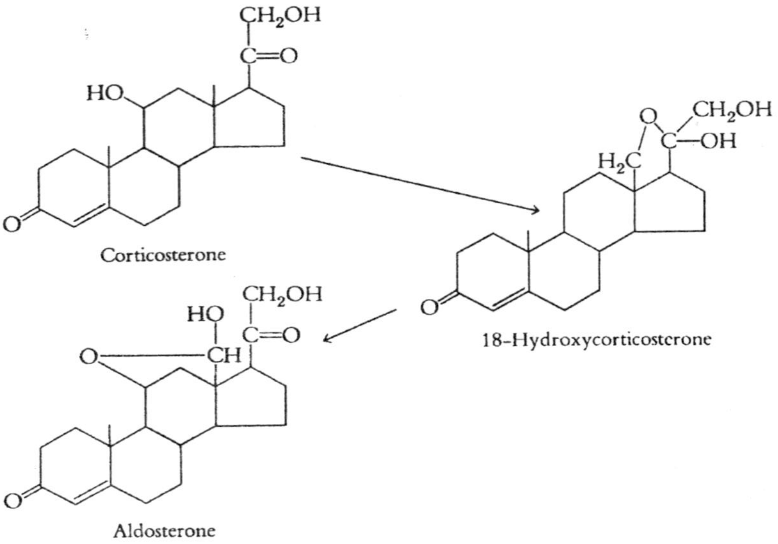

Corticosterone

18-Hydroxycorticosterone

Aldosterone

Fig. 11.4. Synthesis of aldosterone from corticosterone.

convoluted tubule control renin release. Intrarenal arteriolar pressure is monitored by stretch receptors in the juxtaglomerular body. Renin is released in response to a decrease in this pressure. Sodium concentration in the tubular lumen is monitored by cells of the macula densa, and low sodium levels somehow trigger communication between the macula densa and the juxtaglomerular body, resulting in renin release. Changes in either or both parameters influence renin secretion.

Once renin enters the blood it comes in contact with a plasma protein termed *renin substrate* that was synthesized by the liver. Mammalian renin substrates are large glycoproteins (mol wt 58,000-110,000) and behave like α_1-globulins, α_2-globulins and albumin in humans, herbivores and rodents respectively. Renin causes the enzymatic release of a small peptide (decapeptide) known as *angiotensin I* from renin substrate (angiotensinogen). Angiotensin I facilitates release of norepinephrine from the adrenal medulla and produces direct and indirect pressor effects on the cardiovascular system. One effect is the elevation of intrarenal blood pressure that is an important contributor to pathological renal hypertension. A *converting enzyme* converts angiotensin I to an octapeptide, *angiotensin II*. This octapeptide is also a potent vasoconstricting agent and helps restore blood pressure to normal by decreasing arteriole diameter. Converting enzyme is localized to a number of capillary beds, suggesting that angiotensin II produces

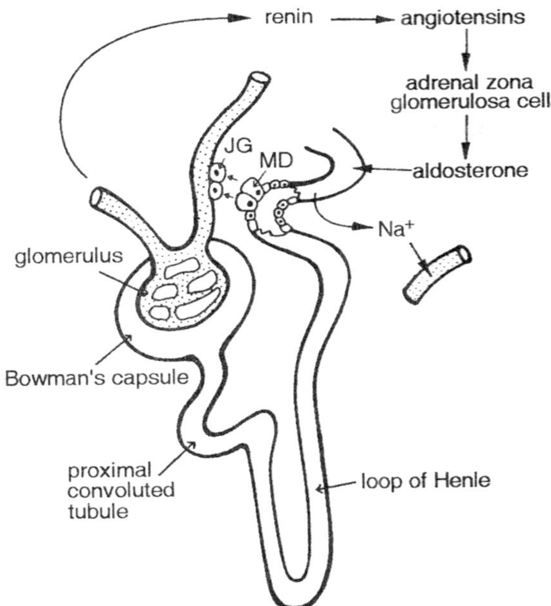

Fig. 11.5. The juxtaglomerular apparatus. JG, afferent arteriole; MD, macula densa.

vascular effects in a number of tissues. The second important action of angiotensin II is the causation of increased synthesis and release of aldosterone from cells of the zona glomerulosa. Angiotensin II may be further metabolized to a heptapeptide known as ***angiotensin III*** that may also stimulate aldosterone release. The importance of this latter

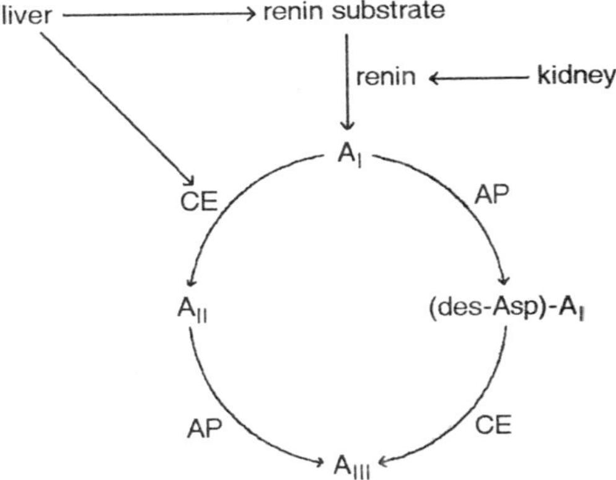

Fig. 11.6. Synthesis of angiotensins.

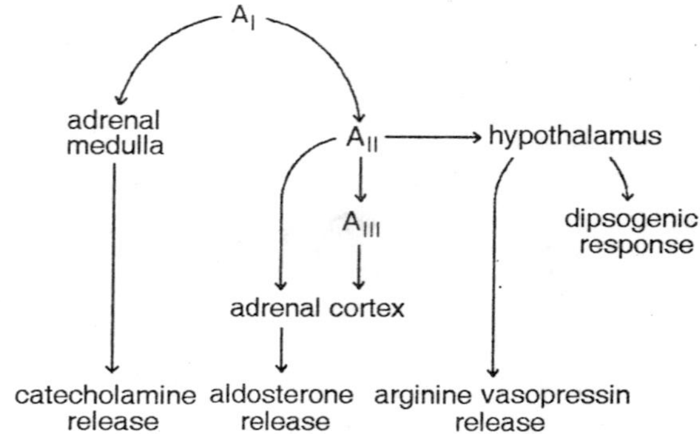

Fig. 11.7. Action of angiotensins.

conversion is not clear but may be related to stepwise hydrolysis and inactivation. Aldosterone stimulates increased reabsorption of sodium and some increased excretion of potassium by cells of the distal convoluted portion of the nephron as well as increased sodium reabsorption by the proximal convoluted portion. This increased sodium reabsorption aids in water retention, reduces urine volume and helps to restore fluid volume. The renin-angiotensin-aldosterone mechanism is closely linked functionally to the role of vasopressins (arginine and lysine vasopressins) in maintaining normal blood osmotic conditions. Angiotensin II stimulates drinking (dipsogenic response) and causes release of vasopressins through actions on the hypothalamus. These two effects contribute to the maintenance of normal fluid balance.

This elegant but rather circuitous mechanism for regulation of extracellular fluid volume and sodium levels may be relatively unimportant, however, in normal homeostatic regulation of sodium concentrations and extracellular fluid volume. This suggestion is supported by observations that fixed quantities of glucocorticoids and mineralocorticoids are sufficient to maintain normal homeostatic balance in people with Addison's disease regardless of wide variations in salt and water intake. Aldosterone, like its cousins the glucocorticoids, may play a permissive role in the regulation of salt and water balance and other mechanisms may be more important for determining the magnitude and direction of salt and water movements.

Additional factors controlling aldosterone secretion

High levels of potassium in extracellular fluids directly stimulate aldosterone secretion, which, in turn, promotes renal potassium loss. In contrast, extracellular sodium variations do not directly influence aldosterone secretion unless unusually large variations are produced.

One additional extrarenal mechanism has been proposed for controlling aldosterone release. Presumably in response to low circulating sodium levels, the pineal gland secretes a peptide (?) hormone, *adrenoglomerulotropin*, that supposedly stimulates release of aldosterone. However, the importance of this mechanism, not to mention its existence, is questionable.

Mechanism of aldosterone action

Aldosterone stimulates sodium reabsorption and potassium excretion in the distal convoluted tubule of the kidney and possibly enhances some sodium reabsorption in the proximal convoluted portion, intestinal mucosa, salivary glands and sweat glands.

Aldosterone produces the typical steroidal pattern of action on target cells. Increased nuclear RNA synthesis results in production of a specific protein, *aldosterone-induced protein*, which somehow mediates the movement of sodium across the target cells somewhat reminiscent of the mechanism for movement of calcium ions across cells of the duodenal mucosa. The actual mechanism of this sodium pump and its relationship to potassium excretion are unknown.

Pathologies of the Adrenal Axis

Glucocorticoid Hypersecretion

Cushing's syndrome

This syndrome was described by Gushing in 1932. Hypersecretion of ACTH by basophilic pituitary adenoma is the cause of this condition. The adrenal cortices are hypertrophied and plasma cortisol levels are elevated (hypercortisolism). Excessive cortisol levels have adverse effects on metabolism of many tissues including the brain, muscles, skin, vascular tissue, kidney, liver and skeleton. Hypophysectomy alleviates the symptoms but necessitates extensive replacement therapies. Adrenalectomy requires only corticoid therapy.

Cushing's disease

The symptoms of this disease are the same as those for Cushing's syndrome, but the cause is different. Hypercortisolism in Cushing's disease occurs from excessive secretion of cortisol by adrenal adenomas, adrenal carcinomas, gonadal-adrenal rest tumors or from endocrine or nonendocrine cancers that elaborate an ACTH-like peptide.

Corticosteroid-secreting tumors suppress CRH and ACTH which brings about atrophy of the adrenals. Adrenal hypertrophy occurs when the cause is ectopically produced ACTH-like peptides.

Glucocorticoid Hyposecretion

Addison's disease

This disease is characterized by a shortage or absence of cortisol resulting in hypersecretion of ACTH. Hyperpigmentation (excessive darkening of the skin) occurs in most cases because of the innate MSH-activity of this ACTH. Aldosterone and adrenal androgen production are also depressed.

The person with Addison's disease is usually hypoglycemic due to lack of cortisol. Absence of aldosterone causes other symptoms including muscle weakness, water losses, hypotension and salt-craving. The loss of adrenal androgens becomes a problem only for postmenopausal or castrate women who rely on the adrenal as their sole androgen source.

The most common cause of Addison's disease is bilateral atrophy of the adrenals resulting from tuberculosis. It may also occur as a result of drug-induced or congenital deficiencies in steroidogenetic enzymes.

Secondary hypoadrenocorticism

Hypothalamic or pituitary lesions that block production of CRH or ACTH can produce the symptoms of hypoadrenocorticism. People with secondary hypoadrenocorticism lack the hyperpigmentation which usually accompanies Addison's disease but have the other symptoms. This condition can develop as a result of hypophysectomy, autoimmune disorders, viral illnesses, prolonged morphine administration and other causes. The most common origin of the disorder is prolonged therapy with cortisol or related steroids employed as anti-inflammatory agents or immune suppressants.

Disorders of Aldosterone Secretion

Hyperaldosteronism

This condition is characterized by low blood potassium, high blood sodium and muscle weakness. Elevated sodium levels bring about water retention and may produce hypertension. Hyperaldosteronism may result from an adenoma or carcinoma in the zona glomerulosa that autonomously secretes excessive amounts of aldosterone. These symptoms also develop when ACTH or ACTH-like peptides are elevated chronically. Treatment with aldosterone inhibitors can reduce these symptoms until the source of the excessive aldosterone is removed.

Hypoaldosteronism

Loss of the zona glomerulosa through adrenalectomy or Addison's disease, congenital or drug-induced depression of aldosterone production and defects in the renin-angiotensin system can induce hypoaldosteronism. Potassium excretion is reduced, sodium is lost in the urine and water retention is impaired. Imbalances in sodium/ potassium ratios alter muscle and nerve function. These conditions can be alleviated by treatment with aldosterone.

Adrenal Excesses in Androgens

Adrenal androgen production may be elevated in several hyperadrenocorticoid conditions such as Cushing's syndrome. Adrenal tumors may secrete excessive quantities of adrenal androgens. Symptoms of excessive production of adrenal androgens in women include hirsutism, acne, seborrhea, irregular menses and reduced fertility, lowered voice pitch, atrophy of the breasts, possible thinning of hair and recession of the scalp in the temporal region, clitoral enlargement and hypertrophy of skeletal muscles. These same symptoms may be present in males but are not as noticeable. Young adult women that exhibit anorexia nervosa (a disorder characterized by greatly reduced food intake) and an elevated adrenal axis may develop facial hair (hirsutism) as a result of ACTH-induced secretion of adrenal androgens.

Side Effects of Corticosteroids

Corticosteroids are administered in high doses to achieve their therapeutic effects. This is especially true of glucocorticoids. Adverse side effects occur as a consequence of prolonged therapy.

Adverse effects of glucocorticoid therapy

The beneficial therapeutic effects of glucocorticoids are manifest only when applied at doses two to three times physiological levels. Consequently a number of adverse side effects occur with prolonged administration of glucocorticoids, including mild diabetes mellitus, muscle weakness due to extensive protein catabolism (an effect that does not occur with physiological doses), osteoporosis due to destruction of bone substance, reduced activity of the immunological response system and mental depression.

Adverse effects of aldosterone therapy

Excessive doses of mineralocorticoids (or glucocorticoids) cause sodium retention and consequent accumulation of fluids (edema), as evidenced by rapid weight gain following their administration. However, an "escape" phenomenon, due to unknown factors, occurs, and the condition is alleviated before serious complications arise and irreparable damage has occurred. In cases of cardiac failure the escape mechanism also fails.

Aldosterone therapy results in decreased blood potassium and increased urine potassium. Excessive aldosterone causes severe potassium losses that can induce muscle cramps and muscle weakness. These events occur primarily because of adverse effects on cell membrane characteristics and resultant alterations of normal muscle cell physiology.

Comparative Aspects of Adrenocortical Tissue

The anatomical organization of the adrenocortical homologues and chromaffin cells differs markedly with the only obvious uniformity being a tendency for combining both cellular types in one organ, the adrenal gland of amniotes. Although the term chromaffin may be used to designate the catecholamine-secreting cells responsible for elaborating epinephrine and norepinephrine in all vertebrates, several terminologies have been proposed in attempting to deal with the diverse character of the adrenocortical homologues, including *interrenal cells*, *corticosteroidogenic cells* and *adrenocortical cells*. The last term will be used here because even though the adrenals of nonmammals lack the anatomical cortex-medulla relationship (and in many cases no adrenal gland per se is present), "adrenocortical" does denote the functional and evolutionary relationship of this cellular type to those of the mammalian adrenal cortex.

Cytologically the adrenocortical cells of nonmammals resemble the steroidogenic cells of the mammalian zona fasciculata. Adrenocortical cellular types of cyclostomes, teleosts and nonmammalian tetrapods possess a well-developed smooth endoplasmic reticulum, mitochondria with tubular cristae and numerous osmiophilic (lipoidal) inclusions. Following stimulation with ACTH, pituitary extracts or appropriate environmental stimuli, these cells exhibit increased basophilia, increased enzymatic activity (Δ^5,3β-hydroxysteroid dehydrogenase [HSD]) and decreased lipid content. Atrophy of the adrenocortical cells follows hypophysectomy.

Zonation of the adrenocortical cells is suggested cytologically in some anurans, reptiles and birds, and two separate cellular types have been claimed for the bullfrog *Rana catesbeiana* and for birds. However, too little work has been done to establish firmly the existence of more than one type of adrenocortical cell in most nonmammals.

Adrenocortical cells of fishes differ most from the general mammalian pattern of corticosteroidogenesis with respect to some of the hormones produced. However, the general sequences for corticosteroidogenesis are similar in all the vertebrates with respect

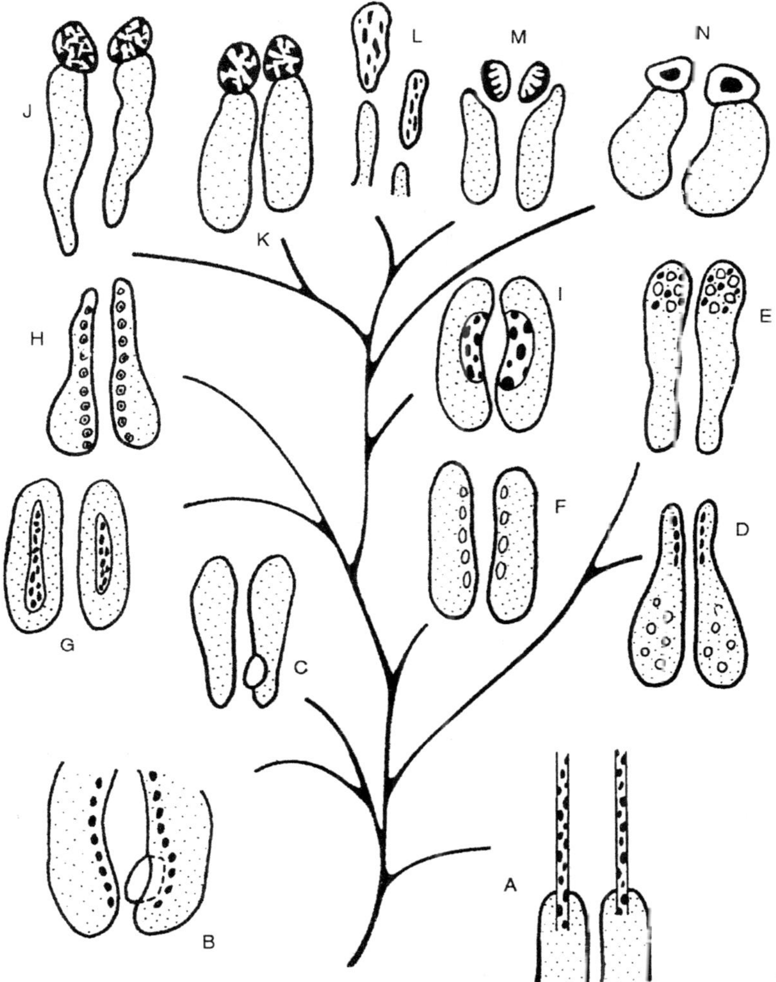

Fig. 11.8. Phylogeny and anatomy of vertebrate adrenal glands. A, Cyclostomata; B, Elasmobranchii; C, Holocephali; D, Holostei; E, Teleostei; F, Dipnoi; G, Anura; H, Caudata; I, Chelonia; J, Aves; K, Crocodilia; L, Squamata (snakes); M, Squamata (lizards); N, Mammalia.

to precursor-product relationships, with many of the same enzymes being involved. Among the nonmammalian tetrapods the nature of corticosteroid secretion is nearly identical, and the pattern of secretion is very similar to that described for cells of the mammalian zona glomerulosa.

Daily rhythms and seasonal variations in corticosteroid secretory patterns occur in nonmammals as well as in mammals. Peak seasonal adrenocortical activity is roughly correlated to periods of reproductive activity, although a cause-effect relationship cannot be categorically applied. Some studies suggest that "stress" may be the critical factor and that stressors associated with reproduction may be only one component involved in stimulating adrenocortical function (albeit a major one). The responses of nonmammals to stresses such as surgery, forced exercise and handling are very much like those described for mammals.

Class Agnatha: Cyclostomata

In lampreys presumed adrenocortical cells have been identified as islands of cells above the pronephric funnels in the kidney as well as in the walls of the large dorsal blood vessels (postcardinal veins) in this same region. Although in-vitro studies employing radioactively labeled steroidal precursors (pregnenolone, progesterone or both) have not demonstrated the ability of these presumed adrenocortical cells in lampreys or hagfishes to produce corticosteroids, cortisol, cortisone, corticosterone and 11-deoxycortisol have been isolated from hagfish and lamprey plasma. Enzymes such as HSD have not been demonstrated in cyclostome adrenocortical cells, however. The site for corticosteroidogenesis has not been established in cyclostomes, and more studies of this group are needed.

No evidence of either renin or a juxtaglomerular apparatus has been found in cyclostomes.

Class Chondrichthyes

Elasmobranchs and holocephalans have one large unpaired gland consisting exclusively of adrenocortical cells and located between the posterior ends of the kidneys (that is, truly interrenal). In addition, small islands of adrenocortical cells may be found on the surface of the kidneys extending anteriorly.

In 1934 Grollman and coworkers used extracts of the interrenal glands from three *Raja* species to maintain adrenalectomized rats, demonstrating the presence of corticosteroids in these glands.

Elasmobranchs produce a unique corticosteroid, 1α-hydroxycorticosterone. The enzyme necessary for synthesizing this unique steroid, *1α-hydroxylase*, is found only in the elasmobranch interrenal gland. Holocephalans lack 1α-hydroxylase and cannot secrete 1α-hydroxycorticosterone. The Pacific ratfish *Hydrolagus colliei* secretes primarily cortisol.

In-vitro studies of shark interrenals (*Scyliorhinus caniculus*) using exogenous pregnenolone as a substrate result in synthesis of primarily corticosterone and 11-deoxycorticosterone with lesser amounts of 1α-hydroxycorticosterone. When endogenous precursors only are involved, the primary product in vitro becomes the anticipated 1α-hydroxycorticosterone. There are no 18- or 17α-hydroxylases present in the shark interrenal, and consequently aldosterone, cortisol, cortisone and 11-deoxycortisol are not synthesized. Data from plasma analyses, however, indicate that not only is 1α-hydroxycorticosterone secreted in elasmobranchs but also cortisol and corticosterone. Such discrepancies between in-vivo plasma levels and in-vitro synthesis occur in other vertebrate classes as well, and one should be extremely cautious in extrapolating from in-vitro capabilities in which various intermediates and products accumulate to in-vivo situations in which the final products are removed (secreted into the blood). Accumulation of intermediates and products in the vicinity of the secretory cells under in-vitro conditions upsets chemical equilibria so that unusual ratios of steroids are observed.

Aldosterone has not been identified in vivo or in vitro in either elasmobranchs or holocephalans, and the capacity to produce this steroid may be lacking. Indeed, it would appear that the renin-angiotensin system is absent also in elasmobranchs although it may be present in holocephalans.

Class Osteichthyes: Actinopterygii

The anatomical arrangement of adrenocortical cells in the Actinopterygii differs markedly from that described for all other fish groups. In the sturgeons and polypterine fishes (Chondrostei) as well as in the ganoids (Holostei) the adrenocortical cells are scattered in small clumps throughout the kidney. The identification of these cells is hampered in the ganoids (*Amia*, *Lepisosteus*) by the presence of large numbers of *corpuscles of Stannius*, which although not steroidogenic do resemble cytologically the adrenocortical cells. The teleostean adrenocortical cells are embedded in the most anterior portion of the kidney, known as the *head kidney*. Frequently these cells are associated with the dorsal posterior cardinal veins as described for cyclostomes. The head kidney has lost its renal function and consists mostly of lymphoid tissue, nonfunctional pronephric tubules and small islands of adrenocortical cells. The teleostean adrenocortical cells are often referred to as *interrenal tissue*, although "intrarenal" would be a more descriptive term. In a few species all of these cells surround the posterior cardinal veins, and none are associated with the kidney. Because of the diffuse nature of the adrenocortical tissue in teleosts it is not possible to remove these cells surgically, and one must resort to the use of selective inhibitors of corticosteroidal synthesis such as metyrapone.

The principal circulating steroid in the Chondrostei, Holostei and Teleostei is cortisol, with corticosterone, aldosterone and some others present in minor quantities in the teleosts. Bony fishes lack the 1α-hydroxylase of elasmobranchs and consequently do not synthesize 1α-hydroxycorticosterone. In-vitro studies with teleostean adrenocortical cells indicate that they convert pregnenolone preferentially to cortisol. When progesterone is supplied as a precursor the principal product is corticosterone. Nevertheless, cortisol primarily is produced in vivo.

Renin-angiotensin system

Renin activity is present in all of the spiny-rayed fish groups. Histological and cytological identification of renal cells exhibiting renin granules has not been verified in any of the nonteleostean actinopterygian fishes, however.

Corpuscles of stannius

The corpuscles of Stannius embedded in the kidney of actinopterygian fishes were once thought to be steroidogenic. However, they lack the necessary steroidogenic enzymes, and it is unlikely that they play any significant role in the endogenous synthesis of corticosteroids. The major function of the corpuscles of Stannius appears to be related to calcium regulation.

Class Osteichthyes: Sarcopterygii

The Dipnoi have been of special interest to comparative endocrinologists seeking to understand the evolution of corticosteroids since they represent close relatives to both the fish and tetrapod lines. In dipnoan fishes the adrenocortical cells are found as small cords located between renal and perirenal tissues adjacent to branches of the postcardinal veins. Adrenocortical cells from estivating *Protopterus* synthesize corticosterone in vitro from progesterone. However, only cortisol was identified in the plasma of the aquatic phase, suggesting a tetrapod-like secretion (corticosterone) during its moist, air-breathing phase and a teleostean-like secretion (cortisol) during its aquatic phase. Although it is tempting to speculate on the evolutionary significance of these data, it would be premature to do so without further investigation.

Aldosterone, cortisol, corticosterone and a trace of 11-deoxycortisol are found in the blood of the South American lungfish *Lepidosiren paradoxa*, which is more dependent on remaining in the water than its African cousin *Protopterus*. The levels of both aldosterone and cortisol were high (about 6 μg/dl for each), whereas corticosterone levels were much lower (0.16 μg/dl). The aquatic Australian lungfish *Neoceratodus fosteri* secretes aldosterone and deoxycorticosterone, but in very small amounts (1.2 and 4.4 ng/dl respectively in males and 5.0 and 31.9 ng/dl respectively in females). These data would suggest that the lungfishes are intermediate between the tetrapod condition (aldosterone and corticosterone) and the actinopterygian fishes (cortisol) with respect to which corticoids are prominent. It would be interesting to know which corticosteroids are secreted by adrenocortical cells of the coelacanth fish *Latimeria* (Crossopterygii).

Renin-angiotensin system

Renin activity and the presence of renal cells containing renin granules have been observed in the coelacanth *Latimeria chalumnae* and in two genera of lungfishes. No experimental evidence for either a vascular effect or an effect on adrenocortical cells has been reported, however.

Class Amphibia

The adrenocortical cells of the Amphibia are extrarenal and extremely variable with respect to their location. In anurans, adrenocortical tissue is found in irregular nodules organized loosely into a pair of interrenal glands on the ventral surface of the kidneys. However, in *Xenopus laevis* the adrenocortical tissue is organized as small islets on the ventral surface of the kidney. Each of these adrenocortical islets also contains two or three chromaffin cells. In most anurans, some chromaffin cells are associated with the interrenal glands, and in one anuran, *Rana hexadactyla*, there are more chromaffin cells than adrenocortical cells in the interrenal glands.

In addition to the adrenocortical cells and chromaffin cells, a third cellular type, the *summer* or *Stilling cell*, has been found in the interrenal glands of ranid frogs. This Stilling cell appears in summer and regresses in winter frogs. It is an eosinophilic cell and resembles a mast cell (histamine-producing cell). The functional significance of the Stilling cell is unknown. It has been suggested to produce renin, but this has not been confirmed.

Adrenocortical cells of both apodans and urodeles occur in scattered islands on the ventral surface of the kidney. This anatomical arrangement in part explains the virtual absence of synthetic studies employing apodan or urodele adrenocortical cells in vitro.

Studies with adult anuran adrenocortical tissue in vitro have shown that the major corticoids synthesized are aldosterone and corticosterone, and both hormones have been identified in adult amphibian plasma. In addition, in-vitro syntheses result in production of a large quantity of *18-hydroxycorticosterone*, which can be a precursor for aldosterone. The ratio of aldosterone: 18-hydroxycorticosterone:corticosterone in vitro is 6:3:1. The high levels of this aldosterone precursor may simply be an artifact of in-vitro conditions, as mentioned previously. Ovarian production of significant quantities of 11-deoxycorticosteroids has been reported, and this may be an important source for corticosteroids in sexually mature females.

Although corticosterone is the dominant corticoid reported for terrestrial amphibians, cortisol has been reported to be the major corticoid in metamorphosing ranid tadpoles and in the aquatic frog *X. laevis*, as well as in the permanently aquatic urodele *Amphiuma*. Aquatic-phase adult newts, *Notophthalmus viridescens*, produce substantial amounts of cortisol. These observations would support the hypothesis that cortisol is important for maintaining sodium balance in freshwater amphibians, as reported for fishes, and that corticosterone becomes more important following metamorphosis to a terrestrial-phase amphibian. Corticosterone is also the major corticoid in reptiles and birds. It may be that there is a transition from cortisol to corticosterone secretion that takes place during or following metamorphosis that is possibly controlled by thyroid hormones.

Renin-angiotensin system

Renin activity is present in amphibian renal tissue, although the secretory renin-containing granules differ morphologically from those of mammals. Several studies report failure to locate a macula densa in Amphibia, but a macula densa-like structure has been reported for a toad. Substantiation of this claim is of considerable importance since a macula densa has not been reported for reptiles.

Class Reptilia

Chelonians, crocodilians and most snakes have paired suprarenally positioned adrenal glands exhibiting a variable degree of intermingling chromaffin cell cords within a mass of adrenocortical cells. In lizards and some snakes the adrenocortical cells are partially encapsulated by chromaffin cells, resulting in a "cortex" homologous to the mammalian medulla. Some chromaffin cells are also found within the central mass of adrenocortical cells. The chromaffin cells of *Sphenodon* (Rhynchocephalia) surround the dorsal aspect of the gland as well as form islets within the mass of adrenocortical cells.

Although few species have been studied, all of those examined (including turtles, lizards, snakes and alligator) synthesize aldosterone and corticosterone as the major corticoids under in-vitro conditions and what appears to be 18-hydroxycorticosterone. Corticosterone synthesis predominates in vitro with the amounts of 18-hydroxycorticosterone exceeding the levels of aldosterone.

Most investigators have failed to show any effect of ACTH on in-vitro corticosteroidogenesis in snakes (*Natrix natrix*, *Naja naja*), turtles (*Pseudemys scripta elegans*, *Chrysemys p. picta*) and the alligator (*Alligator mississipiensis*). However, adrenocortical cells from turtles (*C. picta*) secrete corticosterone in vitro when mammalian ACTH or crude extracts of avian, chelonian or anuran pituitaries are added to the culture medium, and ACTH stimulates adrenocorticoid synthesis in the cobra, *Naja naja*. Corticosterone levels in vivo and in vitro are elevated following administration of ACTH to *Caimen crocodilus* or *C. sclerops*. Corticotropin also elevates corticosterone levels in two lizard species but does not influence aldosterone levels.

One extraadrenalocortical source of corticosteroids has been reported. Isolated ovaries from the night lizard *Xantusia vigilis* synthesize 11-deoxycorticosterone, but the importance of this observation to either reptilian corticosteroidogenesis or to reproductive function is uncertain.

Renin-angiotensin system

Renin activity has been reported for turtles, lizards and snakes, but no macula densa has been described. This is curious considering the possibility of a macula-like structure in amphibians and evidence for a macula in birds and mammals.

Class Aves

In birds the adrenal glands are organized in the same manner as described for turtles, crocodilians and most snakes. The relative quantities of chromaffin with respect to adrenocortical cells varies, however. There appears to be some zonation of cellular types on histochemical and cytological bases similar to that seen in mammals.

The major corticosteroids synthesized by adrenocortical cells taken from domestic species and incubated in vitro are corticosterone, aldosterone and 18-hydroxycorticosterone. This sequence of steroidogenesis (corticosterone → 18-hydroxycorticosterone → aldosterone) characteristic of birds as well as anuran amphibians and reptiles is essentially the same pattern found in cells of the mammalian zona glomerulosa.

Studies with duck adrenal slices in vitro suggest that corticosterone is the immediate precursor for 18-hydroxycorticosterone and that the latter is converted to aldosterone. However, only corticosterone and aldosterone have been found in avian plasma with the exception of 11-deoxycorticosterone in the herring gull. The absence of 18-hydroxycorticosterone in plasma further supports the conclusion that it is a precursor for aldosterone synthesis, and its accumulation in vitro is an artifact.

Although mammalian ACTH stimulates adrenocorticoid secretion (corticosterone) in chickens, hypophysectomy does not cause complete cessation of corticosteroidogenesis. It has been proposed that melanophore-stimulating hormone produced and released by the hypothalamus maintains steroidogenesis in the hypophysectomized chicken. This suggestion needs further investigation because of its implication for other studies with hypophysectomized animals.

Renin-angiotensin system

Renin activity is present in birds, and both the juxtaglomerular apparatus and a macula densa have been described. Cells of the avian macula densa are similar to those of mammals, with only some minor differences.

Physiological Roles for Corticoids in Nonmammalian Vertebrates

The major function of corticoids investigated in nonmammals has been the effects on salt transport, particularly sodium, that is, mineralocorticoid activity. Unlike the mammalian condition, aldosterone, cortisol or corticosterone may possess mineralocorticoid activity when tested in nonmammals.

It has been hypothesized that the renin-angiotensin system evolved as a mechanism for regulating blood pressure. Control over mineralocorticoid synthesis and regulation of sodium/potassium balance was acquired later.

Class Agnatha: Cyclostomata

The blood of myxinoids (hagfishes) is isosmotic to sea water, but there are some minor differences in concentrations of specific ions. Therefore, although osmotic balance per se is no problem, distribution of certain ions must be maintained actively. Injections of aldosterone or deoxycorticosterone acetate alter electrolyte composition of the body fluids with respect to sodium ions, but cortisone has no effect.

Lampreys are either freshwater organisms or migrate between fresh water and the sea. While lampreys are in fresh water their body fluids are hyperosmotic to their surroundings, and they produce a large quantity of dilute urine. Sea lampreys (*Petromyzon marinus*) do not secrete corticosteroids when in sea water, but when they are in fresh water corticosteroids can be identified in the circulation. Aldosterone treatment can reduce renal and extrarenal sodium losses from lampreys held in fresh water. The significance of this latter observation to osmoregulatory control in lampreys must await demonstration of aldosterone in lampreys.

Class Chondrichthyes

Elasmobranchs have plasma that is hyperosmotic to sea water and hence do not have the osmoregulatory problems exhibited by most marine fishes. This hyperosmotic condition is achieved by maintaining high circulating levels of urea and trimethylamine oxide. There is no clearly defined role for any corticosteroids, including 1α-hydroxycorticosterone, in elasmobranchs. Although corticoids stimulate salt excretion by the rectal gland, this salt-secreting gland probably plays only a minor role in iono-osmotic homeostasis. A possible role for corticoids in carbohydrate metabolism of elasmobranchs has been suggested. There are no studies of corticoids in holocephalans.

Class Osteichthyes

Among the bony fishes, corticosteroids have been investigated with respect to function only in the teleosts. In general, corticosteroids (especially cortisol) stimulate sodium transport across gills (both influx and efflux), the mucosa of the gut and the kidney of freshwater fish. Cortisol appears to be the major corticoid in bony fishes, and it appears to regulate sodium fluxes.

Freshwater-adapted eels, *Anguilla* spp, exhibit greater renin activity in sea water than in fresh water. Furthermore, there are gradual changes in plasma renin activity during adaptation of eels to either fresh water or sea water. Administration of either ACTH or renin causes elevation of circulating cortisol levels in eels, suggesting that both mechanisms for regulating corticoid levels are present in bony fishes, infusion of isosmotic sodium chloride solution into the lungfish *Neoceratodus fosteri* causes reduction in plasma renin activity, suggesting the presence of a functional renin mechanism in the Dipnoi as well.

The problems of endocrine regulation of iono-osmotic balance in teleosts are discussed specifically with respect to adaptation of fishes to different salinities and the interactions of cortisol with various hormones, such as prolactin, on salt and water balance.

Seasonal and daily rhythms in circulating levels of cortisol have been reported for several species. Maximum levels of corticosteroids in rainbow trout maintained on long photoperiod are observed at night (2400 hours). In the spring, plasma corticoids rise in juvenile salmonid fishes prior to their migration from fresh water to the ocean or while adapting to sea water. Adult salmonids exhibit elevated corticosteroid levels during their spawning migration from the ocean to fresh water and during spawning.

Comparisons of absolute corticoid levels can not be made among different species, or in some cases even within the same species, since it is not possible to assess the role of stress in most cases. Stress produces marked elevations of corticosteroids in fishes. Just bringing wild trout into the laboratory can increase levels of plasma cortisol 2 to 5 times. Elevated corticosteroids in migrating juvenile and adult salmonids probably reflect a response of the adrenal axis to chronic stress. Importance must be placed upon relative levels within a species under carefully described conditions.

Class Amphibia

Hypophysectomy or exposure of frogs to sea water induced atrophy of the adrenocortical tissue, presumably because of reduced requirements for corticosteroids. Interrenalectomy of frogs causes a decrease in plasma sodium and an increase in plasma potassium similar to that observed in mammals. Winter frogs usually survive interrenalectomy, but summer frogs die; possibly death is correlated to the yet unknown importance of the Stilling cells in summer frogs.

Aldosterone seems to be the important salt-regulating hormone acting on the skin and urinary bladder, increasing sodium influx and retention respectively. The action of aldosterone on the urinary bladder involves synthesis of new protein and would appear to be analogous to the production of aldosterone-induced protein in the mammalian distal convoluted tubule. Corticotropin elevates aldosterone, levels in *Rana esculenta*, implying a direct action of ACTH on aldosterone, a condition very different from that described for mammals. Salt-depleted frogs exhibit elevated renal levels of renin, suggesting the presence of both mechanisms for regulating aldosterone production, although plasma renin levels do not appear to differ between distilled-water-adapted toads and saline-adapted toads.

Aldosterone treatment reversed the depression of plasma sodium caused by aminoglutethamide in the tiger salamander. Higher doses of corticosterone than aldosterone were needed to bring plasma sodium back to normal, supporting a physiological role for aldosterone as a mineralocorticoid in amphibians.

Seasonal and daily rhythms of cortisol levels have been reported in plasma samples from *Rana esculenta* and *Bufo americanus*. The time of the daily maximum varied in *R. esculenta* from 2400 hours in May to 1900 hours in July and 0800 hours in November. Greatest secretion of corticoids coincided with reproduction in the spring. Two peaks of corticoid secretion were observed in captive and free *B. americanus*. One peak coincided with reproduction in the spring but the second peak occurred in the fall at the time of pre-hibernating migrations. Daily maxima corresponded to increased locomotor activity.

Stress causes an elevation in plasma corticosteroids. Placing freshly collected bullfrogs in sacks for up to 24 hours doubles their corticosterone levels. Stress also can alter secretion of other hormones. The secretion of PRL, gonadotropin and androgen all decrease in amphibians following capture. The failure of some species to breed in captivity and the stimulus for some larvae to undergo metamorphosis may be consequences of stress.

Class Reptilia

A general role for corticosteroids in reptiles has not been demonstrated although one functional role has been suggested. Reptiles have nasal (orbital) salt-excreting glands similar to those found in aquatic birds. Corticosteroids appear to regulate the secretion of salt by these glands, but definitive studies are needed. Injections of concentrated sodium chloride solutions (salt-loading) depress plasma aldosterone levels in several species of lizard. Similar observations are reported for the tortoise, *Testudo hermanni*. Salt-loaded lizards also have elevated levels of AVT. Sodium depletion apparently does not affect renin activity in turtles.

Class Aves

Corticosteroids have been implicated in salt regulation by birds. The nasal salt-excreting glands of certain aquatic birds secrete a hypertonic NaCl solution. Salt secretion is enhanced by treatment with

ACTH or corticosterone, and corticosterone uptake by nasal salt glands is followed by an increase in salt excretion. Adrenalectomized ducks cannot excrete a salt load, but corticoid therapy restores this salt-excreting ability. Metyrapone, which disrupts corticosterone synthesis, blocks nasal gland secretion.

Increases in environmental salinity are correlated with increases in Na-K-dependent adenosinetriphosphatase in salt gland cells, and this event is possibly related to increased protein synthesis caused by corticosteroids. Marine birds have larger adrenal glands than do freshwater or terrestrial birds, which supports the role for corticosterone in nasal salt gland regulation. Birds inhabiting brackish water have intermediate-sized adrenals.

Renin activity and angiotensin activity in blood plasma of ducks and pigeons increase following hemorrhage, establishing a physiological role similar to that described for mammals. The control of corticosterone by ACTH and the importance of corticosterone in salt regulation suggest that avian corticosteroidogenesis may not be affected by the renin-angiotensin system. Studies in species that lack the dependence of nasal salt gland excretion are needed to clarify this situation.

Circadian rhythms for corticosteroids have been described in some birds. The time of the daily corticosteroids peak is believed to determine the behavioural response following injection of prolactin in white-crowned sparrows. These observations have led to hypotheses concerning the control of migratory behaviour by the phase relationships between rhythms of corticosterone and PRL secretion.

Stress activates the hypothalamo-adrenal axis in birds. This involves the ACTH-corticosterone axis and release of epinephrine from the adrenal medulla. Activation of the endocrine stress mechanism also influences other endocrine glands. Chasing (stress) of male zebra finches for 15 minutes results in a significant depression in plasma androgens measured 2 hours later. Isolation of males in small cages for 12 hours virtually obliterates testosterone in the circulation.

Mammalian Adrenal Medulla

The medullary portion of the mammalian adrenal consists of sympathetic preganglionic neuronal endings (cholinergic) and modified cells derived from neural crest and homologous to postganglionic sympathetic neurons (adrenergic). These adrenal cells secrete either norepinephrine or epinephrine directly into the blood. In other words, the adrenal medulla is a modified sympathetic ganglion. Both epinephrine and norepinephrine (as well as small quantities of dopamine) can be extracted from the adrenal medulla, but the ratio in adult mammals strongly favours epinephrine. This proportion of norepinephrine to epinephrine varies throughout life, however. Fetal and neonatal adrenals secrete predominantly norepinephrine followed by a gradual increase in the proportion of epinephrine that eventually dominates in adult mammals. Whales are an apparent exception in that the adult adrenal consists of about 83% norepinephrine.

Treatment of adrenal medullary cells with potassium dichromate or chromic acid results in formation of a yellowish or brown oxidation product, the *chromaffin reaction*. Cells that exhibit a positive chromaffin reaction are termed chromaffin cells. The catecholamine-secreting cells of the adrenal medulla show a positive chromaffin reaction, but so do other cells in the body (for example, in the intestinal epithelium and skin). Cells containing the tryptophan derivative 5-hydroxytryptamine (serotonin) also exhibit a positive chromaffin reaction. Norepinephrine-secreting cells can be distinguished from epinephrine-secreting cells by the formaldehyde treatment devised by Hillarp and Falck. Formaldehyde combines chemically with norepinephrine storage granules and the resulting complex will fluoresce.

Synthesis and Metabolism of Adrenal Catecholamines

Two distinct cellular types in the mammalian adrenal medulla are related to production of norepinephrine and epinephrine respectively. Both epinephrine and norepinephrine are synthesized from the amino acid tyrosine and employ the same biochemical pathway. However, only one cellular type possesses the critical enzyme phenyl-ethanolamine N-methyltransferase (PNMT) necessary for converting norepinephrine to epinephrine through addition of a methyl group donated by S-adenosylmethionine.

Norepinephrine and epinephrine may be found circulating free in the plasma or as conjugates with sulfate or glucuronide. Most of the circulating epinephrine is bound to plasma proteins, especially albumin. Norepinephrine binds to plasma proteins to a much lesser degree than does epinephrine.

Circulating catecholamines have a short biological half-life and are rapidly excreted via the urine in either free or conjugated forms. The biological half-life for epinephrine is about 5 minutes. The most common metabolic pathway for inactivation of catecholamines involves methylation by the enzyme catechol-O-methyltransferase and subsequent

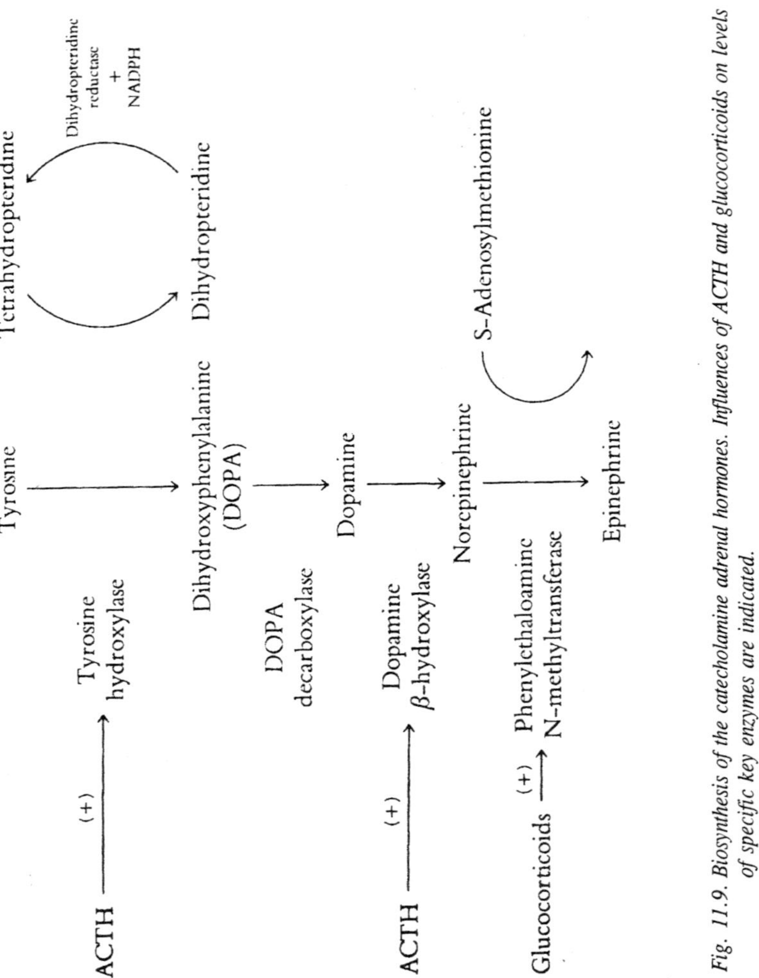

Fig. 11.9. Biosynthesis of the catecholamine adrenal hormones. Influences of ACTH and glucocorticoids on levels of specific key enzymes are indicated.

deamination and oxidation (by amine oxidase) to inactive metabolites that appear in the urine. The most common urinary metabolites are 3-methoxy-4-hydroxymandelic acid, metanephrine and normetanephrine.

Regulation of Catecholamine Secretion

Central nervous pathway

Stimulation of the cholinergic sympathetic fibers innervating the medulla causes local release of acetylcholine (ACh), which in turn stimulates release of norepinephrine, epinephrine or both from the

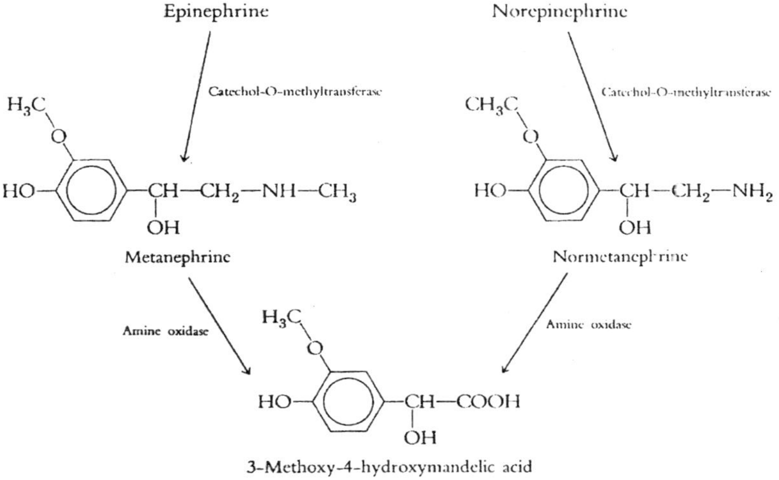

Epinephrine

Catechol-O-methyltransferase

Metanephrine

Norepinephrine

Catechol-O-methyltransferase

Normetanephrine

Amine oxidase

Amine oxidase

3-Methoxy-4-hydroxymandelic acid

Fig. 11.10. Interactive metabolic products of epinephrine and norepinephrine.

chromaffin cells. This action of ACh involves Ca^{++} uptake by the chromaffin cells. Apparently separate control centers for release of epinephrine and norepinephrine are located in the anterior and medial hypothalamus, and selective release of these medullary hormones occurs. Norepinephrine and epinephrine may be released separately under differing physiological conditions, and these two catecholamines have independent physiological roles in homeostasis.

Effects of physiological factors on catecholamine release

Walter B. Cannon first formulated an emergency reaction hypothesis involving secretions of the adrenal medulla and activity of other portions of the sympathetic nervous system. These emergency responses include increased heart rate, vasodilation of arterioles in skeletal muscle, general venoconstriction, relaxation of bronchiolar muscles, pupillary dilatation, piloerection and mobilization of liver glycogen and free fatty acids. All of these responses contribute to increased efficiency of operation so that the organism can best respond to whatever emergency has arisen. This type of response to short-term stress may be distinguished from the response to chronic stress associated with the glucocorticoids and the general adaptation syndrome of Selye. However, the emergency reaction of Cannon may be thought of as being partly equivalent to the alarm reaction of the Selye hypothesis.

The major physiological actions of adrenal medullary hormones are their effects on metabolism in response to emotional stress (anxiety, apprehension), physical stress (injury, exercise) or what might be distinguished as physiological stress (temperature, pH, oxygen availability, hypotension and hypoglycemia). The actions of adrenal catecholamines on cardiovascular events other than the acceleration of heart rate (which is actually due to metabolic effects of epinephrine on cardiac muscle) are probably secondary to the effects of norepinephrine released from postganglionic sympathetic fibers or, in the case of the skeletal muscle arterioles, the release of ACh from postganglionic sympathetic fibers. This secondary role for adrenal catecholamines is further supported by the relatively low percentages of norepinephrine in the adrenals of most adult, mammals. Sympathetic control mechanisms are not well developed in fetal and neonatal mammals, and this observation could be linked to the high proportion of norepinephrine in their adrenals.

Epinephrine stimulates hydrolysis of liver glycogen to glucose and production of lactate from muscle glycogen stores. Circulating norepinephrine produces a similar effect on liver glycogen, but muscle glycogen stores are not affected by norepinephrine. This would explain the ineffectiveness of exogenous norepinephrine as a cardioacceleratory drug, whereas epinephrine is very potent. The mobilization of lipids and release of free fatty acids from adipose tissue is under neural sympathetic control and is not regulated by adrenal catecholamines.

Emotional and severe physical stress increases circulatory levels of catecholamines via hypothalamic adrenal pathways. However, response to emotional stress such as written examinations involves an increase only in epinephrine, whereas adrenal response to anticipation involves primarily norepinephrine. Exercise causes an increase in norepinephrine levels, presumably from both adrenal and neural sources. Epinephrine secretion is not influenced by moderate exercise, but it is markedly increased during long-distance running. Several physiological factors such as cold and heat stress, alkalosis or acidosis and hypotension do not appear to involve primary actions of adrenal medullary hormones. However, responses to asphyxia or anoxia and to hypoglycemia are major factors influencing epinephrine release in adult mammals. Asphyxia causes an increase in epinephrine release, probably through direct actions of oxygen deprivation on the nervous system. In fetal or neonatal animals, asphyxia directly evokes catecholamine release from the adrenal. Insulin-induced hypoglycemia results in cardiac acceleration through stimulating epinephrine release. Hypoglycemia induces epinephrine release primarily through direct effects on glucose-sensitive centers in the hypothalamus. Epinephrine also retards the insulin-induced decrease in blood sugar through its antagonistic actions on liver glycogen.

Effects of ACTH and Glucocorticoids on Adrenal Catecholamine Secretion

Development of a close anatomical association between adrenocortical and chromaffin tissues during vertebrate evolution has suggested a concomitant development of a functional relationship as well. Studies by Wurtman and his co-workers over the past 15 years or so have demonstrated that ACTH exerts a stimulatory effect on epinephrine secretion through the former's action on circulating glucocorticoid levels. Hypophysectomy reduces adrenal epinephrine levels, and treatment with either ACTH or glucocorticoids restores adrenal levels of epinephrine to normal. Furthermore, glucocorticoids increase the activity of adrenal medullary PNMT, the enzyme responsible for conversion (methylation) of norepinephrine to epinephrine. Some studies indicate that ACTH may have a direct action on the medulla as well. Levels of both tyrosine hydroxylase (tyrosine→DOPA) and dopamine-β-hydroxylase (dopamine→norepinephrine) but not PNMT are increased by ACTH treatment. These observations indicate that chronic stress may influence epinephrine secretion not only during the alarm reaction but also in the later stages of the response. It has been reported that epinephrine can cause release of ACTH through actions at either the hypothalamic or adenohypophysial level, but the physiological significance of these observations is not clear.

Mechanism of Action for Adrenal Catecholamines

The presence of specific receptors for epinephrine was first postulated in 1906 by Sir Henry Dale, who showed that ergot alkaloids (drugs such as ergocornine and ergocryptine) blocked some of the actions of epinephrine. Later studies suggested there are two kinds of *adrenergic receptors* in target cells that are capable of binding adrenal catecholamines: alpha and beta receptors. These receptors also respond to a number of epinephrine-like drugs that have been termed *sympathomimetic drugs* since they mimic actions of sympathetic catecholamines. Two common sympathomimetic drugs are isoproterenol and phenylephrine. Norepinephrine binds mainly to alpha receptors, whereas epinephrine binds to both. When both alpha and beta receptors are present on a target cell that binds epinephrine, the alpha effect predominates unless epinephrine is administered with an alpha-blocking agent (for example, phentolamine).

Detailed studies of the mechanism of adrenal catecholamine actions on target cells have concentrated on the effects of epinephrine in cardiac muscle and liver cells. In fact, it was studies in cardiac cells that led Sutherland and his co-workers to the discovery of the second-messenger role for cyclic adenosine 3′,5′ monophosphate (cAMP) and an eventual Nobel Prize in physiology and medicine. Epinephrine stimulates the breakdown of glycogen to glucose in both liver and muscle cells by first stimulating an increase in intracellular cAMP. The glucose released from liver glycogen tends to enter the general circulation, whereas the glucose liberated from muscle glycogen is utilized for rapid ATP synthesis and production of lactate.

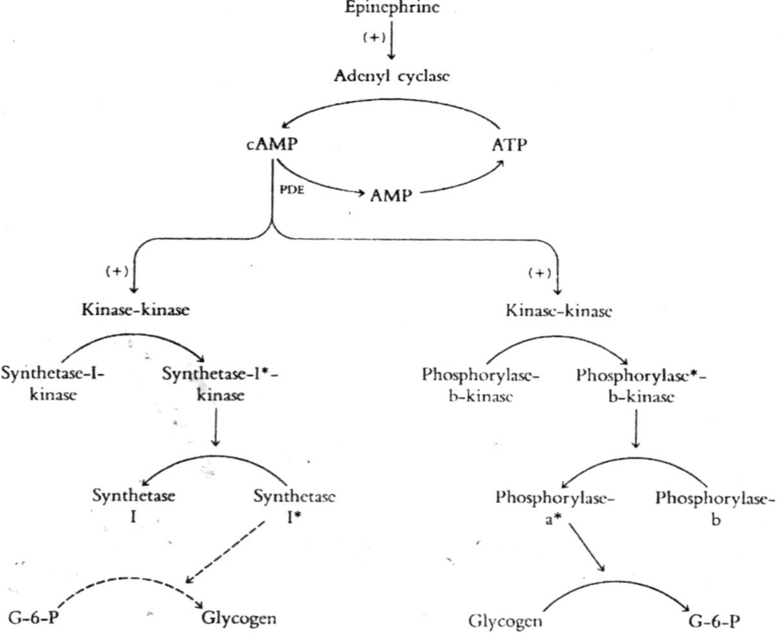

Fig. 11.11. Action of epinephrine on liver or muscle glycogen.

Evolution of Adrenal Medullary Hormones

Morphologically there has been a general trend to develop a close anatomical relationship between chromaffin tissues homologous to the mammalian adrenal medullary cells and the adrenocortical cells. Chromaffin tissue in agnathans is found in association with the posterior cardinal veins as are the separate clusters of adrenocortical cells. In the cartilaginous fishes the chromaffin tissue is more widely separated from the adrenocortical cells than in any other vertebrate group. There is considerable anatomical variation among the teleosts. Chromaffin tissue may be entirely separate from the adrenocortical cell clusters in the head kidney or may be intermingled with it, or in some species both conditions may be found. Clusters of adrenocortical cells (islets) in amphibians typically contain a few chromaffin cells. The anatomical distribution of chromaffin cells in reptiles is varied among the different groups. The adrenal glands of squamates and the tuatara *Sphenodon punctatus* have peripherally located norepinephrine-secreting chromaffin cells and central islets of epinephrine-secreting cells intermingled among the adrenocortical cells. The avian adrenal glands consist of a mixture of cortical and chromaffin cells such that no distinction of "cortex" and "medulla" is possible.

Early studies of the proportion of adrenal catecholamines in nonmammals suggested that the mammalian fetal pattern of high norepinephrine production with increasing production of epinephrine following birth is an example of ontogeny recapitulating phylogeny. Analysis of extractable catecholamines from the shark *Squalus acanthias*, which has entirely separate chromaffin tissue, the frog *Rana temporaria*, which exhibits some mixing of chromaffin and adrenocortical cells; and the rabbit *Oryctolagus cuniculus* suggests an evolutionary change from reliance on norepinephrine to the progressive reliance on epinephrine. Although it seems logical to assume that the methylation step (norepinephrine→epinephrine) was a later evolutionary event, the picture is more complex. Histochemical and chromatographic procedures indicate that norepinephrine predominates in some bird species whereas others produce mainly epinephrine. Still other groups show only a slight preference for one catecholamine over the other. Although norepinephrine predominates in extracts of adrenals prepared from chickens, turkeys and pigeons the major circulating catecholamine is epinephrine. These data point out the dangers of extrapolating too broadly from gland content to secretory activities.

Few comparative functional studies of adrenal catecholamines have been reported. The hyperglycemic action of epinephrine seems to be a primitive action in nonmammalian vertebrates. Norepinephrine, on the other hand, produces hypoglycemia in the ratfish and in 6 to 9-week-old Japanese quail. It is not clear whether these observations represent a true role for norepinephrine.

12

Islets of Langerhans

The mammalian pancreas plays essential roles in digestion and metabolism. The digestive role is accomplished by the exocrine glandular portion of the pancreas which produces digestive enzymes. The endocrine portion of the pancreas influences metabolism by secreting hormones that regulate carbohydrate, lipid and protein metabolism.

The exocrine or acinar pancreas secretes digestive juices into the pancreatic duct whereby they reach the lumen of the small intestine. (An acinus is similar in structure to a follicle except that the lumen of the acinus makes contact with a duct through which the secretion products of the epithelium [acinar cells] are transported.) This pancreatic juice contains digestive enzymes and bicarbonate ions that buffer the acidic material entering the small intestine from the stomach.

The endocrine pancreas consists of small masses or islands or islets of endocrine cells scattered among the acinar tissue. The pancreatic islets secrete *insulin*, *glucagon*, *somatostatin* (growth hormone release-inhibiting hormone) and an additional peptide known as *pancreatic polypeptide*. Insulin is primarily a hypoglycemic agent that lowers blood glucose, whereas glucagon is hyperglycemic. Both of these hormones produce effects on lipid and protein metabolism as well. The roles for pancreatic somatostatin and pancreatic polypeptide are not established.

Recognition for the role of the endocrine pancreas in blood glucose regulation originated from observations of the clinical syndrome *diabetes mellitus*. This disease was first described by the Greek physician Aretaeus in about 1500 B.C. as a condition where "flesh and bones run together" and are siphoned into the urine. The term diabetes, a siphon, comes from the Greek word diabainein: *dia* through + *bainein* to go. Indian physicians reported the sweet taste of the diabetic's urine in the 6th century. The term mellitus referring to the presence of glucose in the urine was added in the 18th century. This condition was always fatal. There is more than one form of diabetes mellitus, but here we refer to the type related to insufficient production of insulin.

Diabetes mellitus is characterized by the production of large quantities of glucose-containing urine. Normally little or no glucose should appear in the urine. It was the presence of glucose in the urine of pancreatectomized dogs that first linked diabetes mellitus to a possible disorder of the pancreas.

The occurrence of glucose in the urine results from excessively high circulating levels of glucose as a result of the failure by the endocrine pancreas to secrete adequate amounts of insulin. Blood levels of glucose become elevated, resulting in more glucose in the glomerular filtrate produced in the kidney. The proximal convoluted portion of each nephron transports all glucose that appears in the glomerular filtrate back into the blood under normal conditions. However, should blood sugar levels be sufficiently high, so much glucose enters the glomerular filtrate that the recovery transport mechanism is saturated, and all of the glucose cannot be returned to the blood. Some glucose remains in the nephron, and its presence in the urine upsets the osmotic balance between urine and blood. Consequently less water can be reabsorbed and a greater volume of urine is produced. Because of the increased water loss associated with the loss of glucose in the urine this syndrome achieved the descriptive title of the pissing evil during the Middle Ages.

Other consequences of the inadequacy of insulin production contribute to water losses as well. Reduction in glucose availability causes increased utilization of fats and proteins for energy. Proteins are hydrolyzed to amino acids that are in turn deaminated to produce carbohydrates plus ammonia. Most of the highly toxic ammonia is converted to urea and excreted through the urine. Urea excretion places an additional demand on water required for removal, contributing to increased urine production. Lipid stores are mobilized to help compensate for carbohydrate shortages in the diabetic. Fats are hydrolyzed to glycerol and fatty acids, which in turn are oxidized directly for energy. In the process of fatty acid oxidation a number of by-products accumulate, including β-hydroxybutyrate, acetone and acetoacetic acid. These substances are known as *ketone bodies*, and their generation is referred to as *ketogenesis*. Ketone bodies increase in the blood (ketonemia) and consequently appear in the urine (ketonuria) and contribute to additional water losses through increased urine production.

The diabetic individual suffers from an important nutrient loss, and metabolic alterations occur in an attempt to compensate for these losses. The loss of glucose into the urine and increased production of nitrogenous wastes and ketone bodies all contribute to a process of desiccation.

Development of the Mammalian Pancreas

The vertebrate pancreas develops from the endodermal lining of the primitive or embryonic gut (endoderm). A dorsal bud from the embryonic intestine fuses with one or two ventral buds to form the definitive pancreas. Usually only the dorsal connection to the intestine is retained as the exocrine pancreatic duct. The exocrine tissue in the developing pancreas can be identified by the formation of small ductules. These ductules coalesce into larger ducts until they form one or more large pancreatic ducts connecting the exocrine pancreas with the lumen of the small intestine. Small buds develop from the ductules and can be identified as solid clumps or islands of cells dispersed among the exocrine tissue. The clumps are referred to as the *islets of Langerhans*. In addition, some of the endocrine cells may be scattered singly or in groups of only a few cells throughout various regions of the pancreas.

Although the islets originate as outgrowths from the pancreatic ductules the actual origin for these cells is not clear. It has been suggested that they arise from mesoderm, from neural crest cells of neuroectodermal origin and from endoderm of the gastrointestinal tract. The islet cells are not derived from neural crest, at least in birds, and it is probable that mammalian islet cells are not derived from either neural crest or other neuroectoderm. Transplantation of quail neural crest cells containing a unique cytological marker into chick embryos confirms their incorporation into the avian pancreas, but they do not give rise to the pancreatic endocrine cells. Instead they differentiate into parasympathetic ganglia. Until this matter is clarified, it might be legitimate to refer to the endocrine pancreas as being of endodermal origin.

Several functional schemes have been proposed for the origins of the endocrine pancreatic cells that are divorced from their possible germ layer origins. It is possible that these endocrine cells represent modified gastrointestinal mucosal cells that initially synthesized "inducers." Exocrine cells in teleosts that are homologous to avian and mammalian islet cells often possess a zymogen mantle, presumably induced by secretions of the adjacent pancreatic endocrine cells. Another suggestion is that these cells originally secreted hydrolytic enzymes related to digestion but that these "enzymes" have lost their catalytic properties and become hormones. In either event these modified mucosal cells presumably lost their contact with the gut mucosa and specialized as centers for internal secretion. The similarity of peptides produced in the islet cells and those of the gastrointestinal tract supports a common origin for these pancreatic and intestinal cells.

Cellular Types in Vertebrate Pancreatic Islets

At least five different cellular types have been identified in the endocrine pancreas: B, A, D, PP and an unnamed type. The so-called clear cell appears to be a fixation artifact since clear cells are most prevalent in Bouin-fixed material but "disappear" with other fixation procedures' (Bouin's fixative is a standard solution used for preparing tissues for histological and cytological observation.) There are four general vertebrate patterns with respect to the dominant cellular types present in the endocrine pancreas. However, careful analyses indicate that it is misleading to characterize classes of vertebrates as exhibiting a particular pattern.

B cells

B cells (β-cells) stain with aldehyde fuchsin (AF+) and pseudoisocyanin (PIC+) following oxidation. The latter staining procedure has been claimed to be specific for insulin granules, but it stains other intracellular structures as well (for example, neurosecretory granules). Immunofluorescent techniques have verified that insulin is produced and stored in the B cell. Several drugs, such as alloxan and streptozotocin, selectively destroy the B cells and impair the ability of the pancreas to secrete insulin. The affected B cells undergo a process of *hydropic degeneration* that is characterized by clumping of nuclear chromosomal material (formation of pyknotic nuclei) and eventual cell death. This technique of chemically induced degeneration is often used in experimental animals to selectively remove the insulin-secreting cells without altering the ability of the islets to secrete other pancreatic peptides.

A cells

The A cells (α_2-cells) of the pancreatic islets are generally acidophilic and argyrophilic (affinity for silver-staining techniques). They do not stain with AF or PIC procedures. The A cell is considered to be the source of the second major pancreatic hormone, glucagon. In some species A cells may contain other peptide hormones in addition to glucagon. The round secretory granules in the cytoplasm of the A cells are morphologically distinct from the angular insulin granules of the B cells, making these cells easy to distinguish with the electron microscope. This difference in granule shape is not found in all mammals, however. Treatment with cobalt selectively impairs the ability of A cells to secrete glucagon, but such treatment does not lead to the destructive degeneration such as alloxan produces in the B cells.

D cells

D cells (α_1-cells or δ-cells) can be cytochemically distinguished from A cells by applying the toluidine blue staining procedure. They can be further distinguished from A cells and B cells by their staining with PIC following methylation but not following oxidation. Immunoreactive somatostatin has been localized in the D cells, although the physiological significance of somatostatin in the pancreas is not clear. The release of both insulin and glucagon is inhibited in paracrine fashion by somatostatin, and the D cell may influence the secretory activities of the A and B cells. An early suggestion for the role of the D cell was the production of pancreatic polypeptide. This substance has been localized in a separate cellular type, however, and is apparently not produced by the D cell.

Pancreatic polypeptide-secreting cells

Pancreatic polypeptide (PP) has been localized by immunofluorescence techniques in cells found at the periphery of the endocrine islets as well as in cells scattered throughout the exocrine pancreas. These cells are distinct cytologically and immunologically from B, A and D cells. The distribution of these PP-secreting cells on the periphery of the islets varies greatly among different species, and it is difficult to generalize for all mammals.

Amphophils

Amphophilic cells have been demonstrated in many mammalian species as well as in sharks, teleosts, amphibians and reptiles, but no conclusions have been generated regarding their functional roles. They may represent either differentiating or degenerating forms, or both, of the other four cellular types that have been implicated in hormonal secretion.

Hormones of the Endocrine Pancreas

The two major pancreatic hormones are the hypoglycemic hormone, insulin, and the hyperglycemic hormone, glucagon. In addition, the pancreatic endocrine cells produce somatostatin and PP. The physiological significance of PP is unknown, but its increase in the circulation following ingestion of a meal suggests a role in postabsorptive metabolism. Somatostatin appears to regulate insulin and glucagon release and may influence release of PP as well.

Insulin: hypoglycemic hormone

Von Mering and Minkowski in 1889 observed the first correlation between the pancreas and blood sugar regulation when they induced diabetes mellitus in dogs following pancreatectomy. The suspected hypoglycemic factor of the pancreas was later named "insuline" to emphasize its origin from the islets of Langerhans. Purified insulin, however, was not isolated until a physician, Frederick Banting, and a graduate student in physiology and biochemistry, Charles Best, teamed up at the University of Toronto in the summer of 1921. Much of their success was due to the collaborative efforts of J.B. Collip and the project's director, J.J.R. Macleod. Banting and Best traveled the poorer sections of Toronto in their Model-T Ford, nicknamed The Pancreas, purchasing dogs at $1.00 each for their experiments.

The presence of proteolytic enzymes in the exocrine pancreas had thwarted earlier attempts to extract insulin from a whole pancreas. Since it had been shown that tying off the pancreatic duct caused degeneration of the acinar pancreatic tissue but did not affect the islet tissue, Banting suggested they employ this technique to avoid contamination of their extracts with digestive enzymes. The hypoglycemic factor they isolated was first named isletin, but they later changed its name to conform to the earlier proposed name. The importance of the successful isolation of insulin and its use to alleviate the fatal symptoms of diabetes mellitus led to the awarding of the Nobel Prize in Physiology and Medicine for 1923 to Banting and Macleod. Best and Collip were not officially recognized in the award, although the recipients acknowledged them for their contributions.

Techniques for measurement of insulin

The standard challenge for the ability of the pancreas to secrete insulin is the *glucose tolerance test*. Following a period of fasting, a glucose load (excessive amount of glucose) is administered orally or intravenously, and its rate of disappearance (clearance) from the blood is measured. The clearance rate for a given glucose load is directly proportional to the secretion of insulin and hence reflects the ability of the pancreas to respond to hyperglycemia with insulin secretion.

Insulin may also be bioassayed in vitro for its abilities to stimulate uptake of glucose from the medium into cells. Muscle cells from the rat diaphragm are often employed in this bioassay. Radioimmunoassay (RIA), however, is routinely employed to measure circulating insulin levels, and many commercial RIA kits are available for performing these analyses. Insulin RIAs are highly specific, and some can detect the difference between porcine and human insulins that vary in only one amino acid substitution.

Chemistry and synthesis of mammalian insulins

The insulin molecule is a small protein composed of two different polypeptide chains linked together by disulfide bridges. The A chain consists usually of 21 amino acids, and the B chain has 30 amino acid residues. Although insulin was the first protein whose primary structure was elucidated, attempts to synthesize the molecule in cell-free systems were unsuccessful because of the inability to induce proper formation of the disulfide bridges between the two polypeptide chains. Later it was discovered that the B cell synthesizes insulin by first forming a much larger single polypeptide chain that is folded by formation of disulfide bonds. This preproinsulin is synthesized at the rough endoplasmic reticulum and cleaved to a smaller proinsulin molecule. Proinsulin is so constructed that A and B chains of insulin are connected by a chain of amino acids called the C peptide. Hydrolysis of proinsulin occurs at the Golgi apparatus as the result of actions of two proteolytic enzymes. One of these enzymes is trypsin-like in its action, and the other behaves like carboxypeptidase B. The resulting secretion granules contain crystals of insulin and C peptide fragments. The insulin fragment forms a complex with available zinc ions to produce a crystalline hexamer within the secretory granule. It is this crystalline structure that imparts the unique structure to insulin secretion granules in the B cell.

When the contents of the secretory granules are released from the B cells by exocytosis, both insulin and the C peptide enter the blood as does a small amount of unhydrolyzed proinsulin. No function is known for either proinsulin or the C peptide once they have entered the general circulation.

Regulation of insulin release

Hyperglycemia can directly stimulate release of insulin through a direct action of glucose on the B cell. It is apparently not the presence

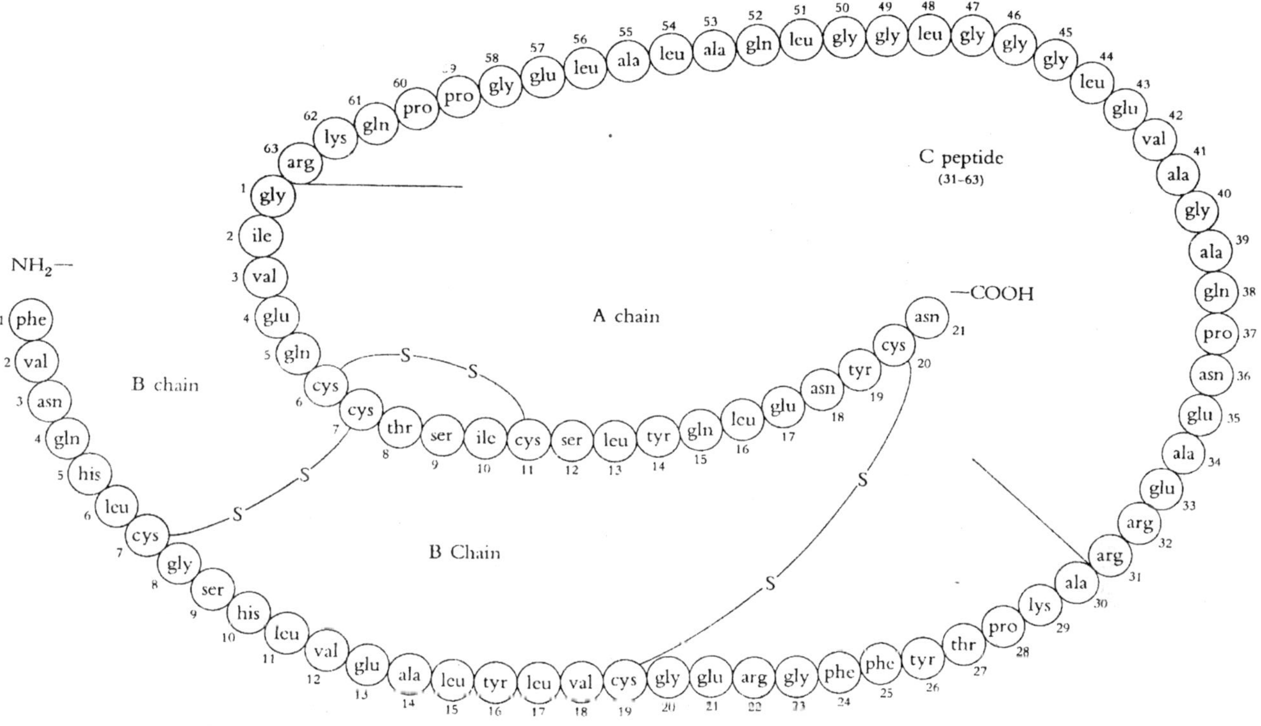

Fig. 12.1. Amino acid sequence of a mammalian proinsulin molecule.

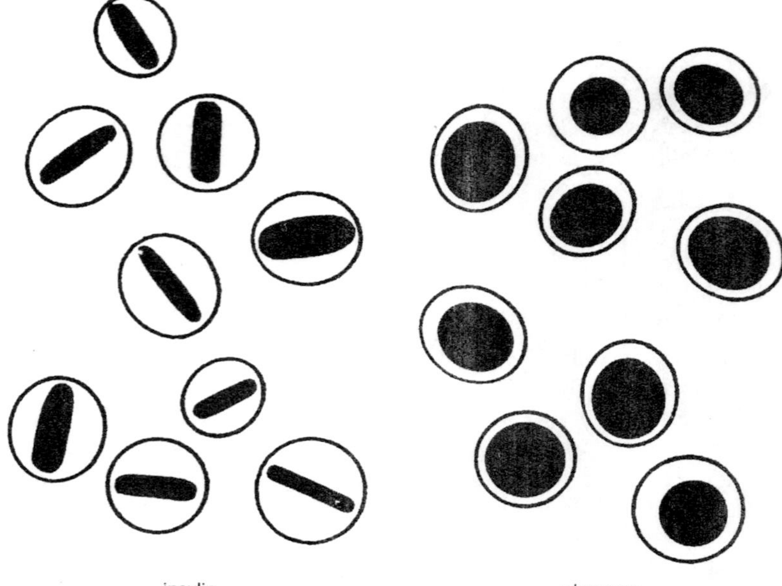

Fig. 12.2. Insulin and glucagon crystals in pancreatic cells.

of glucose but events associated with its metabolism by the B cell that stimulate insulin secretion. Decreases in circulating glucose cause reduction in the secretion of insulin, and it appears that the basic regulatory control mechanism is directed through changes in blood levels of glucose. Any agent capable of elevating circulating glucose levels will also evoke insulin release.

In recent years the importance of nervous regulation of insulin secretion has been demonstrated. Nonmyelinated axonal fibers are present in the islets, and vagal (parasympathetic) stimulation induces insulin release. Similarly, acetylcholine stimulates insulin release, whereas the anticholinergic drug atropine blocks insulin release. Epinephrine, which is a hyperglycemic agent affecting liver, muscle and adipose tissue, also directly blocks the release of insulin. Thus, circulating epinephrine can potentiate its own hyperglycemic action by blocking the normal response of the B cell to increased blood glucose. This action of epinephrine is mediated via β-receptors on the surface of the B cell.

Somatostatin has been shown to directly inhibit release of insulin from the B cell, and it may play a paracrine role in regulating insulin

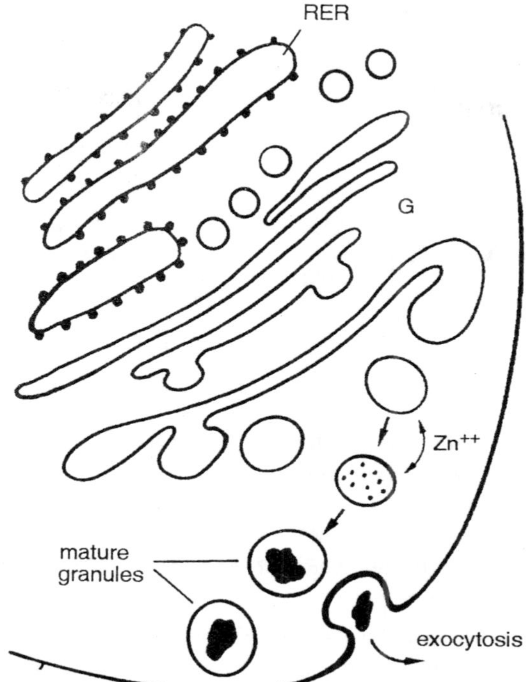

Fig. 12.3. Cellular events associated with the synthesis and release of insulin.

release locally in the pancreas. It is not known what regulates somatostatin release from the pancreatic D cells.

Actions of insulin on target tissues

Insulin promotes anabolic metabolism and growth through three types of action. After binding to membrane receptors there is an immediate increase in the transport of glucose, amino acids, fatty acids, nucleotides and various ions into the target cell. Within a few minutes enhancement of anabolic pathways and a decrease in catabolic pathways occur. Finally some hours later, cellular growth is stimulated. This growth-promoting action is due to an interaction between insulin and certain *insulinlike growth factors*. There are a number of these factors including insulinlike growth factors I and II (IGF I, IGF II), somatomedins and multiplication-stimulating factor (MSA). Most of these factors occur in the blood. MSA is a substance isolated only from calf serum and from cultures of liver cells. These factors are all moderately sized polypeptides (7000-9000 daltons) with about 50% aminc

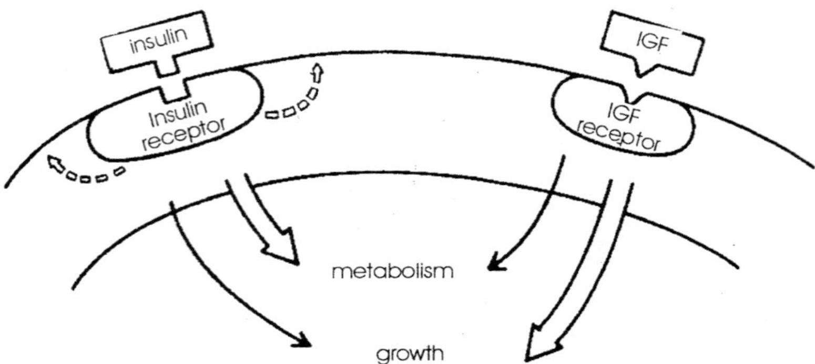

Fig. 12.4. Interaction of insulin and insulinlike growth factors (IGFs) on metabolism and growth.

acid sequence homology to insulin. Nevertheless, these growth factors all have separate membrane receptors from those for insulin. Once insulin binds to its receptor, a proteolytic step is initiated causing formation of glycopeptides which are rapidly internalized. It is the glycopeptides which are believed to regulate the metabolic actions of insulin and to interact with the insulinlike growth factors.

Insulin appears to bring about hypoglycemia through stimulating increased uptake of nutrients from the blood (glucose and amino acids). These nutrients in turn alter metabolic pathways within the responsive cells. Insulin facilitates transport of glucose into muscle and fat cells as well as transport of amino acids into muscle cells. As a consequence, insulin stimulates glucose oxidation and lipogenesis (adipose tissue) and protein synthesis (muscle). Insulin also prevents glycogen breakdown (glycogenolysis) and enhances glycogen synthesis following uptake of glucose in muscle and liver.

Glucose oxidation increases the intracellular pools of precursors for fat synthesis, that is, glycerol, acetylcoenzyme A and fatty acids. In addition to indirectly enhancing lipogenesis, insulin inhibits lipolysis and hence reduces fatty acid oxidation and formation of ketone bodies (ketogenesis), the by-products of fatty acid oxidation. These actions of insulin on adipose tissue and lipid metabolism are marked in carnivores, but there is little effect of insulin on either glucose metabolism or lipogenesis in herbivorous mammals. Consequently carnivorous mammals are more sensitive to pancreatectomy than are herbivores.

The increase in intracellular glucose and amino acids in both liver and muscle cells promotes protein synthesis. These are important actions of insulin in the postabsorptive state, that is, immediately following ingestion of a meal and the entry of large quantities of digestive products such as glucose and amino acids into the blood.

The details of the mechanism of action for insulin in these cells is not clear even after several decades of intensive research. Many of insulin's effects on muscle and adipose tissues can be explained simply on the basis of changes in permeability of the plasmalemma following the binding of insulin. An extension of this hypothesis is the idea that insulin interacts with structural molecules associated with the plasmalemma and initiates a series of molecular rearrangements that are propagated along the membrane and into the interior of the cell. This event could account for all other observed effects of insulin. Changes in enzyme concentrations reported by many investigators may be a secondary result of altered metabolism stemming from permeability changes associated with binding of insulin to the plasmalemma, and these enzymatic changes may not be the primary cause of the altered metabolism observed.

Insulin may enter the target cell itself, and its mechanism of action may relate to its binding to some intracellular site. On the other hand, internalization of insulin may be only a step in its metabolic degradation. It is very clear that insulin does not operate via activation of adenyl cyclase and formation of a second messenger in the form of cyclic adenosine 3´,5´-monophosphate (cAMP). Insulin may influence the intracellular levels of cyclic guanosine 3´,5´-monophosphate, however, and thus secondarily influence cAMP levels.

"Oral insulins"

"Oral insulin" refers to the synthetic hypoglycemic drugs known as *sulfonylureas* and related compounds. Because they are not proteins and thereby not subject to enzymatic degradation in the gut, the sulfonylureas may be taken orally and can relieve the symptoms of diabetes mellitus. Oral insulins thus have considerable therapeutic value in certain types of insulin-deficiency disorders. The effectiveness of these drugs may be related to their abilities in displacing insulin from nonspecific binding and noneffective linkages in serum, connective tissues and islet cells to provide marginally effective quantities of insulin for binding to effective (target) sites. If too high a dose is employed, the oral insulins may also block the effective sites as well, and no hypoglycemic effect is observed. The oral insulins are useful only in disorders related to insulin insufficiency and would not be useful in totally pancreatectomized animals or in humans who can not produce insulin. Furthermore, oral insulins are effective only if diet is controlled.

Carbutamide $H_2N-C_6H_4-SO_2-NH-CO-NH-C_4H_9$

Tolbutamide $H_3C-C_6H_4-SO_2-NH-CO-NH-C_4H_9$

Chlorpropamide $Cl-C_6H_4-SO_2-NH-CO-NH-C_3H_7$

L-Phenethylbiguanide $C_6H_5-CH_2-CH_2-NH-C(=NH)-N(H)-C(=NH)-NH_2$

Synthalin A $H_2N-C(=NH)-NH-(CH_2)_{10}-HN-C(=NH)-NH_2 \cdot 2\,HCl$

Synthalin B $H_2N-C(=NH)-NH-(CH_2)_{12}-HN-C(=NH)-NH_2 \cdot 2\,HCl$

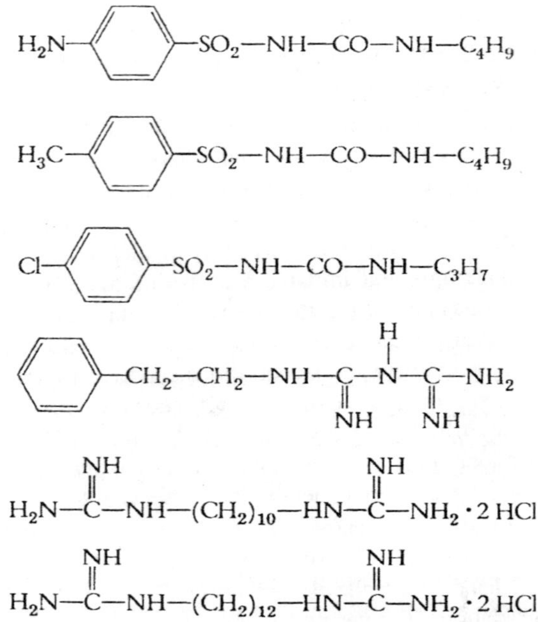

Fig. 12.5. Structures of some "oral insulins."

Glucagon: a hyperglycemic hormone

The discovery of glucagon was not marked by great publicity or special prizes as was that of insulin, and this second pancreatic hormone has always been overshadowed by the clinical importance of insulin. Glucagon is only one of several hyperglycemic factors, and more attention has been given to growth hormone and epinephrine as hyperglycemic hormones. Glucagon becomes more important in fasting carnivores and in herbivores than it appears to be in fed humans or carnivores.

Bioassay of glucagon

Related to its role in fasting mammals, glucagon is often bioassayed in the fasting cat in which it induces hyperglycemia. As in the case of insulin, however, rapid RIAs have been developed for measuring blood glucagon levels, and these are employed routinely.

Chemistry of glucagon

Glucagon is a straight-chain polypeptide hormone consisting of 29 amino acid residues and with a molecular weight of 3485 daltons. The circulating form of glucagon is derived from a prohormone by hydrolysis prior to release. Structurally, glucagon is similar to the family of gastrointestinal hormones that includes secretin and pancreozymin-cholecystokinin.

Actions of glucagon on target tissues

Glucagon promotes glycogenolysis in liver cells, its primary target with respect to raising circulating glucose levels. This effect appears to be mediated through stimulation of adenyl cyclase and production of intracellular cAMP. Increased glycogenolysis accompanied by decreased intracellular oxidation of glucose directs the movement of glucose into the blood.

Lipolysis is stimulated by glucagon in both liver and adipose tissue of fasting animals. This increase in fatty acids following lipolysis results in fatty acid oxidation accompanied by ketogenesis. Glucagon-induced lipolysis depresses glucose metabolism and adds to the hyperglycemic effect.

Extrapancreatic glucagon

There is one cellular type in the duodenal region of the intestine that stains very much like the pancreatic A cell and is ultrastructurally identical to the A cell. This cell produces an immunologically reactive polypeptide that is very similar but slightly larger than pancreatic glucagon (mol wt about 6000-10,000 daltons). After a meal this *enteroglucagon* (*enteron* gut) increases in the blood and may even exceed levels of pancreatic glucagon. Apparently calcium ions in the lumen of the gut stimulate release of enteroglucagon, which in turn elicits release of calcitonin from the parafollicular cells (C cells) of the thyroid. A role for enteroglucagon in carbohydrate or lipid metabolism, however, has not been demonstrated.

Another extrapancreatic glucagon-like hyperglycemic factor has been characterized in extracts of submaxillary salivary glands obtained from rats, mice, rabbits and humans. Release of this *salivary glucagon* is influenced in vitro in the same manner as pancreatic glucagon. Somatostatin, however, does not block release of salivary glucagon.

The biological importance of these extra-pancreatic glucagons is not clear, although they might explain the presence of immunoreactive glucagon in the circulation of pancreatectomized and even pancreatectomized-eviscerated rats. To avoid confusion, the term glucagon will henceforth be used to indicate glucagon of pancreatic source. Extrapancreatic glucagons will always be designated according to their source as enteroglucagon or salivary glucagon.

Somatostatin: regulator of pancreatic hormone release?

The D cell is believed to be the source of pancreatic somatostatin that is chemically identical with hypothalamic somatostatin. Levels of pancreatic somatostatin in rats have been reported to be equivalent to hypothalamic levels' Somatostatin is believed to inhibit the release of insulin, glucagon and pancreatic polypeptide in a paracrine fashion.

Pancreatic polypeptide: a third metabolic hormone?

Pancreatic polypeptide has been purified from ovine, bovine, porcine and human pancreatic tissue. The average molecular weight for PP is 4200, and all peptides isolated consist of 36 amino acid residues. The physiological role of PP is unknown, although there are several reasons for giving PP hormonal status. In man, ingestion of a protein meal such as ground beef causes an increase in plasma levels of PP from preingestion levels of 57 pg/ml plasma to 229 pg/ml, 400 pg/ml and 580 pg/ml respectively in 5, 10 and 240 minutes. Infusion or ingestion of glucose does not evoke this sort of increase in hPP, and it appears that the rapid increase in plasma hPP following ingestion of protein is at least in part a neural response mediated via the vagus nerve. This release of hPP can be blocked by somatostatin, although it is not known whether this latter additional pancreatic agent is involved in the normal regulation of PP release.

Bovine PP is reported to be a potent inhibitor of pancreatic exocrine secretion, but it does not seem to influence carbohydrate (glucose) metabolism. Avian PP has been reported to be a potent stimulator of gastric acid and pepsin secretion, but this effect has not been reported in mammals. It is too early to decide whether these reports are related to the physiological roles of PP or not.

Clinical Aspects of Pancreatic Function

Diabetes Mellitus

Diabetes mellitus is a complex, heterogeneous assemblage of disorders having the common feature of the appearance of glucose in the urine. Diabetes mellitus is always fatal if not treated, and even if treated, serious complications may develop including blindness, kidney disease, and circulatory problems. Approximately 5% of the U.S. population is afflicted with diabetes mellitus, and the incidence of the disease is increasing at the rate of 6% a year. Since the availability of insulin, relatively few deaths in diabetics are directly related to ketoacidosis of the blood and coma. Prior to this, ketoacidosis was the most common cause of death. Nevertheless, diabetes mellitus is responsible directly for about 38,000 deaths annually. Cardio-renal-vascular complications resulting from diabetes mellitus are responsible for an additional 300,000 deaths annually, making diabetes a leading cause of death in the United States.

Most cases of diabetes mellitus can be characterized as either *maturity onset diabetes* (ketosis-resistant; insulin-independent) or as *juvenile onset diabetes* (ketosis-prone; insulin-dependent). "Maturity" and "juvenile" are not entirely appropriate terms since the former can occur occasionally in children and adolescents, whereas the latter can occur with low frequency in middle-aged or older adults. There are some additional forms of diabetes now recognized, but these are rare and are not discussed here.

Maturity onset diabetes (MOD), accounting for approximately 90 to 95% of the reported cases, typically occurs in overweight people (80% of cases) who are over 40 years of age. It has a slow onset with no obvious pancreatic disease, and can usually be controlled by manipulating the diet. Insulin levels in MOD are usually normal or in excess of normal. Apparently, the target cells become insensitive to insulin, possibly due to a decrease in receptor populations or to a deficiency in the mechanism of action of insulin after it is bound to the cell. Another possible cause might be an increase in nonspecific binding of insulin to non-target cells. Whatever the defect, dietary regulation and the removal of excess weight will bring these patients back into normal carbohydrate balance, and the regulation of carbohydrate metabolism will return to normal. Oral hypoglycemic agents (e.g., sulfonylureas) generally are effective at reducing blood glucose levels in these patients, provided dietary intake is controlled rigidly.

There is evidence to support a familial or genetic component in MOD. A series of studies on twins indicates that if one twin develops MOD after age 50, the other twin develops MOD in almost every case.

Juvenile onset diabetes (JOD) usually develops in persons under 20 years of age and is characterized by an abrupt onset of symptoms. A small percentage of people over 40 who contract diabetes mellitus exhibit this form of the disease. Typically, the pancreatic B-cells are reduced markedly (usually to less than 10% of normal), and insulin levels are very low or absent. Glycogen stores are depleted readily and fatty acid metabolism is accelerated, resulting in accumulation of ketone bodies in the blood (e.g., acetone, β-hydroxybutyrate). This accumulation of ketone bodies produces ketoacidosis in about 48 hours and eventually leads to coma and death unless insulin is administered. Patients with JOD have an absolute requirement for insulin for the rest of their lives. Because of this reliance on exogenous insulin and difficulties in administering the appropriate dose when needed, the diabetic must always be prepared for the onset of hypoglycemia following excessive insulin. This usually can be counteracted by consuming a high glucose source.

No strong evidence for a genetic link has been offered for JOD; however, there is some evidence to suggest a role of infection. The symptoms of JOD appear abruptly, and there is a higher incidence of onset during the fall and winter months when infectious diseases also increase. In certain genetic strains of mice, JOD has been linked to a specific virus. There also appears to be a correlation between the occurrence of certain viral diseases (e.g., mumps, rubella *in utero*) and the later appearance of JOD.

Dietary restrictions are similar for both MOD and JOD patients although a reduction in total caloric intake is necessary for overweight MOD patients. Specific diets must be determined individually for every patient. Carbohydrates are still essential to the diabetic diet but need to be in the form of polysaccharides (e.g., starch). Mono- and disaccharides are usually avoided because of rapid uptake and resultant hyperglycemia soon after ingestion. The ingestion of simple sugars is less a problem for the MOD patient as the pancreas can respond with some insulin release.

Comparative Aspects of the Endocrine Pancreas

Five anatomical arrangements for pancreatic islet systems can be distinguished readily by examination of the major vertebrate groupings.

1. *Cyclostome type*. The cyclostome type is composed of aggregations of islet cellular types with no obvious relationship to the homologous intraintestinal equivalent of mammalian acinar cells. In hagfishes islets surround the base of the common bile duct, and in lampreys they are located on the surfaces of the intestine (epi-intestinal), within the intestinal mucosa (intraintestinal) and even within the liver (intrahepatic). The lamprey pattern of islet distribution is considered to be the most primitive pattern invertebrates. Exocrine pancreatic elements are embedded in the intestinal lining.
2. *Primitive gnathostome type*. Islet tissue of elasmobranchs, holocephalans and the sarcopterygian coelacanth (*Latimeria*) occurs as layers surrounding the smaller ducts of a compact, extraintestinal

pancreas as well as some scattered cells throughout the exocrine pancreas.

3. *Actinopterygian type*. Usually the exocrine pancreatic cells are scattered (diffused) along the bile ducts, abdominal blood vessels and on the outer surfaces of the gastrointestinal tract, gallbladder and liver. An intrahepatic pancreas has been observed in several species. The islet tissue often accumulates as clumps near the common bile duct but is not usually separated from the acinar tissues.
4. *Lungfish type*. The pancreas of lungfishes is a compact, intraintestinal structure with a number of encapsulated islets.
5. *Tetrapod type*. Tetrapods typically exhibit a compact, extraintestinal pancreas containing scattered islets and occasionally some scattered individual endocrine cells. In some tetrapods (for example, birds and the toad Bufo), islets may be concentrated in particular lobes of the pancreas and not evenly distributed.

There are some generalizations concerning the origin of the various components of the pancreas and their relationship to the distribution of islets. The dorsal pancreatic bud gives rise to the tail and body of the pancreas (also known as the splenic portion) and the ventral buds give rise to the head or duodenal portion. Islets associated with the splenic lobe typically are larger and consist mainly of B, A and D cells. The islets of the duodenal pancreas are smaller and may contain many PP cells.

Comparative Structure of Insulin

Insulin is a conservative protein in that few amino acid substitutions have occurred at the critical sites related to its structure and biological activity. The C peptide of proinsulin "tolerates" a greater number of substitutions, yet portions of this peptide also remain rather conservative, for example, the glycine-rich central core of the C peptide. Single amino acid substitutions may have profound effects on the resultant molecule. One substitution in hagfish insulin is believed responsible for the observation that hagfish insulin does not bind Zn^{++} and therefore does not form the typical hexamer crystals characteristic of most vertebrate insulins. The number of amino acids in the B chain varies when insulins from different vertebrate classes (especially reptiles) are compared.

Comparative Structure of Glucagon

Mammalian glucagons that have been characterized chemically exhibit no amino acid substitutions. Turkey glucagon differs from mammalian glucagons only at position 28 where serine is substituted for asparagine. Chicken glucagon is identical to turkey glucagon, whereas duck glucagon has threonine at position 16 instead of serine. Glucagon has been isolated from the spiny dogfish shark *Squalus acanthias*, but the primary structure of this glucagon has not been reported.

Occurrence of Vertebrate Pancreatic Hormones in Invertebrates

B cells and insulin activity have been identified in the guts or digestive organs of a number of nonchordate invertebrate animals. Mammalian insulin reduces blood sugar levels and promotes glycogen synthesis in a gastropod mollusc, *Strophocheilus oblongus*. Insulinlike molecules have even been found in protozoans, bacteria and fungi. Insulin is therefore a widely distributed factor that is capable of regulating carbohydrate metabolism in a great variety of organisms and possibly does so. The evolution of insulin must involve a long and interesting history, most of which remains to be discovered.

Immunoreactive glucagon has been localized in the digestive glands of a crab, *Cancer pagurus*, and two gastropod molluscs, *Patella caerulae* and *Helix pomatia*. A glucagon-like molecule is present in extracts of invertebrate chordates, including tunicates and cephalochordates.

Class Agnatha: Cyclostomata

Hagfish "islet" tissue consists almost exclusively of B cells plus some "light" cells that could represent precursor or "spent" secretory cells. Some B cells occur in the gut mucosa as do immunologically reactive A cells. Lamprey islets exhibit insulin-secreting B cells as well as four acidophilic cellular types for which functions have not been identified. Cyclostomes do not have A cells and do not respond to injections of glucagon.

Insulin has been purified following the extraction of islet tissue obtained from several thousand hagfishes. Hagfish B cells synthesize insulin through processes that appear to be identical to those described for mammals. Insulin has also been extracted from islet tissue of the river lamprey *Lampetra fluviatilis*.

Hagfish insulin behaves similarly to other insulins in mammalian assays. The biological activity of purified hagfish insulin in mammals, however, is only about 4- to 7% of that reported for mammalian insulins, indicating the evolution of greater specificity in target cells as well as some changes in the insulin molecule itself. Both mammalian and hagfish insulins cause hypoglycemia when injected into hagfish. Hagfish insulin stimulates both glycogen and protein synthesis in hagfish skeletal muscle but only stimulates protein synthesis in liver. Amino acids are known to stimulate insulin release in lampreys providing additional support for insulin's role as a regulator of protein metabolism. Known insulin antagonists, such as epinephrine, glucagon, T_4, corticotropin and prolactin (containing growth hormone as a contaminant) were all ineffective in altering blood sugar levels. Curiously, isletectomy of *Myxine* does not cause hyperglycemia, and the response to a glucose load in isletectomized hagfish is normal. Hypophysectomy has no effect on islet cytology or blood sugar levels. There must be an explanation for the presence of insulin and the demonstration of its hypoglycemic action in the hagfish with the number of negative observations just mentioned, but it will require more research to reconcile these contradictory data.

Class Chondrichthyes

The endocrine pancreas of a number of species of selachians and holocephalans has been examined. The selachian pancreatic islets contain B, A, D, and PP cells. Both insulin and glucagon have been extracted from the pancreas of the spiny dogfish *Squalus acanthias*. In the ratfish *Hydrolagus colliei*, approximately 50% of the islet cells are of a type referred to as *X cells*. These cells resemble both A and D cells but also possess some unique cytological features of their own. It is not clear whether these X cells represent a transient stage between known cellular types or perform some unique function yet unknown.

It appears that blood sugar regulation is not very precise in these fishes, and blood sugar may vary considerably even within a species. Administration of mammalian insulin evokes hypoglycemia in these fishes, and glucagon has been shown to induce hyperglycemia in some species. Epinephrine also elevates the blood glucose level, and the effect of injected epinephrine occurs very rapidly. On a superficial basis it appears that regulation of carbohydrate metabolism is similar to the mammalian pattern.

There are many unusual features of metabolism in these fishes that require further study. Although liver glycogen levels in dogfish sharks conform with predicted responses to fasting and force feeding, the liver of holocephalans is apparently incapable of storing glycogen. Injection of a glucose load does not alter tissue glycogen even though circulating glucose is increased. These observations may explain the inability of glucagon to increase blood glucose in the ratfish. Protein and lipid metabolism with respect to the actions of pancreatic hormones have not been studied. Lipid metabolism in these fishes should be of special interest since hepatic fat content may be as high as 70% in the dogfish and 80% in the ratfish.

Class Osteichthyes

Most species possess isolated masses of pancreatic islet tissue called *Brockmann bodies*. A given species may have a number of Brockmann bodies, or the islet tissue may be concentrated largely in one "principal islet" as in the Atlantic eel *Anguilla anguilla* or the bullhead *Ictalurus* spp. Brockmann bodies of actinopterygian fishes contain A, B, and D cells, but PP cells occur only in Brockmann bodies located in the pyloric region.

Insulin and glucagon are produced by the B and A cells respectively. Somatostatin is apparently synthesized by the D cells of the pancreas of the angler fish, implying that the role of somatostatin in pancreatic function is not unique to mammals. A fourth cellular type of undefined function has been reported in catfishes and definite PP cells have been reported in *Xiphophorus helleri*.

The lungfish (*Protopterus annectens*) pancreas is located intraintestinally and contains a number of scattered encapsulated islets. The islet tissue of the lungfish possesses A, B, and D cells. There are probably PP cells present, but they have not been identified.

Insulins prepared from several teleosts appear to be similar to mammalian insulins. Glucagon has been isolated from the anglerfish. It consists of 29 amino acid residues that differ somewhat from those of mammalian glucagons. It is produced from a larger proglucagon peptide. Generally insulin is hypoglycemic and glucagon is hyperglycemic in bony fishes. Injections of glucose or the amino acid leucine stimulate insulin secretion in the toadfish *Opsanus tau* and in the European silver eel *A. anguilla*. Purified codfish insulin is a potent hypoglycemic agent in Northern pike, *Esox lucius*. Furthermore, insulin increases uptake and incorporation of labeled glucose and glycine into muscle lipids and protein. Administration of mammalian glucagon increases blood glucose in channel catfish. Glycogenolysis in isolated hepatocytes (liver cells) of goldfish is stimulated by glucagon and epinephrine.

Class Amphibia

The anatomy of the pancreas of anurans and urodeles is similar to that of the other tetrapods. Data are lacking for apodans. The islets are distributed in the acinar tissue, although in at least one anuran, *Bufo arenarum*, many of the islets are concentrated in one lobe of the pancreas. The islets of anuran amphibians resemble those of mammals cytologically with A, B, D, PP and a fifth unidentified cellular type present. Although B cells predominate in amphibians, there are almost equal proportions of A and B cells, whereas A cells comprise only about 10%- 20% of the islets of most mammalian species. D cells are also present. The "islets" of *Necturus maculosus* are not encapsulated and occur as groups of cells. Most urodeles, however, exhibit the typical tetrapod type of islet arrangement.

Mammalian insulin and glucagon produce effects in amphibians similar to those observed in mammals. Insulin induces hypoglycemia in anurans and in urodeles, stimulates incorporation of amino acids into proteins and lowers fatty acid levels. The response to a glucose load in anurans and the effects of pancreatectomy do indicate an endogenous role for insulin. Mammalian glucagon causes hyperglycemia in anurans and in the salamander *Ambystoma annulatum*. Although mammalian glucagon has been found to be ineffective in some amphibians, glucagon treatment is effective in raising blood glucose levels at certain times of year.

Epinephrine is also hyperglycemic in tadpoles and adults of *Xenopus laevis*. Administration of epinephrine or norepinephrine causes reduction in liver and muscle glycogen as well.

Research on the amphibian pancreas has led to a widely employed procedure for research into mammalian diabetes mellitus. It was in the toad *B. arenarum* that the antagonism of pituitary hormones and blood glucose of pancreatectomized or insulin-deficient animals was discovered by Bernardo A. Houssay almost 50 years ago. This discovery led to the widespread use of hypophysectomy to alleviate extreme diabetic conditions in experimental animals, and such doubly operated-on dogs and cats and other laboratory beasts became known commonly as Houssay animals. This action of hypophysectomy in the pancreatic-deficient animal is termed the Houssay effect.

Class Reptilia

Reptiles have a distinct pancreas containing both acinar and islet tissues. There is a predominance of A cells in lizard islets (ratio of 4:1 over B cells), and there is some tendency for the islets to concentrate in the caudal (splenic) lobe of the pancreas. In contrast, the B cells represent about 50% of the islet cellular types in the alligator. D cells are present also in the reptilian pancreas but are not a dominant type. PP cells have not been described.

Reptiles exhibit mammalian-like responses to injected insulin and glucagon and to glucose loading. Lizards, however, are particularly insensitive to insulin, and large quantities of mammalian insulin are required to invoke hypoglycemic responses. It is virtually impossible to induce insulin shock with massive doses. Pancreatectomy does induce a hyperglycemic state in lizards as it does in snakes and alligators. These observations suggest that in spite of the predominance of A cells in lizard islets, insulin may plan an important endogenous role in carbohydrate regulation in lizards as well as in other reptiles.

Glucagon has not been isolated from reptiles, and its role in carbohydrate metabolism has been inferred from observations of the actions of mammalian glucagon administered to reptiles. Mammalian glucagon is ineffective in hypophysectomized snakes, suggesting that pituitary hormones may be more important hyperglycemic agents. Perhaps the effects of injected mammalian glucagons are somehow dependent on pituitary hormones.

Insulin has been purified from rattlesnakes, *Crotalis atrox*, and partially characterized chemically. The amino acid composition of rattlesnake insulin is considerably different from bovine insulin with some uncommon substitutions occurring in the B chain. Comparative analyses of reptilian insulins may shed some light on pancreatic regulation of carbohydrate metabolism in reptiles, especially the apparent insensitivity of many species to mammalian insulins. Cyclostome insulin is more like mammalian insulins than is rattlesnake insulin, which points out the difficulties of inferring evolutionary events from molecular structures.

Of particular interest is the absence of studies in which lipid and protein metabolism of reptiles were examined. Insulin and other "metabolic" hormones have not been examined for their roles in lipid or protein metabolism. Until all aspects of the hormonal regulation of metabolism in reptiles are integrated with the observations of effects of insulin and glucagon, a complete picture for the roles of reptilian pancreatic hormones will not emerge.

Class Aves

The avian pancreas consists of dorsal, ventral and splenic lobes. In some species a fourth lobe occurs (inappropriately called the "third" lobe). The dorsal embryonic pancreatic bud gives rise to the splenic lobe (and "third" lobe when present), whereas the ventral buds produce the dorsal and ventral lobes. Unlike the condition for other vertebrates, there appear to be two cytologically distinct types of islets in the avian pancreas. The first type consists of A and D cells. These islets are called A islets. The B islets are composed primarily of B and D cells. Both types of islets are distributed throughout the pancreas although the splenic lobe contains predominantly A islets. Hyperglycemia initially observed following pancreatectomy is now believed to have been a result of overlooking the splenic lobe during pancreatectomy. Total pancreatectomy results in hypoglycemia rather than hyperglycemia and suggests that glucagon plays a major role in avian blood sugar regulation. This suggestion is further supported by observations that a high glucose load is required to induce insulin release in the duck and by the relative insensitivity of avian tissues to insulin. Elevated glucose levels depress glucagon release, whereas high circulating levels of free fatty acids induce glucagon released. PP cells are uncommon in islets of the splenic lobe but occur in islets of ventral and dorsal lobes.

Avian tissues respond with hyperglycemia, glycogenolysis and lipolysis following administration of mammalian glucagon. The avian pancreas contains about five to ten times the extractable glucagon per gram of pancreas as compared to mammals, emphasizing its importance to metabolic regulation in birds. In one of the few studies performed in wild birds, mammalian glucagon stimulates both plasma levels of free fatty acids and glucose in penguins, whereas insulin causes a reduction in blood glucose. Additional studies with wild avian species representing diverse taxonomic groupings would be welcome in order to confirm or alter generalizations based upon observations of domestic species.

Insulin has been extracted from the avian pancreas, and its structure has been examined. Chicken insulin is more effective in inducing hypoglycemia in birds than it is in mammals, and it is more potent in birds than is mammalian insulin. Glucagon has been extracted and its primary structure determined for duck, turkey and chicken hormones. These hormones are almost identical structurally to mammalian glucagon.

In the process of extraction and purification of insulin from chickens, a third biologically active peptide was isolated and named avian PP. Like mammalian PP, aPP consists of 36 amino acid residues and has a molecular weight of approximately 4200 daltons. Avian PP, unlike its mammalian counterpart, appears to have a strong stimulatory influence on gastric acid secretion in chickens. The physiological role for aPP, however, is uncertain at this time.

13

Reproductive Endocrinology

The oversimplified way in which even biologists sometimes view reproduction is exemplified by the following quotation from a textbook on fish biology: "the reproductive function in fishes is primarily the job of the reproductive system." In actuality, the "reproductive system" includes not only the gonads and their associated structures but also the complex hypothalamo-hypophysial axis as well as both neural and endocrine input to the hypothalamus. Furthermore, environmental factors (chemical, visual, photic, thermal and tactile stimuli) frequently determine reproductive events through their effects on sensory organs and receptors. Environmental factors contribute ultimately to both neural and endocrine input to the hypothalamus. When one thinks of reproduction, one should be considering a process of evolutionary adaptation that extends far beyond the inclusion of what is usually termed the reproductive system.

The major difficulty that arises when generalizations are made about reproduction and reproductive mechanisms stems from the central importance of these processes with respect to past and future evolutionary events. The reproductive system is possibly the central focus for selective agents since reproductive success is the major determinant of evolutionary success. Consequently the reproductive system has been highly responsive to selective forces throughout the long evolutionary history of vertebrates, and many features of extant species are more examples of specific adaptations to solve common environmental problems than they are representative of progressive "improvements" in the basic system that "culminated" in the placental mammals. A case in point would be the achievement of viviparity in all but two extant vertebrate classes, the Agnatha and Aves, as specific adaptations that, coupled with varying degrees of parental care, result in greater percentage survival of a small number of offspring. Viviparity represents only one solution, however, to similar selective pressures that confront all species. In spite of the problems of environmental adaptations that tend to confuse evolutionary relationships, there remain numerous conservative features in regulatory mechanisms of reproductive biology, and it is these that are emphasized in this chapter.

Reproduction is closely regulated through the hypothalamo-hypophysial axis, which coordinates specific gonadal events through regulation of circulating gonadotropins.

In mammals the release of luteinizing hormone (LH) and follicle-stimulating hormone (FSH) from the adenohypophysis appears to be under the control of a single hypothalamic hormone, the gonadotropin-releasing hormone GnRH. Release of this stimulatory factor is regulated by two separate regions of the hypothalamus: the *tonic center*, which maintains a relatively constant circulating level of both LH and FSH, and the *cyclic center*, responsible for the midcycle LH surge observed in sexually mature female mammals. Secretion of GnRH is under stimulatory control of adrenergic neurons and can be inhibited by endogenous opioids. One hypothesis suggests that combined increases in adrenergic activity and an inhibition of opioid secreting neurons allows the ovulatory LH surge to occur.

These gonadotropins stimulate gamete maturation in both males and females as well as steroidogenesis and release of estrogens, androgens and progestogens into the general circulation. Gametogenesis (oogenesis, spermatogenesis) appears to be controlled primarily by FSH, whereas LH is primarily responsible for controlling steroidogenesis in both sexes as well as induction of ovulation and formation of corpora lutea in the ovaries of females. Steroidogenesis can also be influenced by FSH in both males and females. The details of these events are discussed later.

The gonadal steroids released through the action of gonadotropins control differentiation and maintenance of many primary and secondary sexual characters such as secretion by the prostate gland and beard growth in men. This was recognized in the 12th century and probably much earlier by the Chinese who used gonadal (and placental) preparations routinely to treat disorders ranging from impotence to the inability of a woman to bear sons. The interstitial cells of the testes and the cells comprising the growing follicle in the ovary are the sources of these steroids. The hypothalamic centers regulating gonadotropin release are sensitive to circulating steroids, which generally produce a negative effect on GnRH release. The notable exception to this pattern is the stimulation of LH and prolactin (PRL) release by estrogens. Additional evidence has been presented for a gonadal factor produced under the influence of FSH in both testes and ovaries that selectively inhibits FSH release with no effect on LH release. This factor is possibly a peptide, called *inhibin*. Alternatively, the ratio of testosterone to estradiol in the circulation can influence differential release of FSH and LH and could produce the same results as claimed for inhibin. The inhibin hypothesis is favoured at this time.

Several other hormones are involved in mammalian reproduction in addition to those of the hypothalamo-hypophysial-gonadal axis. Prolactin and, to a lesser extent, corticotropin (ACTH) from the adenohypophysis, oxytocin from the pars nervosa, and thyroid hormones all influence reproductive events. The placenta of eutherian mammals has assumed an endocrine role in pregnant females, producing both steroids (primarily estrogens and progesterone) and polypeptide hormones (chorionic gonadotropins, chorionic somatomammotropin, chorionic corticotropin, chorionic thyrotropin). In addition, the endocrine glands of the fetus may influence reproductive events (for example, contribution of the adrenal cortex to steroidogenesis by the placenta). Finally there are numerous reports of chemical agents termed *pheromones* that are produced by one sex to influence reproductive physiology, behaviour or both of the opposite sex. In the next section the generalized events of reproduction in eutherian mammals are presented, followed by a brief description of the reproductive cycle of the ewe, human female and 4-day cycling female rat. Separate discussions of the monotremes and some of the special features of marsupials complete consideration of mammalian reproduction. The final portion of this chapter is devoted to nonmammalian vertebrates.

Embryogenesis of Gonads and Their Accessory Ducts

The primordia of the mammalian gonads arise from the intermediate mesoderm as a genital ridge on either side of the midline in close association with the mesonephric kidney. Numerous derivatives of the mesonephric kidney and duct system are retained as functional portions of the vertebrate reproductive system. The gonadal primordium consists of an outer cortex derived from peritoneum and an inner medulla. Differentiation primarily of the medullary component results in a testis, whereas the cortical component gives rise to an ovary. Primordial germ cells do not arise within the gonadal primordium itself but migrate from their site of origin in the yolk sac endoderm to whichever component is genetically destined to develop. Initially the medullary component in males and females differentiates into primary sex cords that in males become the seminiferous tubules of the testis. The cortical components degenerate in males. The interstitial cells arise from medullary tissue surrounding the primary sex cords. Primordial germ cells that have migrated into the primary sex cords give rise to spermatogonia. In females the primary sex cords degenerate, and secondary sex cords differentiate from the cortical region. These secondary sex cords become the definitive ovary, and the primordial germ cells give rise to oogonia.

In males the central portion of the differentiating testis forms a network of tubules, known as the *rete testis*, that do not contain

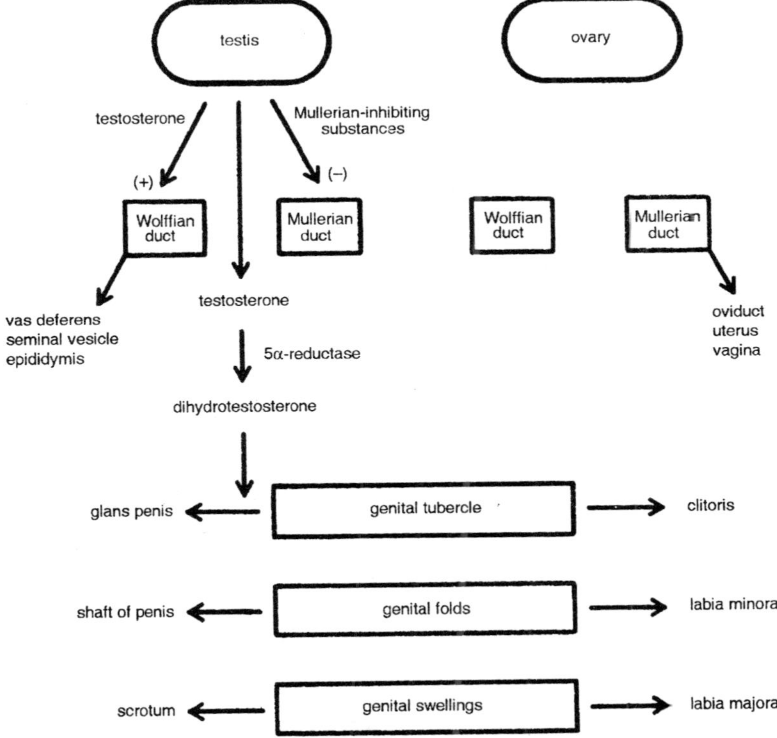

Fig. 13.1. Development of mullerian ducts, wolffian ducts and homologous external genitalia in mammals.

seminiferous elements. The rete testis connects the seminiferous tubules with a portion of the mesonephric kidney duct called the *wolffian duct*, which under the influence of testosterone, differentiates into the *vas deferens* and conducts spermatozoa to the urethra. Some of the anterior mesonephric kidney tubules do not completely degenerate, as does most of the mesonephric kidney in mammals. In the presence of testosterone, this tissue together with a portion of the mesonephric duct forms the epididymis and seminal vesicle.

A second pair of longitudinal ducts develops from the mesial wall of the mesonephric duct and lies parallel to the wolffian ducts. These structures are known as the *Mullerian ducts*. In genetic females the mullerian ducts develop into the oviducts, uterus and part of the vagina, usually fusing together to form a common vagina and, in some species, a single uterus as well. The wolffian ducts degenerate in females. In males it is the mullerian ducts that are suppressed in favour of wolffian duct development. *Mullerian-inhibiting substance* (MIS) was first proposed by Jost in the 1940s to explain the inhibitory effect of the testes on development of mullerian ducts in rabbit embryos. Implantation of a testis into a female embryo prevents development of the mullerian ducts. Evidence suggests that MIS is a glycoprotein. It not only blocks mullerian duct development but is capable of inhibiting growth of tumors from ovaries and mullerian duct derivatives. It appears that MIS acts cooperatively with testosterone in producing its effects

Mammalian Life History Patterns

Obvious differences in life cycles can be found by examining the three major mammalian subgroupings: the egg-laying monotremes (Prototheria), the pouched marsupials (Metatheria), and the "placental" mammals (Eutheria). The monotremes have retained the reptilian feature of laying eggs but have added the element of parental care so characteristic of mammals. The duckbill platypus lays its eggs in a nest, but the spiny anteater or echidna places the eggs in a transitory pouch where they develop to hatching. Upon hatching the monotreme appears as a tiny, fetus-like creature with only a few well-developed features that enable it to attach itself to the mother and obtain nourishment. The degree of development at hatching is similar to that of marsupials at birth.

There are about 230 species of marsupials distributed among Australia and North and South America. A primitive type of placenta does develop in marsupials, but it apparently has no endocrine function. The period of pregnancy or gestation is very short in marsupials, and the young marsupial is born in an extremely immature condition. Among the macropodids (kangaroos) the "joey" must find its way essentially unaided to the mother's pouch or *marsupium* where it permanently attaches to the nipple of a mammary gland. The joey continues its development as an "exteriorized fetus." After a long period of pouch development (about 200 days in the red kangaroo), the young marsupial disengages itself from the teat and ventures outside of the pouch, returning first at regular and later at irregular intervals for milk.

Eutherian mammals employ the placenta not only as an endocrine organ to maintain gestation but also as a replacement for the mammary glands to supply nutrition to the fetus. Consequently birth (parturition) is delayed considerably, and the newborn placental mammal (neonate) is at a comparable stage of development to the young marsupial when it first leaves the pouch. Like the juvenile marsupial the newborn placental mammal relies at first on the mammary gland as the exclusive source of nourishment but gradually abandons it for other foods.

The patterns of reproduction in eutherian mammals will be discussed in some detail, followed by separate descriptions of three representative females: ewes, women, and 4-day cycling rats. Monotremes and marsupial mammals will be considered, but only major differences with respect to eutherian mammals will be stressed.

Eutherian Reproductive Patterns

Three distinct reproductive patterns occur in sexually mature eutherians, one typically for males and two among females. Males are characterized by continuous secretion of gonadotropins and more or less continuous spermatogenesis. In many species the males are capable of siring offspring at any time of year. Most species, however, exhibit seasonal episodes of spermatogenesis, sexual activity or both such as displayed by nonmammalian vertebrates.

Females exhibit cyclic patterns of gonadotropin secretion that can be traced to a basic rhythmicity within the hypothalamus. There are two types of female cycles: the *estrous cycle* and the *menstrual cycle*. The estrous cycle is typical for most mammals and consists of a repeated series of precisely regulated endocrine events. The periods within the estrous cycle can be readily distinguished. *Proestrus* is characterized by the hormonal changes that bring about ovulation. *Estrus* immediately follows or coincides with ovulation, and it is a short period when the female is receptive to the male and during which mating can occur. It is also the time when fertilization is most likely to lead to pregnancy and successful birth of offspring. The interim between estrus and the onset of hormonal changes characteristic of proestrus in cases in which pregnancy did not result is termed *diestrus*. Carnivores may be classified as *monestrous*. If mating does not occur or if mating occurs but fertilization and implantation are unsuccessful, the female will not return to estrus until the next breeding season. Many mammalian species are *polyestrous*, however, and will return immediately to proestrus if mating does not occur or if mating is unsuccessful. In certain rodents mating without successful fertilization and implantation may result in a short period of simulated or false pregnancy, termed *pseudopregnancy*, after which the female reenters proestrus.

Some primates, including man, exhibit a different sequence of events, known as the menstrual cycle, that is characterized by sloughing of the uterine lining if fertilization does not lead to pregnancy. The sloughing of the uterine lining results in a vaginal discharge of blood and uterine epithelial cells. This stage is known as the *menses* or period of menstrual flow. The onset of menses precedes the *preovulatory* or *follicular phase* (phase of the growing follicle of the cycle) and marks the end of the *postovulatory* or *luteal phase* (the phase of the corpus luteum). Most primates exhibit estrous behaviour to some degree at about the time of ovulation, including species characterized as having menstrual cycles, although a well-defined estrus is not observed in the human female. Possession of menstrual or estrous cycles should not be considered as alternative cycles.

Some mammals such as the cow and bitch discharge blood prior to ovulation and the onset of estrous behaviour. This discharge is estrogen induced and is not similar to uterine breakdown and the menses of primates.

Many mammals ovulate following coitus and are termed *induced ovulators*. Several carnivores (for example, ferret, mink, raccoon, cat), rodents (for example, *Microtus californicus*), lagomorphs (for example, cottontail and domestic rabbits), one bat (lump-nosed bat) and several insectivores (for example, hedgehog, common shrew) have been proven to be induced ovulators. Some other species are suspected to be induced ovulators (the elephant seal, nutria and a marsupial, the long-nosed kangaroo rat). Induced ovulators do not exhibit a cyclic pattern of gonadotropin release, because the LH surge occurs only after copulation takes place. Most mammals are believed to be so-called *spontaneous ovulators* in that coitus does not usually cause ovulation. However, even some spontaneous ovulators can be induced to ovulate following copulation under special conditions. For example, ovulation can be induced in the rat by appropriate hormone treatment followed by copulation.

Although reproductive endocrinologists consistently focus upon estrous and menstrual cycles, it is important to remember that the normal sequel to ovulation is pregnancy. In nature, it is unusual for a female to enter estrous and not become pregnant. The observation that lactating females usually do not enter estrous supports the hypothesis that elevated levels of PRL can block pulsatile LH release as well as the LH surge required for ovulation.

Endocrine regulation of male reproductive events

Maleness in eutherian mammals is dependent upon secretion of androgens from the testis. In the absence of androgens the male genotype (XY) will develop as a female phenotype. Conversely, treatment of newborn females with androgens destroys the cyclic secretory pattern of the hypothalamo-hypophysial system and replaces it with a non-cyclic pattern like that of males. It may be that androgen treatment destroys only the activities of the cyclic hypothalamic center whereas the tonic center of females is unaffected.

In some species postpubertal males are capable of copulating with a female whenever she is receptive. Secretion of GnRH and hence of gonadotropins is more or less continuous but with daily fluctuations

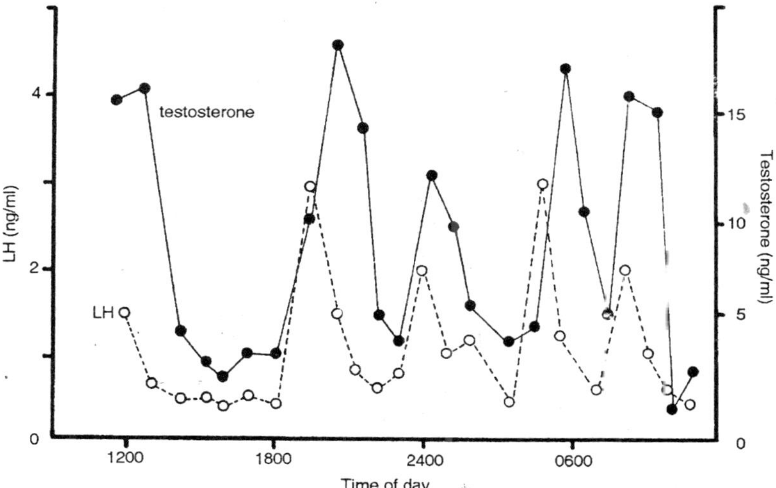

Fig. 13.2. Diurnal variations in blood levels of luteinizing hormone (LH) and testosterone in a bull.

occurring in circulating levels of some gonadotropins. Daily secretory patterns for gonadotropins show considerable variation among different species. Hourly fluctuations of LH have been reported in bulls, and these variations in LH are correlated with following surges in circulating testosterone. However, in human males, FSH shows no cyclic variation in blood levels although LH and testosterone exhibit clear daily patterns with peak levels occurring during early morning hours and minimum values reported for the afternoon. Wild mammals exhibit distinct seasonal breeding, and spermatogenesis may be restricted to a few months or less.

Spermatogenesis

The testis develops primarily from the medullary portion of the embryonic gonadal blastema. Differentiation of the medullary portion with concomitant regression of the cortical components (progenitor of the ovary) appears to be controlled by embryonic androgen secretion. The medullary cords differentiate into *seminiferous tubules* and interspersed masses of *interstitial* or *Leydig cells*. These interstitial cells are steroidogenic and are located between the seminiferous tubules. Androgens are synthesized and released into the circulation by interstitial cells. The seminiferous tubules consist of large *Sertoli* or *sustentacular cells*, *spermatogonia* and cells derived from the latter. The seminiferous tubules are surrounded by *peritubular cells* that are believed to be responsible for contractile activity of the seminiferous tubules. The Sertoli cell has an extensive cytoplasm extending from the outer edge to the lumen of the tubule. The nucleus of the Sertoli cell is located at the outer edge. In addition to the Sertoli cells, the primordial germ cells that give rise to spermatogonia are present along the outer margins of the tubules. Spermatogonial cells proliferate mitotically under the influence of FSH and eventually undergo differentiation characterized by nuclear enlargement to become *primary spermatocytes*. These cells undergo the first meiotic division to give rise to two smaller *secondary spermatocytes*, which are rarely observed in histological preparations because, once formed, they enter the second meiotic division to yield four haploid *spermatids*. Testicular androgens are somehow necessary for meiosis. These spermatids are transformed to *spermatozoa* (spermiogenesis) by concentrating the chromatin material into the spermatozoan head and by elimination of the majority of the cytoplasm. A given histological section of a seminiferous tubule may show varying numbers of spermatogonia, primary spermatocytes, spermatids and spermatozoa in sequence from the outer margin to the lumen. The tails of the spermatozoa extend into the lumen, and the heads of the spermatozoa are typically surrounded by highly folded margins of the Sertoli cells.

Millions of mature spermatozoa may be sloughed off into the lumina of the seminiferous tubules each day. This process is termed *spermiation*. These spermatozoa pass along through the tubules, which eventually coalesce into larger ducts and eventually form the *epididymis* associated with each testis. Vast numbers of mature spermatozoa are stored in the epididymis. Under the influence of androgens the epididymis secretes materials into its lumen where the spermatozoa are being held. Included in this secretion are protein-bound sialic acids (sialomucoproteins), glycerylphosphoryl-choline and carnitine. These particular substances are involved directly in maintaining spermatozoa in viable condition until ejaculation. Androgens and androgen-binding protein (ABP) produced by sustentacular cells in the seminiferous tubules are released along with spermatozoa and travel to the epididymis. Androgen freed from ABP in the lumen or androgen-ABP complexes or both are absorbed by the epididymal cells. These androgens stimulate epididymal cells to secrete materials involved in maintenance of the spermatozoa.

As a consequence of the forcible ejection (ejaculation) that occurs during copulation the spermatozoa leave the epididymis, enter the vas deferens and travel to the urethra. The spermatozoa traverse the length of the penis via the urethra and are deposited in the female's vagina during coitus. Various glands such as the prostate add their fluid secretions to the spermatozoa and epididymal secretions to form a watery mixture of spermatozoa and various organic and inorganic substances known as *semen*. The entire ejaculatory event may be induced by the release of oxytocin from the pars nervosa in response to a neural reflex initiated by mechanical stimulation of the penis.

Roles for gonadotropins in male mammals

Relatively separate roles have been defined for LH and FSH in males. Spermatogenesis is initiated by FSH, which stimulates proliferation of spermatogonia and formation of primary spermatocytes. In addition, androgens are thought to initiate meiotic divisions of primary spermatocytes, resulting eventually in formation of spermatids. The Sertoli cell of the seminiferous tubule also appears to be a target for FSH. The production of ABP by the Sertoli cell is stimulated by FSH. This ABP is released into the testicular fluid where it binds androgens.

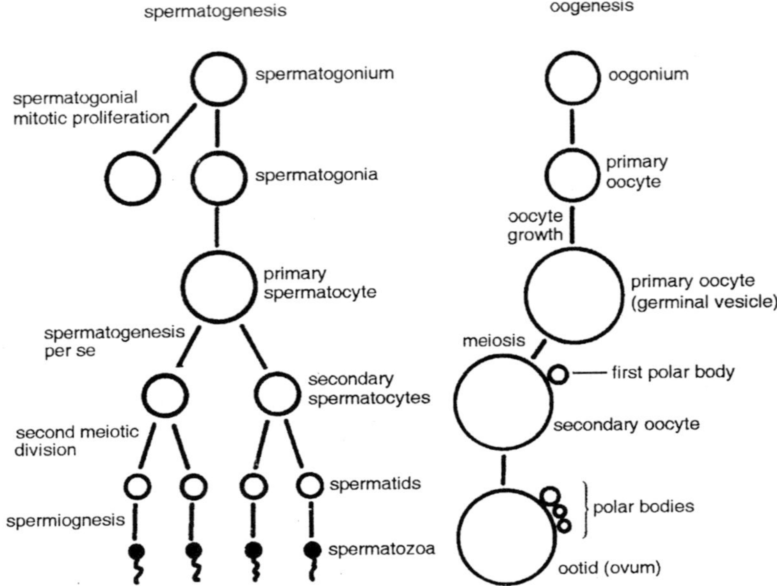

Fig. 13.3. Gametogenesis in testis and ovary.

Spermatogenesis is apparently a temperature-sensitive process, and high temperatures such as found within the body cavity of most eutherians can impair normal spermatogenesis and produce temporary sterility. Consequently, at some time prior to the attainment of sexual maturity or prior to the annual breeding season, the testes descend into the scrotum where spermatogenesis can proceed at a slightly lower temperature. The failure of the testes to descend, a condition known as cryptorchidism, may cause irreparable damage to the seminiferous epithelium in most species. Some mammals lack a scrotum (for example, the elephant, whale, seal), and the testes are permanently located within the abdominal cavity. In such species spermatogenesis obviously does not exhibit the same temperature sensitivity characteristic for scrotal species. The male elephant, for example, is capable of producing viable sperm and copulating with a female at any time of year.

The synthesis and release of circulating androgens by the interstitial cells is controlled by LH. Testosterone is the major circulating androgen, although other androgens such as androstenedione or 5α-hydroxytestosterone (DHT) may circulate in significant amounts. Prior to attainment of puberty in bulls, androstenedione is the principal

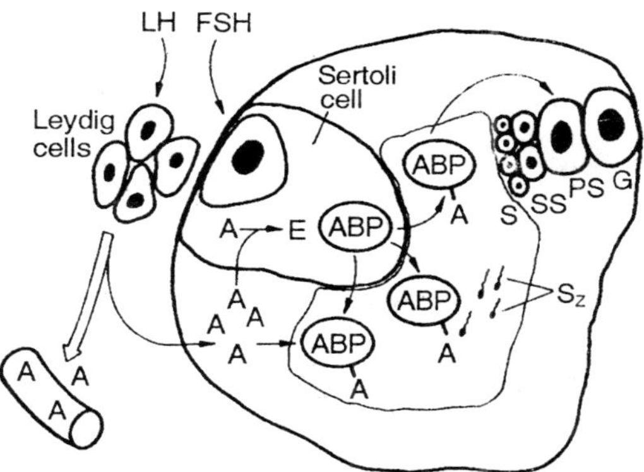

Fig. 13.4. Action of gonadotropins on testicular cells. A, androgens; E, estrogens; ABP, androgen-binding protein; G, gonia; PS, primary spermatocyte; SS, secondary spermatocyte; S, spermatid; Sz, spermatozoa.

circulating androgen, but it is gradually replaced by testosterone at puberty. There is evidence for a separate FSH-sensitive site in the testis for androgen synthesis, and there are data to suggest the Sertoli cell may be that site.

The interstitial cells of the testis synthesize and release estrogens as well as androgens. Testicular estrogens reach dramatic levels in the stallion. Females, on the other hand, secrete small quantities of androgens in addition to estrogens. Maleness or femaleness may be only an expression of the relative proportions of androgens to estrogens in the circulation, although transitory alterations in these ratios may occur normally. Estrogens may have definite physiological roles in males. Estradiol can block androgen synthesis by interstitial cells and can influence the responsiveness of these cells to gonadotropins. The ratio of testosterone to estradiol in the general circulation may alter the ratios of FSH and LH being released from the pituitary.

Interstitial cell function is reduced in humans as a consequence of long term treatment with GnRH. This appears to be a direct action of GnRH on the interstitial cell and not an indirect effect on the pituitary. Receptors for GnRH are found on interstitial cells, and there is evidence that the testis produces a GnRH-like substance. The significance of these observations for normal testicular function is not clear.

Action and metabolism of androgens in males

Circulating androgens influence development and maintenance of several glands and related structures associated with the male genital tract, such as the prostate gland and seminal vesicles, and induce development of certain secondary sexual characters such as growth of the beard in man. Androgens also exert a negative feedback effect upon the secretion of gonadotropins primarily through actions at the level of the hypothalamus.

The action of testosterone in some of its target cells is believed to involve first its conversion to 5α-dihydrotestosterone by the enzyme 5α-reductase. It is DHT which then binds to the cytoplasmic androgen receptor. Development of prostate and bulbourethral glands, the penis and scrotum are dependent upon conversion of testosterone to DHT. Many androgenic responses, however, are not mediated by DHT, and this conversion is not necessary for testosterone to produce its effects. In some target tissues, androgens have been shown to undergo conversion to estrogens through aromatization of the A ring and removal of the C_{19} carbon atom. This process seems to occur, for example, in nervous tissue, but it is not clear whether aromatization is essential to the mechanism of androgen action or is related to metabolism of the androgen (degradation). Development from each wolffian duct of vas deferens, epididymis, seminal vesicle and ejaculatory duct is regulated by testosterone and does not require conversion of testosterone to DHT. Sertoli cells also convert androgens to estrogens, and this aromatization of androgens is stimulated by FSH. This action of FSH on aromatization of androgen is similar to that proposed for the synthesis of estrogens in the ovarian follicle. The role of these testicular estrogens is uncertain, however, but may be related to the action of FSH on the synthesis of ABP by the Sertoli cell.

Aromatization of androgens such as testosterone to estrogens occurs in several areas of the brain, and this conversion is necessary for some of the behavioural effects of androgens. Aromatization to an estrogen is not requisite for all behavioural effects induced by androgens, however. Induction of some male behaviours in castrates requires aromatization and cannot be induced by nonaromatizable androgens such as DHT, whereas others may be induced by either aromatizable androgens or by DHT.

A separate testicular product, *inhibin*, has been proposed to regulate FSH release from the adenohypophysis. This activity has been reported for rete testis fluid, seminal plasma, testicular extracts and ejaculate. Sertoli cells of the rat testis produce a water-soluble, heat-labile *Sertoli cell factor* that selectively inhibits synthesis and release of pituitary FSH without affecting LH. This Sertoli cell factor may be the inhibin molecule originally proposed by McCullagh.

Female cycles

Although marked differences can be pointed out between endocrine events related to estrous and menstrual cycles, there are probably greater differences among species exhibiting one of these types than in comparison of the two types. Nevertheless there are several distinctive similarities that characterize females, regardless of whether they exhibit estrous or menstrual cycles.

Follicular phase

The basis for the cyclical nature of female reproductive events resides in the hypothalamus and is a genetically determined female characteristic. During the *follicular phase* of the ovarian cycle the tonic hypothalamic center releases small quantities of GnRH into the portal circulation and relatively low but rather constant circulating levels of FSH, LH or both are maintained. In general, prior to puberty, which is characterized by increased gonadotropin levels, the ovary contains primary oocytes invested with modified stromal cells (the primary follicle). In most mammals there are no oogonia in the ovary because all of them differentiated into primary oocytes prior to or shortly after birth. The arrival of FSH at the ovary stimulates primary follicles to begin to enlarge and differentiate. The growing oocyte becomes surrounded by two distinct layers of cells, the inner *granulosa cells* and the outer *thecal cells*. Thecal cells further differentiate into an inner and an outer theca (theca interna and theca externa). The granulosa cells are separated from the thecal layers by connective tissue, and there is no penetration of capillaries into the granulosa. As the follicle grows the granulosa cells secrete fluid contributing to the *liquor folliculi* that is primarily an ultrafiltrate of blood plasma. Increasing production of liquor folliculi results in formation and progressive enlargement of a fluid-filled cavity within the follicle, the *antrum*. Most follicles degenerate to form *corpora atretica*: only a few will ever ovulate.

Under the influence of LH, FSH-primed follicular cells synthesize and release estrogens, predominantly estradiol-17β, into the general circulation. The synthesis of estrogens in the ovary appears to be a cooperative effort between cells of the theca interna and the granulosa. Recent studies support the interpretation that LH stimulates the thecal cells to produce androgens that are aromatized by the granulosa cells to form estrogens. Furthermore, the conversion of androgens to estrogens by the granulosa cells appears to be stimulated by FSH. Consequently synthesis of ovarian estrogens may involve two

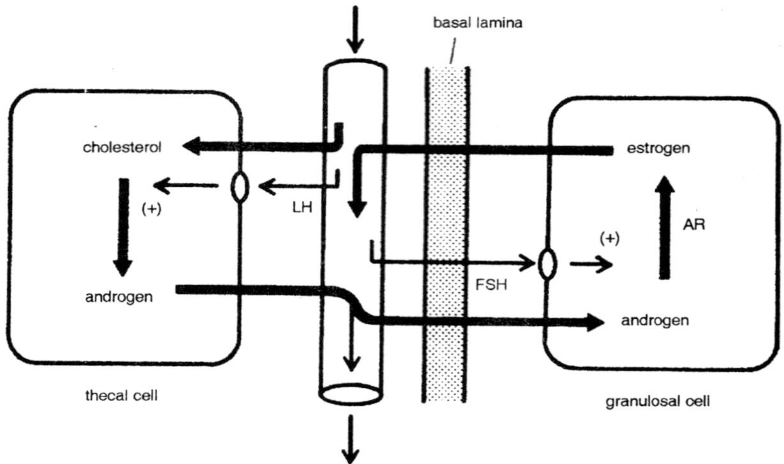

Fig. 13.5. Synthesis of steroids by the ovarian follicle. AR, aromatase.

gonadotropins, each producing its primary effect on a different cellular type in the growing follicle.

Estradiol-17β stimulates differentiation and proliferation of the uterine lining or *endometrium* in preparation for implantation of the *blastocyst*, a blastula formed from the first series of cellular divisions following fertilization. The blastocyst consists of an outer extraembryonic layer of cells, the *trophoblast*, which will form the fetal component of the placenta, and an *inner cell mass*, which will become the embryo proper.

Ovulation

The levels of circulating estradiol-17β increase progressively with the growth of the follicles and the increase in thecal and granulosa cells. A maximum or critical estrogen level activates the cyclic center in the hypothalamus, which releases a large pulse of GnRH. The action of estrogens on GnRH release appears to be mediated via norepinephrine-secreting hypothalamic neurons. The pulse of GnRH released results in a dramatic increase in circulating LH with simultaneous release of FSH.

This so-called LH surge causes ovulation of one or more follicles within a matter of hours. The number of follicles ovulating is species specific, varying from one in women to a dozen or more in the sow. The mechanism by which LH causes the mature follicle to rupture and release the mature oocyte is not known. The LH surge induces granulosa cells as well as some theca interna cells to differentiate

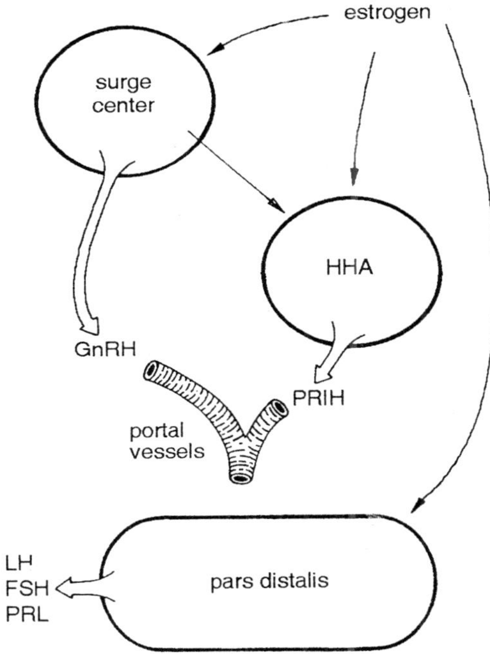

Fig. 13.6. Hypothalamic actions of estrogens in female mammals.

into the *corpus luteum*. This process is known as *luteinization*, and the resulting corpus luteum functions as an endocrine gland, secreting both estrogens and progesterone into the general circulation. One corpus luteum will form from each ovulated follicle. In addition, other developing follicles may undergo luteinization and function as *accessory corpora lutea* during pregnancy. The importance of the smaller FSH surge may be related to initiation of follicle development in the next cycle.

In lower vertebrates, meiosis in the oocyte usually is not completed until after fertilization. Prior to fertilization the ovulated cell is still termed an oocyte. If meiosis was completed prior to ovulation this cell is termed an ovum. The situation in mammals apparently varies from ovulation of oocytes to ova, but here the ovulated cell in every case will be referred to as an ovum for purposes of simplifying discussion.

Luteal phase

Ovulation marks the onset of the luteal phase of the ovarian cycle. The corpus luteum begins secreting large quantities of progesterone,

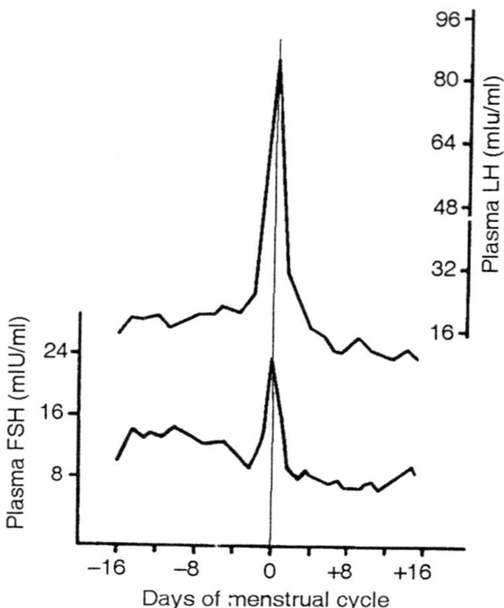

Fig. 13.7. Midcycle gonadotropin surge in normal women.

estradiol-17β and other estrogens and progestogens. Progesterone maintains the proliferated uterine endometrium in the *secretory phase* during which the uterus secretes a fluid sometimes called *uterine milk* or *embryotroph*. Uterine milk is believed to be a source of nourishment for unimplanted blastocysts. The endometrium of marsupials produces a similar secretion. Progesterone also maintains the highly vascularized state of the uterus necessary for implantation and development of the embryo. Muscle layers of the uterus become desensitized by progesterone with respect to responding to certain stimuli with rhythmic contractions. This uterine quiescence is necessary for maintaining a successful pregnancy to term (birth or parturition). Circulating progesterone and estrogens inhibit both the tonic and cyclic hypothalamic GnRH centers during the luteal phase and during pregnancy so that follicular development is arrested and a second ovulatory episode is prevented.

Depending on the species, regulation of corpus luteum function may require LH or be independent of LH once it has formed. In sheep, prolactin (PRL) together with LH might stimulate steroid secretion by the corpus luteum. Only PRL can maintain the activity of the rat corpus luteum. Preovulatory estradiol-17β can produce a surge of PRL release in several species that might be related to corpus luteum function.

The corpus luteum secretes steroids for only a relatively short period in most species (5 to 8 days in man) after which it begins to degenerate. As the corpus luteum undergoes degeneration, steroidogenesis declines, and the uterus enters a regressive phase unless the animal is pregnant. In some species the corpus luteum is relatively long lived, especially in monestrous species like the dog, which will not reenter estrus until the next breeding season. Corpora lutea of the bitch continue to be active for about 63 days after ovulation which is equal to the normal gestation period.

The predetermined life span for the functional corpus luteum has provided one of the most intriguing mysteries of the ovarian cycle. Apparently the corpus luteum sows the seeds of its own destruction. In

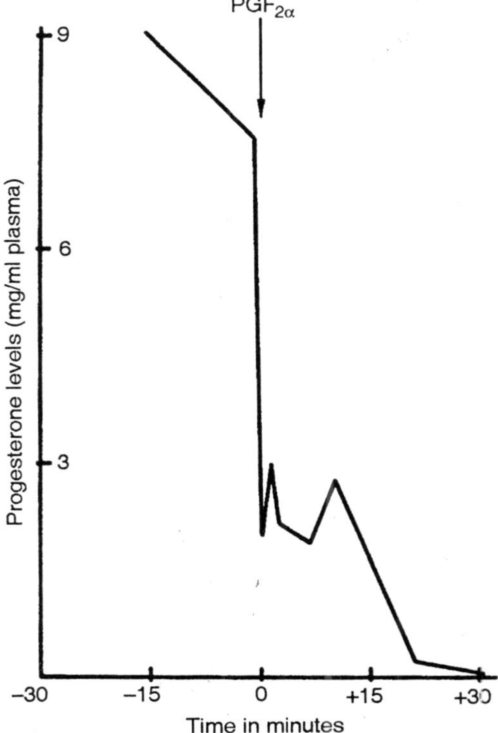

Fig. 13.8. Effect of an intrauterine infusion of prostaglandins ($PGF_{2\alpha}$) on midluteal plasma progesterone levels in the cow.

female rats, mice, hamsters, rabbits, guinea pigs and ewes, progesterone from the corpus luteum stimulates the synthesis and release of molecules known as prostaglandins (PGFs) from the uterine endometrium. These PGFs, especially $PGF_{2\alpha}$, are luteolytic, that is, they cause destruction of the corpus luteum. The mechanism of their luteolytic activity is not clear, although it may relate to an influence on the integrity of the blood vascular supply to the corpus luteum. In primates the destruction of the corpus luteum toward the end of the luteal phase is not influenced by the uterus but appears to be caused locally by a luteolytic estrogen, estrone, produced by the corpus luteum itself. If fertilization has taken place, there are several possible mechanisms by which corpus luteum degeneration is prevented.

Should implantation not occur, the corpus luteum of most eutherian mammals will rapidly degenerate, resulting in a marked decrease in circulating levels of progesterone and estrogens. This decrease in circulating ovarian steroid brings about two major events. The endometrium undergoes regressive changes following steroid withdrawal, becoming less secretory and less capable of supporting implantation of a blastocyst should it occur. In higher primates the exposed layers of the endometrium slough so that considerable rebuilding of the endometrium must occur during the next follicular phase to prepare for implantation of blastocysts resulting from the next ovulation. The second event is the freeing of the hypothalamic GnRH centers from the inhibitory influence of estrogens and progesterone, resulting in a moderate increase in circulating gonadotropins and consequently renewal of follicular development. In fact, increased FSH release occurs during the later stages of the luteal phase in polyestrous species so that follicular growth has resumed even before regressive uterine events become obvious.

It should be emphasized that in wild sexually mature mammals, fertilization and pregnancy are normal events, and coitus occurs frequently during estrus. The character of the ovarian cycle and its rapid resumption in many species if fertilization and successful implantation do not occur ensure either rapid reentry into estrus or rapid appearance of one or more new ova or both and a second opportunity to produce offspring during that season.

The importance of the corpus luteum in maintaining pregnancy varies considerably as does the role of pituitary hormones in stimulating corpus luteum function. In rats PRL is necessary for maintaining the first half of gestation through actions on the corpus luteum, whereas in pigs the corpora lutea secrete progesterone to maintain the uterine secretory phase during the early portion of the gestation period without aid of any adenohypophysial hormones. In ewes, both LH and PRL are necessary for maintaining corpus luteum function during the first third of pregnancy, but maintenance of pregnancy actually resides in the ability of the conceptus to neutralize the uterine luteolytic factor ($PGF_{2\alpha}$). Estrogens of placental origin apparently are responsible for prolonging the life span of corpora lutea in rabbits as well as promoting progesterone synthesis. If the estrogen-secreting placental cells are damaged (for example, by roentgen rays), pregnancy is abruptly terminated.

Actions of estrogens and progesterone on the uterus

Estradiol-17β may have two independent targets in the uterus. One target is the epithelial cell, which responds to estrogens with new RNA and protein synthesis such as that characterized in the mechanism of action for steroids exemplified in the Jensen hypothesis. A second target may be the uterine eosinophil, which is a white blood cell that has infiltrated the uterine lining. These eosinophils supposedly possess specific cytosol receptors for estradiol-17β and appear to be responsible for the rapid uptake of water and release of histamine that characterize uterine responses to estrogens.

The mechanisms for maintenance of the secretory phase in the uterus by progesterone are not understood. However, progesterone acts as a local anesthetic to uterine smooth muscle and thus thwarts premature births, at least in some species.

Placental types

Mammalian placental types differ considerably in terms of the degree of fetal-maternal fusion and with respect to the intimacy of the vascular relationship between the fetus and the mother. Consequently several terminologies have been devised to describe placentas structurally and functionally.

Monotremes have no placenta and are termed *aplacental*. The placenta of marsupials involves the fusion of an embryonic endodermal membrane (the yolk sac) with the chorion (a covering sac derived from the embryonic trophoblast) to form a transitory *yolk-sac placenta*. In a few marsupials, the allantois (another fetal membrane derived from the endoderm like the yolk sac) fuses with the chorion to form a chorioallantoic relationship like that of eutherian mammals.

In the true placental mammals (Eutheria), the chorion produces vascular tufts or villi that form an intimate relation with the uterine endometrium. The allantois fuses with the chorion to produce the *chorioallantoic placenta*. When the allantois is absent the unit is simply termed a *chorionic placenta*.

The true placenta, or *placenta vera*, is fused to the uterine tissues so that some of the uterine tissues are stripped off at birth, causing uterine bleeding (deciduate placentation). If there is no fusion of chorion or chorioallantois with the endometrium the unit is termed a *semiplacenta*. Animals with this type of placenta will not exhibit bleeding and shedding of uterine tissues at birth (nondeciduate placentation).

General shape is often used to classify placentas. Diffuse placentas (lemur, pig, horse) retain chorionic villi over the entire surface. Ungulates (cattle, sheep, deer) possess small tufts or rosettes, known as cotyledons, regularly spaced over the smooth chorion (cotyledonary placenta). Carnivores exhibit a zonal type in which there is a girdle-like band of villi around the middle of the chorionic sac. One or two discoid or disk-shaped masses of villi on the smooth chorionic sac are typical of insectivores, bats, rodents and most primates.

Placentas are also categorized according to the degree of intimacy between the fetal and maternal tissues. The degree of intimacy is directly proportional to the efficiency of transport of materials to and from fetal tissues. The *epitheliochorial placenta* (lemur, pig, horse) is a relationship of apposition, with chorionic tissue simply fitting into grooves and depressions of the maternal tissues. This type of placenta is nondeciduate. A second type of nondeciduate placenta is the *syndesmochorial placenta* typical of ruminant ungulates. The rosettes of chorionic villi occupy deeper pits in the uterine mucosa, resulting in localized destruction of the uterine epithelium so that the chorionic ectoderm comes in direct contact with vascular maternal tissue. At parturition the villi are merely withdrawn, and there is no tearing or bleeding the maternal tissues (nondeciduate).

Carnivores exhibit considerably more intimacy between chorionic villi and the maternal tissues than do ruminants. Erosion of the uterine tissues is more extensive and exposes the endothelium (inner lining) of maternal blood vessels to the chorionic cells. This *endotheliochorial placenta* is deciduate.

Anthropoids, insectivores, bats and lower rodents have what is termed a *hemochorial placenta* in which the endothelium of the maternal blood vessels is lost and the maternal blood directly bathes the chorionic villi. In higher rodents (rat, guinea pig) and lagomorphs (rabbits) the greatest degree of known intimacy is found in the *hemoendothelial placenta*. In this type of placenta the chorionic villi partly degenerate so that the endothelium of chorionic blood vessels is bathed directly by the maternal blood. Both hemochorial and hemoendothelial placentas result in considerable sloughing of uterine tissues and bleeding at birth (deciduate).

Generally all of these factors are taken into account in describing the type of placentation exhibited by a given species. The human, for example, exhibits deciduate placentation with a discoid-shaped placenta of the hemochorial type.

Pregnancy cycle

The ovum still surrounded by some of the granulosa cells (the corona radiata) enters the upper end of the fluid-filled oviduct and is propelled toward the uterus by the action of cilia lining the oviduct. The possible role of muscular contractions of the oviduct wall in transport of the ovum has been suggested but not verified. Fertilization typically occurs in the upper third of the oviduct by spermatozoa that were deposited in the vagina by the male during coitus. These spermatozoa are transported by peristalsis through the uterus and into the oviduct in which recently ovulated ova are descending. Cleavage begins soon after fertilization, and the zygote or fertilized egg becomes a minute, multicellular blastocyst that is capable of eroding and settling in the uterine endometrium for development. The gestation period may be as short as 12 days in the opossum or as long as 22 months in elephants.

Should fertilization and successful implantation occur in eutherian mammals, in which the pregnancy period exceeds the normal life of the corpus luteum, the life-span of the corpus luteum will be prolonged until the placenta can synthesize sufficient progesterone and estrogens to maintain gestation.

A central question that has puzzled reproductive physiologists for many years is related to the mechanism whereby a mammal "knows" it is pregnant soon enough to prolong corpus luteal function. In the mare only fertilized ova ever reach the uterus, implying some sort of early chemical recognition that fertilization has occurred. The signal for prolongation of corpus luteum function in some species is the synthesis of an LH-like hormone called *chorionic gonadotropin* (CG) by the trophoblast of the blastocyst. This trophoblast will develop into the fetal component of the placenta after implantation and will continue to secrete CG. In pregnant mares the chorionic hormone is known as *pregnant mare serum gonadotropin* (PMSG) because it appears in such large amounts circulating in the blood. These placental gonadotropins are structurally very similar to the pituitary gonadotropins and generally produce LH-like effects. Their synthesis and release, however, are not influenced in a negative way by steroids in the manner of the steroidal feedback on pituitary gonadotropins. Such pregnancy-recognition mechanisms are not necessary in some species such as the opossum and carnivores where the length of the normal luteal phase and corpus luteum function are identical to the gestation period.

Secretion of ovarian steroids brought about by extension of the life span of the corpus luteum or during the normal luteal phase of the ovarian cycle of carnivores inhibits hypothalamic centers controlling gonadotropin release so that follicular development and ovulation are blocked in pregnant animals. The adaptive value of not having several embryos in the uterus at different stages of development should be obvious not only with respect to the sequestering of limited nutritional resources by the fetus but also to the destructive effects of parturition on the younger embryos and fetuses. The latter is probably more important since many species produce several offspring at one time and have obviously found ways to adapt to simultaneous competition among the fetuses.

During the last third of pregnancy another pituitary-like hormone, *chorionic somatomammotropin* (CS), is secreted by the placenta of a number of species (primates, mice, rats, voles, guinea pigs, sheep, chinchillas and hamsters but not bitches or rabbits). This placental hormone has both growth hormone (GH)-like and PRL-like activities, and antibodies to CS will cross-react with both GH and PRL in at least some of these species. The major roles for CS appear to be effects on metabolism (GH-like) and stimulation of the mammary gland to begin milk synthesis during the later stages of pregnancy.

The human placenta secretes PRL which is identical to pituitary PRL. Prolactin accumulates in the amniotic fluid during pregnancy where it is thought to regulate volume and ionic composition of amniotic fluid. Levels of amniotic PRL are not affected by drugs that block pituitary PRL release or even by hypophysectomy of the mother.

In humans two additional adenohypophysial-like hormones have been identified in placental tissues. *Chorionic thyrotropin* (hCT) and *chorionic corticotropin* (hCC) have been found, but their functions are not clear. Perhaps they too serve to provide essential adenohypophysial hormones during pregnancy when general adenohypophysial function is inhibited and these essential tropic hormones might otherwise be lacking. The placenta also synthesizes GnRH, but the significance of this is not clear.

A new pregnancy hormone, *relaxin* was named in 1932. It has several possible physiological roles. 18 Relaxin causes relaxation and softening of estrogen-primed pelvic ligaments, allowing the pelvis to stretch and expand (relax) so that the relatively large head of the mammalian fetus may pass through the pelvis during parturition. It reaches peak levels prior to birth and rapidly disappears from the maternal circulation afterward. Spontaneous motility of the uterus may be inhibited by relaxin in some mammals. Relaxin working with estrogens, progesterone and prostaglandins can alter the structural collagen of the uterine cervix, increasing its distensibility at parturition. There are also data supporting an action of relaxin in combination with steroids and PRL. These hormones stimulate growth of the mammary gland and the onset of lactation.

The corpus luteum is the major source of relaxin in species where the corpus luteum is retained throughout gestation (pig, rat, carnivores). Relaxin is produced by the human corpus luteum during early gestation and to some extent by the placenta. A little relaxin is found in placentae of sheep, rats, cows and rabbits, but in horses the placenta is a major source of relaxin. In humans the ovary continues to be the major site for relaxin synthesis after death of the corpus luteum.

Relaxins have been purified from pig, rat and shark. They all consist of two short A chains (22-24 amino acids) and a longer B chain (26-35 amino acids) joined together by disulfide bonds. The positioning of the disulfide bonds is the same as for insulin and the insulin-like growth factors although there are many differences in amino acids. Rat and pig relaxins exhibit numerous amino acid substitutions although insulins from these species are similar. It has been suggested that the relaxin gene arose by duplication from the insulin gene. Among mammals there has been considerable divergence in the relaxin genes. Shark relaxin actually resembles mammalian insulins more closely than it resembles mammalian relaxins. Why has the shark relaxin gene changed so little? Perhaps the answer lies in the acquisition of new functions for relaxin in mammals and intense selection pressure on the relaxin gene products.

Delayed implantation

Several mammals, such as mink, bats and skunks, have evolved a fascinating mechanism known as *delayed implantation* whereby development of the blastocyst is arrested and the unimplanted blastocyst remains in the oviduct or uterus for an extended period prior to implantation. Among some eutherian mammals, delayed implantation appears to be an adaptation allowing copulation to occur at a particular time that is especially advantageous to the parent yet ensuring that the young are born at the most favourable time for their survival. Neither the basis for causing the blastocyst to remain in a healthy, arrested state nor the stimulus to bring about implantation is known. A similar phenomenon occurs in macropodid marsupials, however, and its continuation is related to the presence of a young suckling on a teat. This is clearly not the mechanism involved in eutherians.

Lactation

The development of mammary glands, their synthesis of milk and the ejection of milk to the suckling offspring are all regulated by hormones. Mammary glands in eutherian mammals usually occur as paired structures, from 2 to 18, and may be located on the thorax (man, elephant, bat), along the entire ventral thorax (sow, rabbit), in the inguinal region (horse, ruminants), along the abdomen (whale) or even dorsally (the nutria, a South American rodent). The internal structure includes supporting stromal cells and glandular epithelium that is organized into clusters of minute, sac-like structures called *alveoli*. It is the glandular epithelium that is responsible for synthesis of milk. The alveoli are continuous with ducts and various duct-derived enlargements for storing milk. In addition, there are modified epithelial cells that contain muscle-like myofilaments parallel to the long axis of the cells. The *myoepithelial cells* are capable of contracting and causing ejection of milk from the alveoli into the duct system and out of the gland in the region of the nipple.

Information obtained from the mouse and rat indicates that differentiation of mammary glands from ectoderm involves specific induction by a particular underlying mesenchyme. These glands normally develop with the aid of estrogens during the last third of the gestational period. The fetal ovary is not the source of these estrogens since mammary glands develop in the absence of fetal ovaries. Androgens suppress mammary gland development and are presumably responsible for their altered development in the male fetus.

Postnatal mammary development involves hormones from the pituitary, ovaries and adrenal cortex, at least in mice and rats. Growth of mammary ducts requires estrogens, GH and corticosterone working in concert. However, expansion of the alveoli (lobuloalveolar growth) is dependent upon the direct actions of estrogens, progesterone, PRL, GH, relaxin and corticosteroids.

Lactation can be separated into two basic processes or phases under separate endocrine control mechanisms. The first phase is *milk secretion* or *lactogenesis*. This process is under control of pituitary PRL (or CS) and corticosteroids. In primates, lactogenesis is also stimulated by GH. Lactogenesis involves synthesis of milk fat, milk protein and milk sugar, typically lactose. The synthesis of lactose ultimately depends upon protein synthesis; that is, the enzyme responsible for lactose synthesis, lactose synthetase, must be induced. Lactose synthetase is composed of two protein units, one of which is lactalbumin, which is also found in milk.

Lactose, fat and milk protein (largely casein) are secreted into the lumen of the alveolus. Water and numerous water-soluble substances enter the lumen by osmosis and result in a watery liquid known as milk. Hormones are present in milk including the hypothalamic peptides, TRH and GnRH, as well as TSH, ACTH, PRL, LH, FSH, estradiol-17β, corticosteroids and thyroid hormones. The composition of milk produced by the mammary gland associated with suckling the young is very different at birth from what it will be shortly thereafter. This first milk is known as *colostrum* and is characterized by having a greater concentration of protein and less carbohydrate than does later milk. Colostrum contains substances that serve to protect the neonate against allergies and diseases.

The second phase of lactation is *milk ejection*. A simple reflex mechanism involving the pars nervosa controls milk ejection. Mechanical stimulation of the nipple (suckling) stimulates release of oxytocin from the pars nervosa via a spinohypothalamic neuronal pathway. Release of PRL also occurs when milk is ejected and stimulates further milk synthesis. Oxytocin stimulates contraction of myoepithelial cells that causes milk to be ejected from the alveoli into the ducts and storage channels of the mammary gland. The suckling young strips milk from the gland by expressing it between the tongue and hard palate.

The milk ejection neurohormonal reflex exhibits classical conditioning responses as evidenced by stimulation of milk flow in the cow by sight and sounds of the milking parlor or in women by the cries of their hungry infant. This reflex can be influenced by other neural or chemical inputs to the hypothalamus. For example, stress or physical discomfort can inhibit ejection of milk in the presence of the stimulus that would normally elicit release of oxytocin.

Reproductive cycles in representative female eutherians

Following are detailed accounts of the reproductive cycles known for ewes, women and 4-day cycling rats. These cycles emphasize both the features described previously that are characteristic of eutherian mammals and some of the marked deviations that become obvious when different species are compared. The cycles of the three species described here are among the best known, and it may be argued whether they are truly representative of all mammals. Ewes and rats are polyestrous species whereas women have no seasonal estrous behaviour. Both rats and women are continuous breeders, but ewes are distinct seasonal breeders. Cows and pigs have cycles that are essentially like the ewe's although they differ somewhat in timing of the various events. None of these species exhibits delayed implantation, and all are believed to be spontaneous ovulators except under special conditions for the rat and probably the human.

Ewe

Sheep estrous cycles occur seasonally, and the duration of one complete cycle is 16 days with a return to proestrus if fertilization does not occur. Reproductive cycles can be blocked in ewes by genistein, an antiestrogenic compound found in certain clovers. During the follicular phase (proestrus) there is marked increase in both estrogen and androgen levels, a peak being reached about 24 hours after the

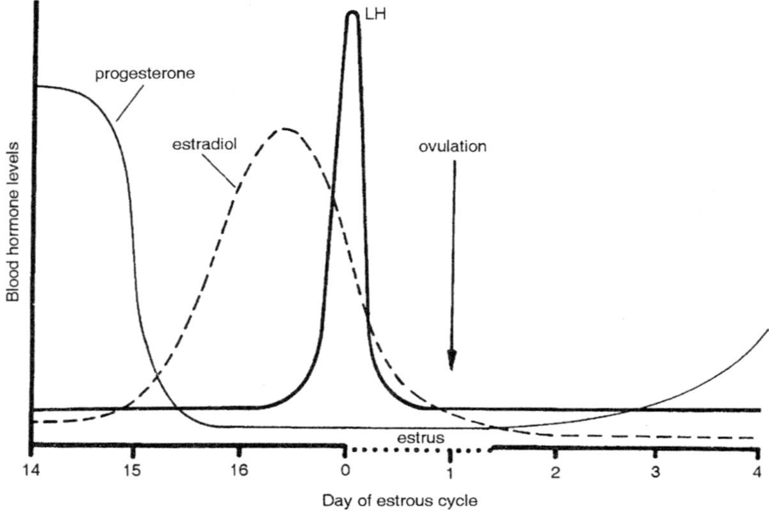

Fig. 13.9. Hormonal events during the estrous cycle of sheep.

onset of proestrus. About 12 hours later a surge of plasma LH occurs caused by the action of estradiol-17β on the cyclic hypothalamic neurosecretory center. The high level of androgen, principally androstenedione, has been suggested to be responsible for inducing estrous behaviour. Usually a single ovulation follows the LH surge by about 24 hours, and a corpus luteum forms from the ruptured follicle under the influence of LH. Low levels of LH following ovulation and the estrogen-induced surge of PRL stimulate the corpus luteum to secrete progesterone. Under the influence of progesterone, the uterine endometrium synthesizes a luteolytic PGF ($PGF_{2\alpha}$) that causes degeneration of the corpus luteum and resumption of proestrus. Fertilization followed by implantation delays degeneration of the corpus luteum. Apparently the conceptus neutralizes the PGFs synthesized under the influence of progesterone. The corpus luteum continues to secrete progresterone until the placenta is capable of producing sufficient steroids to maintain gestation. The placenta also secretes both oCG and oCS

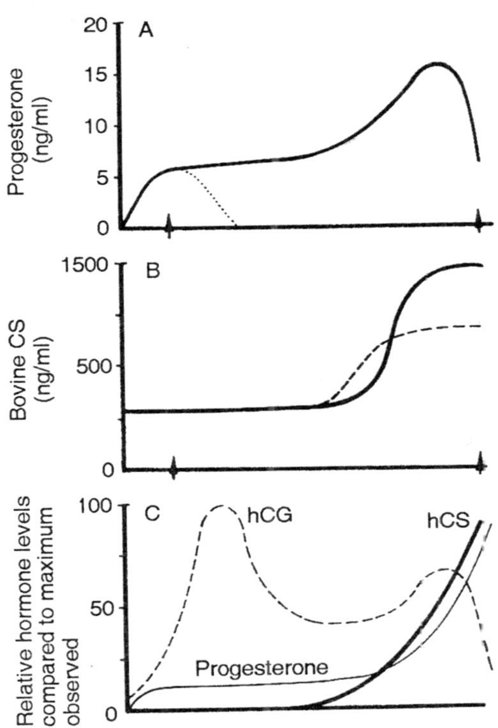

Fig. 13.10. Some hormone levels during pregnancy for ewe, cows and women.

Just prior to birth there is marked reduction in circulating steroid levels. Presumably this marked decrease in progesterone sensitizes the uterus to oxytocin, which was unable to cause contractions in the uterus in the presence of progesterone. The contractions of the uterus initiated by oxytocin result in expulsion of the fetus as well as of the afterbirth (placenta).

It appears that follicle growth, ovulation and luteinization can be brought about by LH alone. However, both LH and PRL are necessary to induce progesterone synthesis by the corpus luteum. Although FSH has been claimed to play no role in ovarian function for the ewe, several studies have reported a stimulatory effect of FSH on follicular development.

Women

The human female exhibits continuous cycling with a mean cycle length of 28 days for most reproductively active women. The rhesus monkey also has a menstrual cycle of 28 days, and there are many parallels in the menstrual cycles of these two primates. Studies of the rhesus monkey have provided insight into factors regulating the human menstrual cycle. Although this cycle is sometimes referred to as a lunar cycle because its periodicity is equivalent to a lunar month, the human menstrual cycle is not correlated with any particular phase of the lunar month and should not be termed lunar. It could be that at one time the menstrual cycle was correlated more closely with moon phases but has become highly modified by numerous environmental and internal factors. Cycles of women can vary from as short as 14 days to as long as 360 days, depending upon both endocrine and psychological factors. Considerable variation can occur in cycle length in a given woman at different times in her life history. Short cycles and irregular cycles are associated with the onset of puberty and with the end of the reproductive life prior to menopause when the ovaries become refractory to pituitary gonadotropins, and both estrogen synthesis and ovulation cease. The absence of circulating steroids releases the hypothalamus from negative feedback, and levels of circulating gonadotropins are exceptionally high in menopausal women.

The menstrual cycle begins at the onset of the menses, which occupies the first 5 days of the cycle. However, as soon as the corpus luteum of a previous cycle begins to regress prior to the onset of menses, there is a moderate elevation in FSH level that initiates growth of new follicles. Typically only one follicle in an ovary will reach maturity in a given cycle, with ovulation often occurring in the

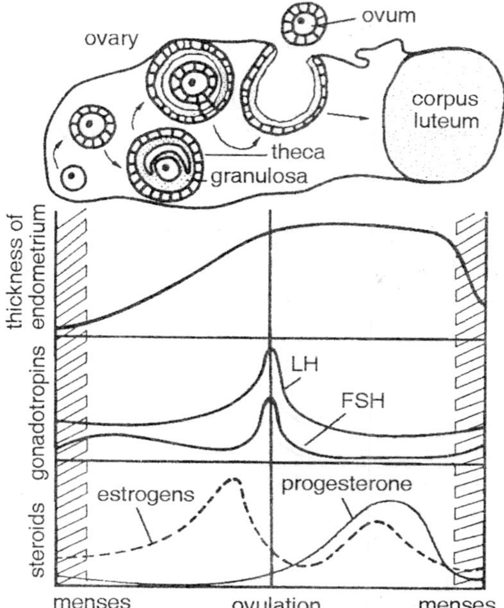

Fig. 13.11. Menstrual cycle of human female.

alternate ovary during the following cycle. The thecal cells of the growing follicles begin to secrete estrogens and progestogens, which peak on about day 14 of the normal cycle. Steroidogenesis is regulated by FSH and LH operating on the thecal and granulosal cells. The peak in estrogen level stimulates GnRH release from the cyclic center, causing a surge of LH with release of less FSH. Ovulation follows the LH surge by about 24 hours, and the granulosa cells undergo luteinization. The resulting corpus luteum is independent of pituitary hormones and begins production of progesterone and estradiol-17β. The hypothalamic centers governing LH and FSH release are inhibited by the high levels of these circulating steroids so that neither follicular growth nor ovulation can occur during the luteal phase of the cycle. The corpus luteum functions for only a few days if the ovum is not fertilized. Unlike the sheep, in women the uterus plays no active role in degeneration of the corpus luteum. Surgical removal of the uterus (hysterectomy) does not affect the duration of the luteal phase. The human corpus luteum may produce its own luteolytic factor as does that of the rhesus monkey. The production of estrone by the corpus luteum of the rhesus monkey increases markedly prior to the onset of luteal degeneration. A similar mechanism may be operating in women. As the production of steroids by the corpus luteum decreases the pituitary is released from steroid-induced inhibition, and a new cycle of follicular growth begins. The outer portion of the uterine lining begins to slough off following the decline in progesterone levels, and the menses begins.

Should fertilization occur, the trophoblast of the blastocyst begins to secrete hCG prior to implantation and the production of hCG continues at an accelerated rate during early pregnancy. Under the influence of hCG, the life span of the corpus luteum may be extended. The corpus luteum secretes both progesterone and relaxin. By about 60 days, it degenerates even in the presence of exogenous hCG. However, the placenta, with the help of the fetal adrenal cortex, is now producing sufficient estrogens and progesterone to continue the inhibition of pituitary gonadotropin release and to maintain the secretory condition of the uterus. Furthermore, the placenta appears to assume the roles for pituitary GH, PRL, ACTH and TSH through production of hCS, hCC and hCT.

The stimulus for birth is not related to a decrease in progesterone levels but to a marked increase in estrogens as well as progesterone. Apparently, high levels of estrogens nullify the anesthetic properties of progesterone on uterine smooth muscle and allow oxytocin to initiate uterine contractions and the onset of labor. This action of oxytocin seems to be a consequence of increased uterine receptors rather than a consequence of elevation in oxytocin levels. Administration of prostaglandins also can induce uterine contractions, and oxytocin may stimulate prostaglandin synthesis in the uterus. Synthetic oxytocin is normally used to induce labor in women.

Four-day cycling rat

The short 4- or 5-day cycle in rats is believed to be a consequence of the failure of the corpora lutea to become functional during the normal luteal phase (diestrus). The rat cycle is closely cued to environmental events, ovulation, for example, occurring invariably shortly after midnight. Furthermore, there is considerable evidence for the involvement of chemical communication between male and female rats that can influence many of the reproductive events.

On the morning of the day prior to estrus, that is, during proestrus, the levels of estrogens in the plasma reach a peak, which stimulates the typical LH surge accompanied by FSH. The surge of gonadotropins occurs on the afternoon of proestrus and is followed rapidly by a

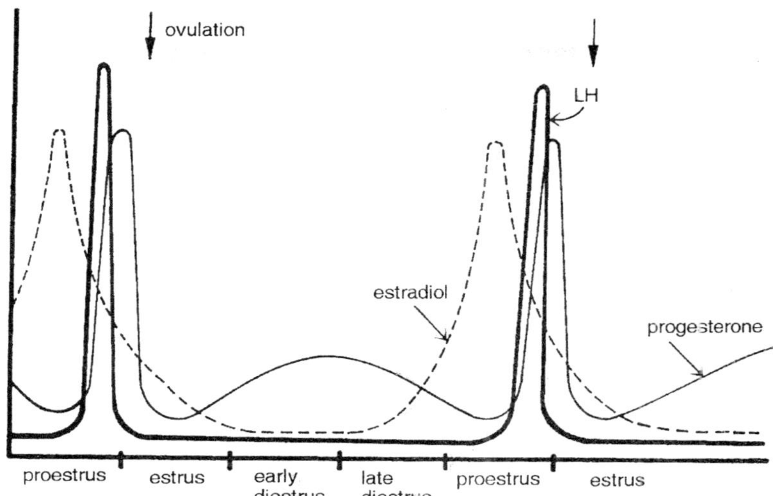

Fig. 13.12 The 4-day cycling female rat.

marked surge of progesterone. Ovulation occurs a few hours after midnight on the day of estrus. Several follicles usually mature simultaneously, and multiple ovulations occur. Estrus lasts about 9 to 15 hours, during which the female is receptive to the male. Ovulation occurs during estrus. Cornified cells, which were produced by the actions of estrogens, appear in the superficial layers of the vagina, and their presence in vaginal smears characterizes estrus.

The third and fourth day of the cycle are termed diestrus I and diestrus II. Vaginal smears prepared during diestrus are characterized by the absence of cornified cells and a predominance of leukocytes in the smear. There is limited secretory function by the corpus luteum, as evidenced by a slight increase in plasma progesterone. Much of this progesterone is produced by ovarian interstitial cells, which constitute what has been called a permanent corpus luteum in the rat ovary. This interstitial tissue responds to LH by secreting both progesterone and 20α-dihydroprogesterone.

Many researchers recognize a transitional period between estrus and diestrus termed *metestrus*. The female is no longer receptive to the male, but some cornified cells still appear in smears prepared from the vaginal mucosa.

Mating may stimulate gonadotropin release, and consequently the corpora lutea begin to secrete significant amounts of progestogens.

This increased production of progestogens will inhibit hypothalamo-hypophysial function and delay the onset of the next cycle. If fertilization and implantation do not occur from this mating, the rat will return to proestrus within a few days. This pseudopregnancy occurs in laboratory rodents and sometimes in domestic mammals. The condition of pseudopregnancy is often accompanied by PRL-like effects on lactation and behaviour presumably due to hypophysial release of PRL.

If implantation does occur the corpora lutea continue to secrete progestogens under the influence of placental CG and begin to secrete relaxin. The rat placenta also produces CS, which contributes to stimulation of mammary gland development and lactogenesis prior to birth.

Monotreme Reproductive Patterns

The egg-laying aplacental monotremes are reproductively more like reptiles than like other mammals, although they do possess some strictly mammalian features. The developing monotreme follicle does not form an antrum. Prior to ovulation, however, a thin layer of antral fluid is secreted by the follicular cells between the zona pellucida and the follicular cells. There is a well-developed theca interna that persists after ovulation and formation of corpora lutea. The function of the corpus luteum is not clear, but it has been suggested that it stimulates secretion of nutrient substances by the uterine glands. These nutritive substances are incorporated by the oocyte. Monotreme eggs have leathery shells like reptilian eggs. The duckbill platypus lays its eggs in a nest, but the female spiny echidna places her eggs in a brood pouch that develops seasonally. Formation of the pouch is probably dependent upon estrogens, although no experimental data are available to support this contention. The eggs develop and hatch within the pouch, and after the breeding season the pouch regresses.

Corpora atretica may form from either large or small follicles in the monotreme ovary. The contents of a large follicle typically are extruded by rupture of the follicle prior to formation of the corpus atreticum. The extruded materials enter the peritoneal cavity or the prominent intraovarian lymph spaces that characterize the monotreme ovary.

The mammary glands of monotremes lack teats. Consequently milk is secreted onto a special area, the *areola*. The newly hatched monotreme must attach itself to its mother with its forelimbs and suck or lick milk from the areola.

Marsupial Reproductive Patterns

Marsupials possess an estrous cycle with ovarian events similar to those described for eutherians, but the period of gestation is much shorter in relation to the estrous cycle. Proestrus in females is characterized by follicular enlargement and estrogen-dependent uterine proliferation and increase in size of elements of the vaginal complex. Peak uterine and vaginal development coincides with estrus and copulation. Ovulation occurs spontaneously one to several days after estrus, and postovulatory follicles transform into corpora lutea that maintain a secretory uterine condition. Progesterone continues the secretory uterine phase in castrates and is undoubtedly the hormone responsible for maintaining gestation as in eutherian mammals.

The luteal phase is the same in mated and nonmated females, and no "pregnancy-recognition signal" is necessary. Pregnancy does not affect ovarian function, and both pregnant and unmated females return to proestrus at about the same time in most species. Equivalent mammary gland development occurs during postestrus in both pregnant and nonpregnant females, and newborn foster young will develop normally if attached to virgin or nonlactating females at the equivalent postestrous state to the time of parturition. Circulating progesterone and urinary pregnanediol levels are similar in pregnant, unmated and luteal phase females. There are no endocrine differences between the pregnant and nonpregnant postestrous phase, which leads to the conclusion that the marsupial placenta is not an endocrine organ.

Another unusual feature of marsupials is the ability to simultaneously produce two kinds of milk. As in eutherian mammals the first milk differs markedly in composition from that produced later during lactation. Macropodids which may have both a newborn joey and one that has already detached itself from a teat, will produce early and late milk simultaneously in the respective glands. The developmental state of a particular mammary gland would seem to be independent of endocrine conditions and strongly influenced by external conditions; that is, the joey.

A major developmental difference between eutherians and marsupials has had a profound influence on reproductive patterns in the latter group. Marsupials exhibit a primitive reptilian pattern of wolffian and mullerian duct origins. Instead of developing medially to the kidneys and ureters as in eutherians, these ducts develop laterally. Consequently it is not possible for left and right mullerian ducts of females to fuse in the midline without placing considerable strain on the ureters. It is believed that the short gestation period of marsupials is a consequence of separate uteri and vaginas and limited space for uterine hypertrophy. A special birth canal must be formed so that parturition can occur, and in some species it forms anew each season. This development of separate vaginas has influenced evolution of the male reproductive system as well. Males of some species have a bifid penis, with left and right prongs apparently being inserted into separate vaginas during copulation.

Embryonic diapause

Macropodid marsupials have developed a type of delayed implantation that has been termed *embryonic diapause* to distinguish it from delayed implantation, described earlier for eutherian mammals. Embryonic diapause has been reported for 14 macropodid species but does not occur in at least one species, the western gray kangaroo. One major difference from delayed implantation occurring in eutherian mammals is the condition of the resting blastocyst. The macropodid blastocyst consists of about 70 to 100 cells of a uniform type termed *protoderm*. It has not differentiated into embryonic and extraembryonic regions like that of the eutherians. The blastocyst is surrounded by a shell membrane and an albumin layer.

Presence of a joey suckling on a teat presumably evokes release of oxytocin from the pars nervosa. Oxytocin is believed to arrest corpus luteum functions while allowing lactation to occur. Removal of the suckling joey will allow the resting blastocyst to implant. Ovariectomy following ovulation induces diapause, but if ovariectomy is performed during diapause there is no effect on the duration of diapause. Progesterone administered to either intact or ovariectomized females stimulates cessation of diapause and reinstates blastocyst development. Estrogen is also effective, but continued embryonic development is not as successful as following progesterone treatment.

In the red kangaroo, embryonic diapause is an adaptation to renew pregnancy immediately following the death of the joey living in the pouch. The gestation period for the red kangaroo is 33 days. After birth the newborn must find its way to the pouch virtually unaided. When it reaches the pouch the joey attaches itself permanently to a teat and continues development as an exteriorized fetus. Soon after parturition the mother kangaroo enters estrus again and mates. The presence of one joey in the pouch inhibits implantation of the blastocyst resulting from the second mating. The blastocyst remains in a suspended state of development for about 200 days, at which time the first joey disengages itself from the teat and ventures into the outside world as a juvenile kangaroo. The newly liberated kangaroo will return at intervals to the teat to which it was formerly attached for nourishment. Mean- while the detachment of the first joey from the teat either releases an inhibition to implantation or somehow provides a stimulus for implantation of the waiting blastocyst. In about 4 weeks the gestation period terminates in birth of the second joey, which enters the pouch and attaches to a teat. The mother kangaroo again enters estrus and mates, and another blastocyst enters embryonic diapause. Thus a female red kangaroo may have a young juvenile that requires occasional nourishment, a joey attached to a teat and a blastocyst "waiting in the wings." It has been suggested that during extensive periods of drought an older joey requiring considerable amounts of milk from the mother would be allowed to die, and implantation of the waiting blastocyst soon provides another joey whose demands upon the mother's stored reserves and water reservoir would be small in comparison to the demands of the larger joey.

Major Endocrine Disorders Related to Reproduction

Precocity

Precocity is defined as the appearance of any one indicator of puberty at an age earlier than 2.5 to 3 standard deviations below the mean age at which the indicator normally appears in that population. The sequence and mean age of appearance for these indicators should be considered only as a guide. Implied precocity may not be evidence of an endocrine disorder, and variations in the sequence of these events is normal. Major deviations in a number of indicators may signal precocial endocrine activity of a pathological nature. Isosexual precocity involves early appearance of the genetically determined sex. It is termed heterosexual precocity if male features develop precocially in a female or if female features appear precocially in a male.

Precocity with normal endocrinology

Idiopathic precocity (of unknown cause) may be familial. Sexual development and body growth appear normal but are accelerated. Reproduction may be possible at an early age. For example, the youngest mother on record was 5 years 8 months of age at delivery!

Precocity and pineal tumors

Pineal tumors are not common but occur most frequently in young males. Precocious sexual maturation occurs in about one third of these cases. Pineal tumors may impair release of antigonadotropic factors from the pineal and allow sexual maturation to occur prematurely. Pineal tumors have been related to delayed puberty in a few cases. These tumors may secrete more antigonadotropic factors than the normal gland.

Precocity from ectopic gonadotropins or gonadal steroids

Rare pituitary tumors may secrete excessive amounts of gonadotropins. Sometimes nonpituitary tumors secrete chorionic gonadotropin. Certain ovarian or testicular tumors produce sufficient steroids to cause external evidence of puberty. Heterosexual precocity can occur from feminizing tumors in testes or androgen-secreting tumors of the adrenals or ovaries.

Delayed Puberty

Males

Prevalence of undescended testes or cryptorchidism is common at birth (10%) but is reduced to only 1% of males by age 1. Only about 0.3% of adult males exhibit cryptorchidism, and the case of only one undescended testis is much more common than the bilateral condition. Because of higher temperatures experienced by an undescended testis, the spermatogenetic tissue degenerates at about the time when spermatogenesis would normally begin (at about age 10). Androgen production is normal or may be reduced. The external testis of a unilateral cryptorchid develops normally, and these males are fertile.

There are several other causes for hypogonadism in males including insufficient levels of LH and FSH due to hypothalamic or pituitary dysfunction. In some cases the interstitial cells may be unresponsive to gonadotropins.

Females

Primary amenorrhea is the failure for menarche to occur at the normal time. This condition can be related to many different causes including disorders of the hypothalamus, pituitary and ovaries. Poor nutrition, stress or rigorous athletic training programs can delay puberty through inhibitory actions on the hypothalamo-hypophysial-gonadal axis. For example, levels of LH and FSH are depressed in women suffering from anorexia nervosa, a disorder in which food intake is greatly reduced. Simple weight loss can depress FSH levels for a time but does not inhibit LH secretion.

Secondary amenorrhea occurs after menarche and can result from many endocrine disorders including thyrotoxicosis, drug therapy, premature menopause and a variety of hypothalamic, pituitary and gonadal disorders. Two of the more common gonadal disorders associated with secondary amenorrhea are *polycystic ovarian syndrome* (POS) and *luteinization of atretic follicles* (LAF). The name for POS is derived from the general thickening and luteinization of ovarian follicles resulting in formation of numerous cysts in the ovaries. These cysts develop from thecal cells, and there is a loss of granulosa cells. Progesterone and estrogen production are diminished and gonadotropin secretion consequently is elevated. Androgenic steroids are produced but can not be aromatized to estrogens. Uterine abnormalities and infertility result. Hirsuitism, occasional balding and obesity may accompany POS.

The LAF syndrome results from premature luteinization of ovarian follicles prior to formation of the cumulus oophorus. Gonadotropin levels are elevated, but masculinization usually does not happen. Numerous small ovarian cysts may be present, but these are easily distinguished from the large cysts that characterize POS.

Hereditary Disorders

Hermaphroditism and pseudohermaphroditism

Occasionally people are born with a combination of ovarian and testicular tissue. Most of these people possess an ovotestis on one or both sides of the body. Rarely, an individual is found who has an ovary and attendant mullerian derivatives on one side and a testis with its wolffian duct derivatives on the other. Pseudohermaphrodites have gonads of one sex but externally resemble the other sex. For example, a person with *testicular feminization syndrome* is a genetic male. Although the testis is normal, the external appearance is that of a woman because of the congenital absence of androgen receptors in the tissues.

A most unusual example of pseudohermaphroditism is the apparent shift of sex at puberty that occurs in a few people from several small villages of the Dominican Republic. This disorder is the result of a genetic deficiency for the ability to synthesize 5α-reductase. Males with this defect are born with undescended testes that synthesize testosterone like normal testes. However, these males can not convert testosterone to DHT without 5α-reductase. Testosterone-dependent structures develop normally (such as the vasa deferentia, epididymi) but DHT-dependent structures such as the prostate gland, penis and scrotum do not develop. These males understandably are raised as girls until they begin to synthesize DHT at puberty. The presence of

Sa-reductase allows the penis to enlarge and facial hair to appear. Although these men change social roles after puberty, they are infertile.

Klinefelter's syndrome

Person's with Klinefelter's syndrome are born with an. abnormal number of sex chromosomes: XXY. This familial disorder occurs at fertilization and is present in about 0.2 to 0.3% of males. Klinefelter's syndrome can exist without obvious somatic abnormalities, although these persons are infertile and exhibit differing degrees of mental retardation. Similar syndromes have been described with additional sex chromosomes (XXYY, XXXY, etc.). Severity of the symptoms increases with the number of X chromosomes present.

Turner's syndrome

Another disorder arising at fertilization is the loss of one sex chromosome so that the resulting genotype is XO (one X chromosome and no Y or no second X chromosome). Sometimes this condition occurs when a twin is found to exhibit Klinefelter's syndrome. Individuals with Turner's syndrome have a female phenotype but are infertile. They also exhibit a number of anatomical defects as well as cardiovascular and kidney disorders.

Galactorrhea

Secretion of a lactescent (milky) fluid from the breasts of either sex is called *galactorrhea*. It is usually caused by excessive secretion of PRL. Breast enlargement is not prerequisite for its appearance. Galactorrhea frequently occurs in severe hypothyroidism characterized by elevated circulating levels of TSH and TRH. Prolactin release is evoked by the high TRH levels.

COMPARATIVE ASPECTS OF REPRODUCTION IN NONMAMMALIAN VERTEBRATES

It is difficult to generalize about nonmammalian vertebrate reproductive patterns, as was demonstrated for mammals in the preceding pages. Attainment of sexual maturity occurs at a definite time characteristic for individual species and is followed by a series of reproductive cycles closely attuned to certain environmental factors. Depending on the species, sexual maturity may occur during the first year of life (many teleosts), after more than 15 years of juvenile existence (Atlantic eel, sturgeon) or at some intermediate period. Some animals have no reproductive cycle in that they breed only once after attaining sexual maturity and die soon afterward (for example, Pacific salmon, *Oncorhynchus* spp), whereas most species exhibit many reproductive cycles over many years. Some of these may produce successive broods in a given year or season or may exhibit only one or two cycles per year.

Environmental factors, such as temperature and photoperiod and presence of suitable breeding or nesting sites, operate through the central nervous system and the hypothalamo-hypophysial axis to regulate gonadal maturation and secretion of sex hormones. Steroid hormones, adenohypophysial hormones or both determine development of various sex-dependent characters and influence courtship, breeding and possibly parental behaviours.

Like mammals, nonmammalian species may be viviparous or oviparous, with the exception of the cyclostomes and birds, which are exclusively oviparous. The term viviparity is used here to designate situations in which fertilized eggs are incubated in utero, and nourishment for development is supplied at least in part by the maternal organism. Young are born alive. Ovoviviparous species retain eggs for variable periods of time during which development procedes. The eggs are usually laid in a more advanced stage of development. In some cases the egg may hatch in utero. If young are born alive, they have received no additional nourishment (other than yolk) from the maternal organism. Use of the term "viviparous" here will indicate live-bearing species and actually may include some ovoviviparous species. Ovoviviparous species that lay eggs will be included in "oviparous". Oviparous species all lay eggs with protective coverings from which the larval or juvenile form will later hatch.

Ovarian structures and events occurring in the gonads of nonmammalian vertebrates are related to the mammalian condition. Oocyte development is regulated by pituitary gonadotropins. The process of yolk formation is called *vitellogenesis*. Synthesis of yolk precursors or *vitellogenins* occurs in the liver and is stimulated by estrogens. When released into the blood, these lipoproteins bind calcium ions and are responsible for an elevation of total blood calcium in females undergoing vitellogenesis. Thus, marked increases in blood calcium can be used as an indicator of vitellogenesis. Incorporation of vitellogenins by growing oocytes and their conversion to yolk proteins are controlled by gonadotropins.

There is a major difference in the structure of testes in most anamniotes and amniote vertebrates. Whereas testes of mammals, birds, reptiles and anurans exhibit the tubular pattern of seminiferous elements with interspersed clumps of interstitial cells, the testes of urodele amphibians and fishes consist of *lobes* or *lobules*, each of which is composed of large *cellular cysts*. Each cyst is derived from a *spermatogonial nest*, and all of the cells within a cyst and usually all the cysts within a lobule will be in the same stage of spermatogenesis. (Seminiferous tubules of anurans also exhibit a pattern of cystic spermatogenesis.) Sustentacular (Sertoli) cells are present in each cyst. The more posterior lobules may be in a more advanced stage of spermatogenesis in repeating breeders than are the more anterior lobules. Spermiation in these anamniotes is usually followed by complete evacuation of spermatozoa from the mature lobules (more posterior ones) and regression of remaining cellular elements. Differentiation of lobules containing new cell nests occurs anteriorly from connective tissue elements and residual germ cells in the covering of the testis (tunica albuginea). In some fish species, however, all lobules develop and discharge sperm more or less simultaneously, and if breeding recurs there must be extensive regeneration of spermatogonial nests and new cysts prior to the next breeding season. True interstitial tissue is lacking in most anamniotes, and the synthesis of androgenic hormones occurs in cells associated with lobule walls. These steroidogenic cells are termed *lobule boundary cells*.

An additional steroidogenic tissue, the *interstitial gland*, may develop in the ovaries of gnathostomes. Interstitial glands develop from thecal cells derived from atretic previtellogenic follicles. It has been suggested that much of the estrogen synthesized during reproductive cycles is from the interstitial gland.

The endocrine factors in nonmammalian vertebrates are similar to and in many cases identical to those already described for mammals. There appear to be two distinct gonadotropins, FSH and LH, in all tetrapods (except teleosts and squamate reptiles), and their release is under stimulatory hypothalamic control. Follicle development in females and spermatogonial mitoses in males are stimulated by FSH, with meiotic events in males being influenced locally by androgens produced in sustentacular cells. Spermiation and ovulation are generally controlled by LH-like gonadotropin. However, in teleostean fishes only an LH-like gonadotropin has been purified, and mammalian LH can regulate all aspects of reproduction in these fishes. Mammalian FSH is ineffective. Even spermatogonial mitoses are under control of LH in teleosts. In contrast, reproduction in squamate reptiles requires only an FSH-like gonadotropin.

The major circulating estrogen in nonmammals is estradiol-17β, and testosterone or a closely related androgen is characteristic for males. Androgen and estrogen levels are higher than in mammals due to high circulating levels of steroid-binding globulin. Furthermore, relative levels of androgens and estrogens are not correlated with sex. For example, females may exhibit levels of androgens at certain times that exceed estrogen levels. The actions of gonadal steroids, including negative feedback effects on the hypothalamo-hypophysial axis, are similar in nonmammals to those described for mammals. Gonaduct differentiation and function, differentiation and maintenance of sex accessory structures and induction of certain behaviours are regulated by gonadal steroids.

As in mammals, PRL exhibits in certain species some specialized functions that are closely linked to reproductive events. The specific involvements of PRL will be discussed in some of the accounts that follow.

In summary, thyroid hormones appear to enhance the onset of gametogenesis, especially in males. It is only in the Amphibia and certain avian species that a negative correlation has been reported between thyroid activity and the onset of sexual maturation.

Internal fertilization requires evolution of a technique for transferring spermatozoa from the male to the female. Some viviparous anurans (for example, *Nectophrynoides*) and birds transfer spermatozoa through cloacal aposition or what has been termed the cloacal kiss.

Aquatic fishes and urodele amphibians, which practice internal fertilization, rely on spermatophores for transfer of spermatozoa. The spermatophore consists of a bundle of spermatozoa that are aggregated and enclosed in a gelatinous substance that will not rapidly dissolve in water. This structure allows the male to directly or indirectly transfer spermatozoa without excessive dilution of the "semen." Elasmobranchs, viviparous teleosts, apodan amphibians, one anuran and many reptiles possess intromittent organs that allow direct transfer of spermatozoa from male to female. In contrast, an aquatic male urodele deposits his spermatophore on the substrate and through a complicated behavioural ritual induces the female to pick it up with her cloaca. Frequently the female receiving a spermatophore has a special storage site (*spermatotheca*) that is capable in some species of storing viable spermatozoa for months. The spermatotheca thus may also possess special mechanisms (enzymatic?) to disperse the bundle of spermatozoa so that they can perform their destined functions.

Class Agnatha: Cyclostomata

Lampreys (Petromyzontidae) are characterized by having no breeding cycle, and all individuals die after spawning only once. In contrast, hagfishes (Myxinoidea) apparently breed more than once. Little is known about the reproductive biology of the hagfishes, and the following account of agnathan reproduction by necessity must emphasize the lampreys.

It appears that the cyclostome gonad arises entirely from the embryonic cortex whether it is destined to be a testis or an ovary. This singular embryonic origin may account for the common observations of what appear to be hermaphroditic gonads among the hagfishes. Males exhibit a single median testis with cystic spermatogenesis. Because of fusion of the paired primordia early in development the female lamprey has a single ovary. The single ovary of myxinoids is due to the failure of one primordium to develop. Steroid-binding proteins are not present in cyclostome blood, and circulating levels of steroids are very low. Gonaducts are absent in cyclostomes, and the gametes are shed into the coelom from which they exit via abdominal pores.

Petromyzontidae

Male lampreys

The mature lamprey testis exhibits the typical piscine pattern of lobules with germinal cysts. When the cysts have completed formation of spermatozoa, they simultaneously rupture and release spermatozoa into the body cavity. The testis of parasitic forms such as *Lampetra fluviatilis* contains only primary spermatocytes at the time of migration to the breeding grounds. These spermatocytes are transformed rapidly near the time of spawning into spermatozoal masses. Typical interstitial cell masses can be identified cytologically between the lobules in testes of migrating lampreys. These cells accumulate cholesterol-positive lipids and have become densely lipoidal by spawning time. Cytologically, interstitial cells appear to be steroidogenic and exhibit maximum hydroxysteroid dehydrogenase (3β-HSD) activity in February and March prior to the time of spawning. Development of lobule boundary cells during the later stages of spermatogenesis has been observed in *L. fluviatilis*. These cells are sensitive to hypophysectomy, and they may be homologous to sustentacular cells, although 3β-HSD activity has not been reported.

Female lampreys

Oogenesis has been carefully examined in the parasitic sea lamprey *Petromyzon marinus* and in the river lamprey *L. fluviatilis*. In *P.*

marinus, oogonia proliferate mitotically in the larvae to form the primary oocytes. By the time of metamorphosis of the larva to the juvenile, there are no oogonia remaining in the ovary. The primary ovarian follicles become more vascularized at this time. During the prolonged parasitic phase of body growth (about 10 to 20 months), the oocytes continue to enlarge slowly. The single follicular cell layer becomes thinner and less vascularized as spawning approaches, and the oocyte enters a period of rapid enlargement to reach the preovulatory condition. The mature follicles rupture immediately before spawning, and the eggs enter the coelom. Follicular atresia occurs throughout the history of ovarian development, and most oocytes undergo atresia, establishing this basic pattern early in the phylogeny of vertebrates. Phagocytes derived from the follicular cells ingest the yolk, and the follicle layers and surrounding stroma collapse into the area formerly occupied by the oocyte.

Oocyte growth in parasitic *L. fluviatilis* accelerates markedly just prior to spawning. The granulosa, which covers only the vegetal pole in close contact with the oocyte, reaches maximal development about 1 month sooner. The thecal cells are greatly reduced and with the aid of the electron microscope can be seen covering the granulosa layer and the animal pole. The theca interna consists of a single layer of cells in which there is a marked increase in smooth endoplasmic reticulum and mitochondrial differentiation during vitellogenesis. These cells show maximum cytological activity prior to the time of most intensive vitellogenesis, following which they undergo progressive regression until the time of ovulation. The theca externa consists of fibroblasts, collagen fibers and capillaries. Hydroxysteroid dehydrogenase activity is apparently confined to the thecal cells where peak activity is observed about 1 month prior to the appearance of secondary sex characters and the acceleration of follicular development.

Vitellogenesis appears to be an estrogen-dependent event in lampreys involving cooperative action of the liver, which produces proteins that are secreted into the blood and are sequestered by the ovary to be incorporated in the developing oocyte. Estrogens stimulate liver hypertrophy and elevate plasma protein-bound calcium, suggesting the presence of a mechanism such as the one that has been documented so carefully in birds and other nonmammals.

In the free-living lampreys that do not feed after metamorphosis the ovarian events occur over a much shorter time and are consequently more dramatic. In brook lampreys the immediate postmetamorphic period is marked by the onset of both vitellogenesis and massive atresia. As many as 70% of the follicles present at metamorphosis may become atretic, and phagocytosis of the yolk may provide an essential nutritional source for growth of the remaining oocytes to maturity.

Endocrine function in lampreys

The endocrine control of reproduction in lampreys has been studied by several investigators. The importance of gonadal steroids to development of sex accessory structures has been demonstrated through classical experiments involving hypophysectomy, gonadectomy and appropriate hormone therapy to either hypophysectomized or castrate animals. Pituitary gonadotropins stimulate gonadal hormone secretions, which in turn stimulate formation of secondary sex characters. The gonads of hypophysectomized *L. fluviatilis* were found to be less developed than those of sham-operated controls, although treatment of immature lampreys with mammalian gonadotropins has no effect on the gonads.

Myxinoidea

The reproductive biology of deep water myxinoids is poorly known, and there is little literature available. The hagfishes are apparently seasonal breeders and, unlike the lampreys, exhibit continuous reproductive cycles. Only *Eptatretus burgeri*, which lives in the shallow coastal waters of Japan, shows a seasonal cycle of gonadal activity. There is a single gonad in adults similar to that described in lampreys. Although a few cytological observations have been reported with respect to follicular development and formation of both preovulatory "corpora lutea" (atretic follicles) and postovulatory "corpora lutea," the possible endocrine functions of such structures are unknown. Male hagfishes apparently lack true interstitial cells, and it is not clear whether steroidogenic lobule boundary cells are present.

Although almost no experimental work is available with respect to the function of the hypothalamo-hypophysial system in hagfishes, hypophysectomy of the Pacific hagfish *Eptatretus stouti* results in testicular degeneration in males, but hypophysectomy of females has no effect on either ovarian structure or circulating steroid hormone levels. Vitellogenesis is stimulated by treatment of *E. stouti* with estradiol.

Class Chondrichthyes

The elasmobranchs have been studied extensively, probably because of the incidence of viviparity in these species (present in ten families of sharks and four families of rays). Unfortunately no data are available on reproduction in ratfishes. Elasmobranchs are characterized by internal fertilization regardless of whether they are viviparous or oviparous. Several different reproductive patterns have been described, extending from species that lay a precise number of eggs in a particular sequence, such as the oviparous clear-nosed skate *Raja eglanteria*, to the viviparous spotted dogfish *Scyliorhinus canicula* that is sexually active throughout the year and may have embryos in different stages of development in utero at the same time.

Male elasmobranchs

Spermatogenesis in paired testes is of the cystic type. Sustentacular cells have been identified, and they possess 3β-HSD activity. These cells become densely lipoidal and cholesterol positive following spermiation and are eventually resorbed. Sustentacular cells for the next cycle differentiate from connective tissue cells (fibroblasts) in the wall of the testis. Nests of spermatogonia proliferate from germ cells in the same regions, and they are responsible for producing spermatozoa utilized during the next breeding period.

Spermatophores produced by elasmobranchs are the result of secretory activities of male accessory ducts. After spermiation, spermatozoa pass through vasa efferentia and enter the coiled tubules of the so-called *Leydig gland*, which is derived from the anterior portion of the mesonephric kidney. Spermatozoa and secretions of the Leydig gland pass on to an expanded region of the vas deferens known as the *ampulla*. Here the spermatozoa are consolidated and receive additional secretory material to form complex spermatophores typical for each species. Fertilization is internal, and spermatophores are transferred to the female by specialized structures termed claspers.

Endocrine factors in male elasmobranchs

The importance of the ventral lobe of the elasmobranch pituitary as the source of gonadotropin controlling spermatogonial proliferation (mitotis) has been demonstrated in the spotted dogfish. Degenerative changes in the testes appear 6 weeks after removal of only the ventral lobe, and 22 months later the testes contain only spermatogonia and mature spermatozoa, indicating that removal of the ventral lobe blocks further differentiation of spermatogonia to spermatocytes, whereas all spermatocytes present at the time of surgery are able to complete meiosis and spermiogenesis. Removal of the rostral or neurointermediate lobes of the pituitary is without observable effect on the testes.

Female elasmobranchs

The elasmobranch ovary is covered by germinal epithelium and may contain a cavity derived from large lymph spaces within the stroma. Elasmobranch follicles are similar to those of mammals in possessing several distinct layers of cells. The connective tissue near a nest of oogonia will differentiate into the theca. As each follicle begins to develop, some epithelial cells undergo hypertrophy and hyperplasia to become the granulosa. In some species the granulosa may consist of only a single layer of cells. These cells are responsible for yolk deposition during oocyte growth as well as for yolk resorption should a given follicle become atretic. Granulosa cells are also thought to be the source of estrogens since they exhibit more 3β-HSD activity than do thecal cells. Most estrogen synthesis occurs in the mature follicle, which has a well developed granulosa. During follicular development a theca interna and theca externa can be discerned. However, both layers largely consist of connective tissue elements, and only a small amount of 3β-HSD activity has been observed in the theca interna cells.

The granulosa cells provide the source of both preovulatory (atretic) and postovulatory corpora lutea. The connective tissue layers surrounding these structures are derived from the theca. Corpora lutea of several species have been shown to possess 3β-HSD activity, and corpora lutea persist during gestation in *Squalus acanthias*, the spiny dogfish. Furthermore, the corpora lutea from pregnant *S. acanthias* produce twice as much progesterone in vitro as do those from nonpregnant females, which possess only preovulatory corpora lutea formed from atretic follicles. These observations strongly support an endocrine role for postovulatory corpora lutea in the viviparous elasmobranchs. Atresia is a common occurrence in elasmobranch ovaries. Depending upon the species being examined, either thecal or granulosa cells may contribute to formation of the preovulatory corpora lutea. Conflicting data have been published with respect to the question whether these atretic follicles are steroidogenic or not, and final resolution awaits further study.

Elasmobranch females have well-developed mullerian ducts that give rise to the oviducts as well as to the uterus of viviparous species. Oviducts have been examined in oviparous species that secrete horny shells to protect the eggs laid in the ocean as well as in viviparous species, and they possess a number of specialized features. *Oviducal* or *nidamental glands* secrete albumen and mucus in oviparious species.

Villus-like structures may develop in the uterine portion of the oviducts of certain viviparous females, and they provide nourishment for their young. The oviductal glands of oviparous species are often differentiated into an anterior albumin-secreting area and a posterior shell-secreting region. An intermediate mucus-secreting zone may be found in some species. In one dogfish species the shell-secreting portion of the oviduct serves as a spermatotheca.

Endocrine factors in female elasmobranchs

Removal of the ventral lobe of the adenohypophysis blocks oviposition in female *S. canicula*, and all follicles containing oocytes larger than 4 mm diameter undergo atresia. As in the male, removal of rostral or neurointermediate lobes has no effects on reproduction. The roles for steroids in reproduction have not been elucidated, although the possible role of progesterone from corpora lutea during gestation was suggested earlier. Additional studies with immature elasmobranchs, employing both gonadectomy and removal of the ventral lobe together with hormonal replacement therapy, are necessary to develop a complete understanding of endocrine factors responsible for reproduction in elasmobranchs.

Class Osteichthyes

The bony fishes, or more specifically the teleosts, exhibit almost every reproductive pattern and strategy known for vertebrates, including some that are unique to these fishes. Most of the account here is based on teleosts, but there are many similarities between teleosts and the other orders of bony fishes. Like that of cyclostomes the teleostean gonad develops only from a cortical primordium. Bony fishes may be dioecious or hermaphroditic. Among the hermaphrodites there are examples of protandry (function first as males and later transform to females) and protogyny (function first as females and later transform to males), and there are simultaneous hermaphrodites that may even exhibit self-fertilization. Fertilization may be external or internal as in the viviparous teleosts and in the viviparous coelacanth, *Latimeria*. In some viviparous teleosts the fertilized egg is known to develop within the ovary. Elaborate patterns of courtship, nest building, parental care and other specific reproductive behaviours have been reported among diverse groups. An incredible account of reproductive variations has been compiled by Breder and Rosen; many endocrine aspects of reproduction have been described.

Breeding is cyclic, with each species exhibiting a well-defined spawning period regulated by environmental factors (seasonal changes in photoperiod, temperature, etc.). Although some species spawn only once and die, others may spawn several times during a single breeding season (for example, *Reprohanus melanochir*, an Australian garfish).

There are many viviparous species of teleostean fishes that may utilize any of several known patterns, not to mention some patterns that may still be unknown to scientists. Some species, such as the guppy *Poecilia reticulata*, have short cycles. Soon after birth of the young, a new batch of oocytes is ready to be fertilized and another brood may be reared. Other species may require a longer "interbrood period" for oocyte maturation and vitellogenesis (*Mollienesia* and *Gambusia*). However, in *Quintana atrizoma*, oocyte development occurs during gestation so that a new batch of eggs can be fertilized as soon as the young are born. It would be most interesting to know the details of the patterns of hormone secretion in these different species, but they have not been studied.

Endogenous seasonal or annual rhythms have not been demonstrated in fishes although seasonal reproductive cycles are clearly evident even in tropical species. The effects of artificial lengthening and decreasing of the photophase may accelerate spawning in spring and fall spawners respectively. A classic demonstration of environmental phasing of reproduction has been demonstrated by transporting a poecilid, *Jenynsia lineata*, from South America, where it normally spawned in January and February, to the northern hemisphere. In the new pond location where photoperiod and seasons were reversed, the fish switched to spawning in July and August. However, the possible importance of the temperature regimen, which was also switched, should not be overlooked. In some species, temperature has been shown to be the critical factor in controlling recrudescence regardless of the light regimen imposed on the fish. In the tropics where photoperiod and temperature show little or no fluctuations, the reproductive cycles of freshwater and brackish water fishes appear to be tuned to the wet and dry seasons, which profoundly influence the aquatic environments. Periodic flooding and drying cause marked changes in water availability and also influence salinity and chemical composition of the aquatic environment. Some tropical fishes, such as many of the cichlid species that live in permanent bodies of water, may exhibit successive breeding over most of the year. It would be most interesting to examine the involvement of endocrines in these species with respect to timing and initiation of each breeding cycle.

Male bony fishes

Spermatogenesis in bony fishes is of the cystic lobular type, except for testes of atheriniform teleosts, such as the guppy, which have a tubular organization. Spermatogenesis in cyclic spawners resumes soon after breeding. Nests of spermatogonia proliferate from germ cells near the margins of the testicular lobules. Sustentacular cells have been described in a number of species, apparently arising from connective tissue elements of the spent lobule walls. Hydroxysteroid dehydrogenase activity has been shown in sustentacular cells, implying a role for these cells in the synthesis of steroids.

The major circulating androgens in teleosts are testosterone and 11-ketotestosterone. In many teleosts the androgen-secreting cells are located in the lobule walls (lobule boundary cells), and true interstitial (interlobular) cells are absent. These lobule boundary cells are believed to differentiate from fibroblasts in the lobule wall. Interstitial cells have been described in some species, and these cells possess 3β-HSD activity in teleosts, lungfishes and in the coelacanth *Latimeria*.

A seasonal cycle of lipid accumulation and depletion has been described in detail for lobule boundary cells of pike, *Esox lucius*, and similar cells have been observed for other teleostean species. Viviparous teleosts, like their distant elasmobranch relatives, produce spermatophores, employing secretions by the male gonaducts. These structures and the endocrine control of their secretory activities have not been examined sufficiently.

Spermatogenesis, 3β-HSD activity in interlobular or lobule boundary cells and spermiation are stimulated by treatment with LH but not FSH. Numerous studies have demonstrated that teleosts would appear to possess only one pituitary gonadotropin and that it is basically LH-like in its action. This is a particularly interesting observation since spermatogenesis in amphibians and amniotes is undoubtedly influenced strongly by FSH rather than LH.

Many species show marked seasonal development in sperm ducts, accessory glands and secondary sexual characters that are presumed to be under androgenic control. For example, testosterone induces formation of nuptial tubercles on the head of fathead minnows, a distinctive male sexual characteristic in several cyprinid species. Sperm ducts are derived from the coelomic walls. There are no wolffian ducts, and the sperm ducts of bony fishes are not homologous to the vasa deferentia of tetrapods or of elasmobranchs.

Female bony fishes

The teleostean ovary has been studied in considerable detail with respect to gonadal differentiation, oogenesis and vitellogenesis and ovulation, both in oviparous and viviparous species. The ovary of most teleosts is hollow, whereas solid ovaries have been reported in most Dipnoi and Chondrostei. Some teleosts also have solid ovaries. Unlike the hollow ovary of elasmobranchs and amphibians, in the ovary of teleosts the cavity as well as the outer surface is lined with germinal epithelium. Each hollow ovary is continuous with an oviduct which is not homologous to mullerian duct derivatives of other vertebrates. Eggs are discharged directly into the oviduct. In species with solid ovaries the eggs are discharged into the body cavity from which they pass to the exterior via oviducts or directly through openings in the body wall.

Salmon gonadotropin, mammalian LH, hCG or PMSG (in large doses) induces vitellogenesis, ovulation and oviposition in several species, including the Indian catfish *Heteropneustes fossilis*, the goldfish *Carassius auratus* and pink salmon *Oncorhynchus gorbuscha*. Synthetic mammalian GnRH also induces ovulation in goldfish. Mammalian FSH has not been effective in stimulating reproductive events in female teleosts. In the carp a peak in plasma gonadotropin occurs during ovulation, and the highest circulating gonadotropin levels in salmonids have been reported for prespawning females.

Evidence has been presented for non-glycoprotein gonadotropin in teleosts. This gonadotropin stimulates uptake by oocytes of yolk precursors from the blood. This function is controlled by LH in other vertebrate groups. Basically the teleostean ovary consists of masses of follicles embedded in a rather sparse stroma. Each follicle begins as a single-layered epithelium surrounding the oocyte. As the follicle grows these epithelial cells undergo hyperplasia and hypertrophy to form the granulosa. Connective elements in the stroma near the follicular nest will differentiate into a theca, which may further differentiate into a theca externa and a theca interna. The granulosa cells are responsible for yolk deposition in the oocyte during follicular growth and resorption of yolk during atresia. Steroidogenesis may occur in granulosa cells based upon measurements of 3β-HSD and 17β-HSD activity, although in some species the theca may exhibit more 3β-HSD activity than does the granulosa. In tilapia, 3β-HSD activity has been reported for both thecal and granulosa cells.

Three types of ovaries can be identified in teleosts. In the *synchronous ovary* all oocytes are in the same stage of development.

Species with a synchronous ovary spawn only once (for example, *Oncorhynchus* spp., *Anguilla* spp.). Species such as rainbow trout and flounder have a *group-synchronous ovary* with at least two populations of oocytes. These species generally spawn once a year during a short breeding season. The last type is the *asynchronous ovary* which has oocytes in all stages of development. These species spawn frequently each year during a prolonged breeding season.

Teleostean ovarian tissue synthesizes estrogens in vitro from radioactive precursors and also synthesizes testosterone, 11-ketotestosterone and deoxycorticosterone. Adrenal androgens may serve as precursors for the synthesis of 11-ketotestosterone. These three steroids have been identified in the peripheral plasma of teleostean species. In several species the levels of testosterone are greater in prespawning females than in males. Ovarian androgens may be precursors for estrogen synthesis. A possible role for DOC in ovulation has been claimed for *H. fossilis*.

All groups of bony fishes develop preovulatory corpora lutea as a result of atresia of developing follicles and develop postovulatory corpora lutea following ovulation. However, a convincing endocrine function of corpora lutea is not yet established among bony fishes. Atretic follicles do not have detectable 3β-HSD activity.

Vitellogenesis by the liver is apparently stimulated by estradiol. When released into the blood, these proteins bind free calcium and cause more calcium to be released from calcium reservoirs to replace the free calcium. Consequently total plasma calcium levels are elevated. Reproductively active oviparous females exhibit significantly greater plasma calcium levels than do males or immature females, presumably because of elevated estradiol levels and vitellogenesis.

Reproductive behaviour in bony fishes

Many aspects of reproductive behaviour have been studied in teleosts, including migration, courtship, nest building, spawning, copulation and parental care. Most of this work has concentrated on roles of testis, testosterone and synthetic androgens in males. Castration of males blocks breeding behaviour and causes reversal to nonbreeding condition of androgen-dependent characters. Some variations have been reported for agonistic (territorial) behaviour, which often accompanies breeding, depending upon the time of castration. In *Gasterosteus aculeatus*, form *trachurus*, castration more than a week before building of the first nest abolishes all related behaviours. If castration is performed within the week prior to building of the first nest, however, agonistic behaviour remains at a high level for 3 to 4 weeks. In some species, castration does not result in a decrease in agonistic behaviour. Androgens may not be required to maintain the behaviour once it has been induced.

The roles of estrogens and androgens in relation to the breeding behaviours of females have been studied less than the roles of androgens in males. Castration of females may result in complete abolition of all reproductive behaviour, loss of only some or only a decrease in intensity. In one case, ovariectomized *C. aculeatus*, form *leiurus*, show more aggressive behaviour than intact females, implying that steroids normally depress aggressive behaviour in females. Estrogens have proven ineffective in inducing female behaviour in females. Possible roles for androgens have not been studied. Spawning behaviour in females appears to be under control of gonadotropins, although in *Fundulus heteroclitus* neurohypophysial preparations or synthetic oxytocin induces reflexive spawning movements in hypophysectomized or castrated females. This spawning reflex is a behaviour not dependent upon shedding of ova. Similar observations have been reported for a few additional species. Spawning behaviour in females may be stimulated by prostaglandins.

Mammalian PRL has been shown to influence certain aspects of parental behaviour and implies a role for endogenous PRL. Fanning behaviour associated with aeration of the eggs can be stimulated in *Symphysodon aequifasciata* and *Pterophyllum scalare* by PRL treatment, whereas similar treatment inhibits fanning behaviour in sticklebacks. Stimulation of a mucous secretion that is fed to young *S. aequifasciata* is a PRL-dependent event. Mucous secretions are also stimulated by PRL in other species. It is not clear whether this behaviour of feeding mucus to the young in *S. aequifasciata* is caused by PRL treatment.

The implication of hormones in migratory behaviour is largely circumstantial. The gonads and their secretions probably do not play a causative role, since gonadal maturation usually occurs during migration. Thyroid hormones have been claimed to be causative factors of migratory behaviour, and increased thyroid activity coincides with migratory behaviour. It is possible that the increased activity of the thyroid gland is correlated to "permissive" effects related to metabolism and osmoregulation. Thyroid hormones may only enhance the physiological states favourable to migration, whereas the behavioural changes are neurally controlled through actions of environmental factors such as photoperiod and temperature or possibly by endogenous rhythmic neural cycles that are regulated by these environmental factors.

Class Amphibia

The origin of modern amphibians from their amphibian progenitors suggests that the primitive reproductive pattern for amphibians involved production of a small number of large, heavily yolked eggs. If direct development is primitive to amphibians, then it follows that the possession of an aquatic larval intermediate is a secondarily derived condition. The importance of the larval stage would seem to be that it allows for an intercalated feeding stage where growth can be optimized without using energy stores or food resources of the parents. If this interpretation is correct the presence of aquatic larvae should not be assumed to be a primitive feature and does not support the notion that study of species exhibiting this life history pattern will provide evolutionary clues. Nevertheless, during metamorphosis of these fishlike larvae to terrestrial-type tetrapods, the same problems are encountered and solved that allowed for the evolution of terrestrial vertebrates.

Within the modern Amphibia several trends in reproductive evolution are evident. All three extant orders show a reduction in the use of the aquatic habitat with a tendency toward terrestrial development. This trend is accompanied by greater reliance on internal fertilization, a secondary reduction in clutch size (number of eggs produced per breeding) and development of simple parental care of eggs and young. Mate selection and courtship patterns have become very elaborate in some species. Finally, oviparity has given rise to viviparity, especially in the anurans and apodans.

Oviparity

Many anuran amphibians are oviparous animals with external fertilization, although internal fertilization occurs in several species. Breeding in oviparous species is tied closely to a seasonal cycle involving photoperiod, temperature or availability of moisture or a combination of these, although a few species are continuous breeders (the Indian frogs *Rana tigrina* and *R. erytrea* and the South American toads *Bufo arenarum* and *B. paracmenis*).

One predominant reproductive pattern is found in oviparous anurans. Spermatogenesis and ovarian follicular development are completed in the fall, and the animals simply "hibernate" until suitable breeding conditions occur in the spring. Many oviparous species lay their eggs in temporary or permanent ponds with the egg developing into a free-swimming larval form. Tadpole larvae are the characteristic fishlike larval form of anurans and differ markedly from the larvae of urodeles, which possess external gills and four limbs. Anurans have internal gills like fishes and obtain their limbs later during metamorphosis. One anuran (*Ascaphus*) is known to lay its eggs in streams, and the tadpole larvae that result have special modifications to keep from being swept downstream. Some anuran and urodele species lay their eggs on land, usually in moist places such as under logs or in the axil of tree branches. Terrestrial eggs that are heavily yolked develop directly into miniature adults. No aquatic larval stage exists except within the egg.

Oviparous urodeles exhibit several reproductive patterns. In some species the pattern is similar to that of anurans (*Triturus cristatus*, *Notophthalmus viridescens*). Spermatogenesis in the hellbender *Cryptobranchus alleganiensis* occurs in July shortly before breeding in August and September. Other species such as the mudpuppy, *Necturus* spp., transfer spermatozoa to the females in the fall, and oviposition occurs the next spring when males are not present.

A number of oviparous apodans have been described, all of which lay terrestrial eggs. In *Ichthyophis* the eggs are laid in a burrow near a stream, and the newly hatched larva must emerge from the burrow and find its way to the stream. Apodans generally produce larger eggs than do the other amphibian groups, and clutch sizes are small.

Viviparity

Two European land salamanders *Salamandra salamandra* and *S. atra*, give birth to live offspring that develop in the posterior portion of the oviducts. In *S. atra*, one young develops and undergoes metamorphosis in each oviduct during a 4-year gestation period. Gestation is shorter in *S. salamandra*, which gives birth to larval salamanders.

Viviparity in anurans typically involves a modification of a pouch that allows the eggs to develop into tadpoles on the body of the maternal animal. The South American tree frogs carry their eggs in a single mass on their back. A fold of skin may develop that completely covers the eggs in a pouch such as that found in the so-called marsupial frogs, *Gastrotheca* spp., of South America. In others, such as the African frog *Pipa pipa*, each egg develops in its own dermal chamber. Oviductal incubation of eggs occurs in *Nectophyrnoides* and *Elutherodactylus*, and at least one species broods its young in its gut.

Viviparity may occur in the majority of apodan species. The contribution of maternal energy through oviductal secretion to support the developing young is considerable. In *Typhlonectes* one female may produce as many as nine larvae, each of which weighs about 40% of the mother's body weight at birth.

Male amphibians

Caudata

Spermatogenesis is of the cystic type in urodeles, and testicular structure and function are very similar to those of fishes. The urodele testis consists of one or more *lobes*, each containing several *ampullae*, which in turn are comprised of several germinal cysts. Germ cells associated with a germinal cyst divide mitotically to produce a cluster of secondary spermatogonia. These cells undergo synchronous differentiation to primary spermatocytes and enter meiosis. All of the cysts within an ampulla develop synchronously although it is typically only the more posterior ampullae that exhibit spermatogenesis prior to a given breeding season. Sustentacular cells develop from fibroblasts in the cyst walls while the spermatogonial divisions are taking place. As the ampullae mature the posterior portion of the testis becomes swollen with spermatozoa, whereas ampullae of the anteriormost portion consist primarily of spermatogonia. The posterior portion of the testis becomes dense and whitish because of masses of spermatozoa. After spermiation occurs the collapsed ampullae that have discharged their spermatozoa into the male ducts are resorbed, and after breeding, spermatogenesis is initiated in the anterior portion of the testis. If spermiation occurs in the fall, spermatogenesis will not be resumed until the next summer. New ampullae differentiate from connective tissue elements and germ cells in the tunica albuginea.

The urodele testis possesses lobule boundary cells in the ampulla walls. These cells exhibit a marked seasonal pattern of lipid accumulation, steroidogenesis and lipid depletion. Lipid accumulation begins when the secondary sexual characters develop or hypertrophy. The lobule boundary cells become intensely lipoidal and cholesterol-positive following breeding when androgen-dependent accessory structures are regressing. Both lobule boundary cells and sustentacular cells possess 3β-HSD activity, and it is likely that both are influenced by pituitary gonadotropins.

Androgen levels generally are greater in urodeles than in anurans. Seasonal patterns of androgen secretion have been reported for *Taricha granulosa* and *Cynops pyrrhogaster*. In both species androgen levels appear to be low during breeding. Failure to find seasonal variations in captive animals of other species may be a response to stress which shuts down androgen secretion.

Testosterone or estrogens can stimulate hypertrophy of the vas deferens, but only testosterone stimulates development of the male-type cloaca in urodeles. This cloacal gland complex is responsible for secretion of various components of the spermatophore. Spermatozoa are stored in the vasa deferentia. Contraction of the vasa deferentia and discharge of spermatozoa are caused by AVT.

Testosterone also stimulates development of nuptial pads in the newt *N. viridescens*, but maximal development is obtained by simultaneous treatment with PRL and testosterone. In urodeles such as *N. viridescens*, *T. cristatus* and *A. tigrinum*, PRL is known to influence the movement of land-phase animals to water for breeding and also induces heightening of the tail fin, which is a male secondary sex character.

Internal fertilization in both aquatic and terrestrial urodeles occurs through the production of an elaborate spermatophore produced through the actions of the cloacal glands of the male. The spermatophore consists of a glycoprotein matrix to which a packet of spermatozoa is attached, the glycoprotein matrix acting as a base upon which the spermatozoan packet rests. Following an elaborate courtship procedure, a female is induced to pick off the spermatozoan packets by her cloacal lips. The spermatozoan packet may then be stored in a specialized portion of the cloaca (spermatotheca) until ovulation occurs.

Anura

The anuran testis is similar in structure to that of amniotes and consists of a mass of seminiferous tubules with permanent germinal epithelium and conspicuous interstitial tissue. The cells of the latter tissue are ultrastructurally like mammalian interstitial cells and possess 3β-HSD activity. The lipid cycle within the interstitial cells and the degree of 3β-HSD activity closely parallel the development of androgen-dependent sex accessory structures such as the enlarged thumb pads of ranids. Interstitial cells of postspawning anurans exhibit considerable lipoidal accumulation but very low 3β-HSD activity. Thumb pads regress in ranids at this time.

During winter months, sustentacular cells of ranid testes lack lipid, but these cells elongate and exhibit small lipoidal granules as the breeding season approaches. Sustentacular cells of breeding animals have a well-developed smooth endoplasmic reticulum, and 3β-HSD activity is detectable. After spermiation the sustentacular cells detach from the tubule wall and degenerate. New cells for the next reproductive period differentiate from fibroblasts in the tubule walls.

Circulating testosterone, FSH and LH vary seasonally, but estradiol levels are very low. Highest values of hormones occur during mating in *Rana catesbeiana* but not in *Rana esculenta*.

Apoda

Male caecilians differ from urodeles and almost all anurans by possessing an elaborate intromittent organ, the *phallodeum*, associated with the posterior part of the cloaca. Consequently fertilization is internal in all apodans. Another unique feature is the retention of the posterior portion of the mullerian ducts that form the *mullerian glands*. These tubular apocrine glands are believed to produce the seminal fluid and would be analogous to the prostate of mammals.

The structure of the lobed testes is similar to that of urodeles, but the cell nests within an ampulla are not all in the same stage of spermatogenesis.

Female amphibians

Anura and Caudata

Amphibian ovaries are hollow, sac-like structures derived from the embryonic cortex and covered by germinal epithelium. A derivative of embryonic medullary tissues forms the inner lining of the ovary. Oogonia are present in the germinal epithelium, and they give rise to nests of oocytes. The follicular epithelium consists of a single layer of granulosa cells throughout the maturation period. A thecal layer does form around the follicle, but it is not easily seen with the light microscope.

Hydroxysteroid dehydrogenase has been demonstrated in granulosa cells as well as in thecal cells, but no 3β-HSD activity was observed in the interstitial cells of the ovary. Follicular cells of vitellogenic and mature oocytes in *Xenopus laevis* have been shown to possess increased 3β-, 17α- and 17β-HSD activity following PMSG treatment. Both thecal and granulosa cells may be sources of circulating ovarian steroids, but cytological evidence favours the granulosa as the major source.

At the end of a breeding season the ovary contains young follicles that will become the next crop of mature oocytes, numerous cell nests that will become the young follicles of the next vitellogenic period and primary germ cells that will give rise to new cell nests. Progression from primary germ cells to mature oocytes may require three breeding seasons for completion.

Ovarian estrogens control development of sex accessory structures such as the hypertrophy of oviducts prior to ovulation. In the marsupial frog *Gastrotheca riobambae*, the development of the brood pouch is also dependent upon estrogens from the preovulatory follicles.

Postnuptial ovaries frequently contain postovulatory corpora lutea, which are short-lived in oviparous species. Granulosa cells hypertrophy after ovulation and accumulate cholesterol-positive lipids. The follicle collapses and becomes a central mass of lipoidal cells surrounded by a fibrous capsule derived from the thecal layer. Postovulatory corpora lutea of *T. cristatus* and *Rana esculenta* possess 3β-HSU activity and may be sources for steroids. No functional endocrine role for postovulatory corpora lutea has been demonstrated in oviparous species.

In viviparous amphibians such as the anuran *Nectophrynoides occidentalis*, postovulatory corpora lutea are more persistent and appear to be functional throughout gestation. Furthermore, postovulatory corpora lutea from this species are capable of converting pregnenolone to progesterone, suggesting they may be endocrine structures. The granulosa-lutein cells of the postovulatory corpora lutea in viviparous *S. salamandra* possess 3β-HSD activity and appear cytologically to be steroidogenic. Thirty or more corpora lutea persist in each ovary during the first 2 years of gestation and gradually decrease in both size and number over the next 2 years until the young are born. In *Nectophrynoides occidentalis* as well as in oviparous *Taricha torosa*, the succeeding crop of follicles begins development only after degeneration of the postovulatory corpora lutea. These latter data suggest an inhibitory action of progesterone produced in the postovulatory corpora lutea on release of gonadotropins from the adenohypophysis.

Atresia occurs frequently during follicular development in amphibians. Granulosa cells are responsible for phagocytosis of yolk and formation of the preovulatory corpora lutea (corpora atretica). Since no 3β-HSD activity has been identified in these structures, they should be termed corpora atretica.

The process of vitellogenesis and yolk deposition in oocytes of oviparous amphibians has been reviewed. In *X. laevis*, gonadotropin stimulates micropinocytosis of vitellogenin by oocytes. Micropinocytotic vesicles of vitellogenin are hydrolyzed enzymatically in the yolk platelets to produce the yolk proteins *phosvitin* and *lipovitellin*. The yolk platelets containing these yolk proteins are utilized as an energy source during early embryogenesis. Estrogens will induce vitellogenin synthesis in both female and male livers when administered in vivo. Circulating vitellogenin binds free calcium ions, resulting in elevated total plasma calcium levels through release of calcium from storage sites.

Hypophysectomy results in atresia of all vitellogenic follicles in excess of about 0.4 mm diameter, indicating the importance of endogenous pituitary gonadotropins in the process of vitellogenesis. Mammalian FSH will augment the growth of vitellogenic follicles and prevent atresia following hypophysectomy. The failure of mammalian gonadotropins to stimulate formation and growth of previtellogenic follicles (less than 0.4 mm diameter) coupled with their apparent insensitivity to hypophysectomy has led to the suggestion that these processes are completely independent of pituitary control. However, experimental studies have not ruled out completely a role for gonadotropin in the development of previtellogenic follicles, and it is possible that low endogenous levels of amphibian gonadotropin are necessary for even the earliest events in gametogenesis.

Amphibian ovaries in vitro produce progesterone, estradiol, estrone, testosterone, deoxycorticosterone and dihydrotestosterone. The high levels of circulating testosterone reported for some anurans may be related to a precursor role for peripheral aromatization to estrogens. The follicular wall and/or interstitial gland is/are the source of these hormones.

Ovulation is under control of an LH-like gonadotropin and progesterone. In-vitro studies of the ovary were initiated by Wright and subsequently elaborated by others. These studies have examined the actions of various pituitary hormones as well as a wide variety of steroids in inducing ovulation and oocyte maturation (completion of meiosis and breakdown of the germinal vesicle) in vitro. Although many hormones are effective in vitro if sufficiently large doses are employed, LH and progesterone are the most potent, and it is reasonable to presume they play a role in normal ovulatory events. Prolactin enhanced the sensitivity of oocytes to gonadotropin or progesterone both in vivo and in vitro. This enhancement can be blocked by simultaneous in vivo treatment with thyroxine prior to examining oocyte maturation and ovulation in vitro. The synthesis of progesterone is under control of LH, and the action of progesterone is believed to be indirect, operating through stimulation of a "maturation-promoting factor. The source of the progesterone "surge" believed to occur just prior to ovulation is not clear, and it has been proposed that progesterone is produced by interrenal cells.

Ovarian maturation typically is completed in autumn, and ovulation is delayed over winter until favourable conditions occur in the spring. The endocrine basis for this diapause is not clear but may involve direct inhibition of ovulation by one or more pituitary hormones. Hypophysectomy of both gravid anurans and at least one urodele results in ovulation and oviposition. Furthermore, hypophysectomy of gravid neotenic tiger salamanders increases their sensitivity to induced ovulation in response to a single injection of hCG. This "reflexive" ovulation could be due to inadvertent release of gonadotropins by the operation itself or to removal of an active inhibitory substance of pituitary origin or to both.

Growth of oviducts is stimulated by estrogens or androgens. In mature females contraction of oviducts is caused by AVT and presumably is the hormonal stimulus in oviposition. Oviducts of breeding animals are more sensitive to AVT than are those of nonbreeding adults. Progesterone induces responsiveness to AVT in immature oviducts, but estrogens are not effective. Possibly the pre- or postovulatory follicle releases sufficient progesterone to alter receptor levels for AVT in the muscles of the oviducts.

Bidder's organ. Among both sexes of bufonids (toads) are found rudimentary ovaries or *Bidder's organs* that form from cortical remnants of the embryonic genital ridge. Histologically Bidder's organ consists of a compact mass of very small oocytes. After castration, Bidder's organ hypertrophies and forms a functional ovary in either males or females. Even prior to castration it is capable of synthesizing steroids. It is not clear whether these structures have functional roles or whether they represent only a curious occurrence.

Fat bodies. Conspicuous masses of adipose tissue called *fat bodies* are located adjacent to the gonads of amphibians. In female anurans and urodeles the size of the fat body is inversely correlated with gonadal weight, and it has been proposed that the lipoidal substances stored in the fat bodies are utilized for oocyte growth. Fat bodies of both male and female newts (*Triturus cristatus*) may be steroidogenic tissues and therefore may influence gonadal function, accessory sex structures or both. In *Rana esculenta*, function of fat bodies appears to be regulated by pituitary gonadotropins.

Apoda

Fertilization normally occurs in the upper portion of the oviduct, and the fertilized eggs are either laid in burrows or, as may prove to be the case for most species, are retained in the oviducts until the developing larvae have completed metamorphosis.

Ovarian development is very similar to that described for anurans and urodeles, but in apodans the eggs tend to be larger and fewer. Postovulatory corpora lutea develop in the ovaries, and they appear to be important for maintaining oviductal secretion (even in oviparous species) and pregnancy. Oviductal secretions provide nutrition for the developing young, and these secretions may be controlled by hormones released from the corpora lutea.

Sufficient numbers of specimens have not been examined to determine the seasonalness of breeding and ovarian cycles in caecilians. The endocrine factors involved in reproductive events can only be inferred at this time from studies on anurans and urodeles.

Reproductive behaviour in amphibians

Numerous aspects of reproductive behaviour have been described for many amphibians, but little is known about its endocrine control. Reproductive behaviour includes migration, calling, courtship, clasping, spawning and parental care. Studies involving castration, hypophysectomy and/or injections of pituitary hormones support the conclusion that testicular hormones are involved in calling, courtship and clasping. However, attempts to stimulate reproductive behaviour with androgens have not been successful. Recent studies support a role for neural peptides (AVT, GnRH, ACTH) in triggering mating behaviour in androgen-primed animals. For example, when a female frog that is not ready to spawn is clasped by a courting male, she croaks to signal her nonreceptivity. A receptive female will not emit this release call. This female has accumulated water that will be used in ovulation and oviposition. Water retention is caused by AVT. Administration of AVT inhibits the release call possibly through effects on the brain.

Class Reptilia

Living reptiles are members of diverse orders, and it is not surprising that considerable differences occur, making it difficult to generalize about reptilian reproduction. In many respects the squamate reptiles possess features unique to their order, whereas the other orders may be more typical of reptiles with respect to exhibition of primitive features. Most reptilian species are oviparous and exhibit well-defined annual reproductive cycles and breeding seasons. In addition, many examples of viviparity are known among snakes and lizards. Only a few, heavily yolked eggs are produced by most species. Clutch sizes may be large in some turtles, crocodilians and snakes. Fertilization is internal in all reptiles. Males have copulatory organs for placing spermatozoa into the cloaca of a female. Mating frequently follows complicated behavioural patterns including male-male interactions. Females of many species can store spermatozoa in the cloaca for months.

Reptiles may produce two distinct gonadotropins, although squamates rely on an FSH-like gonadotropin. There is disagreement with respect to relative roles of LH and FSH in reptilian reproduction, making a definite statement impossible at this time. The hypothalamus regulates gonadotropin release in response to environmental stimuli such as photoperiod and temperature.

Reptiles have been characterized by their lack of parental behaviour and, in some cases, even lack of recognition of their offspring. In recent years, studies have demonstrated that there may be considerable investment in parental care even among oviparous species. Although members of the oldest extant order, the chelonians, typically abandon their nests once the eggs are laid, crocodilian parents participate in the hatching process and in protecting the young. Evidence of nestbuilding and parental care has been unearthed for some extinct dinosaurs as well. Thus, complex parental care did not appear de novo in birds and mammals but evolved in primitive reptiles.

Male reptiles

The reptilian testis is typical of amniotes and consists of convoluted seminiferous tubules, each surrounded by a connective tissue sheath, the *tunica propria*. The entire testis is enclosed by a connective tissue layer, the *tunica albuginea*. Spermatogenesis recurs soon after the breeding season and is completed in most species prior to the onset of winter. Pituitary gonadotropins stimulate spermatogenesis in a variety of reptilian species. Spermatozoa may be stored for up to several months prior to mating during the breeding season. Sustentacular cells are common, and they have been shown to be steroidogenic. After spermiation and testicular collapse, the sustentacular cells fill with cholesterol-positive lipids. Soon there occurs a clearing of the lipid from the cells at the time mitosis resumes in the spermatogonia. Follicle-stimulating hormone causes clearing of these cells and the resumption of spermatogenesis.

Typical interstitial cells have been described in the reptilian testis, and they undergo cyclical changes associated with particular reproductive events. Seasonal changes in lipid content and 3β-HSD activity are correlated with androgen secretion and sexual changes in androgen-dependent sex accessory structures.

In sexually active squamates a portion of the kidney tubules hypertrophy under the influence of androgens. This modified kidney structure is known as the *sexual segment* of the kidney, and it appears to secrete materials that help maintain spermatozoa that are stored in

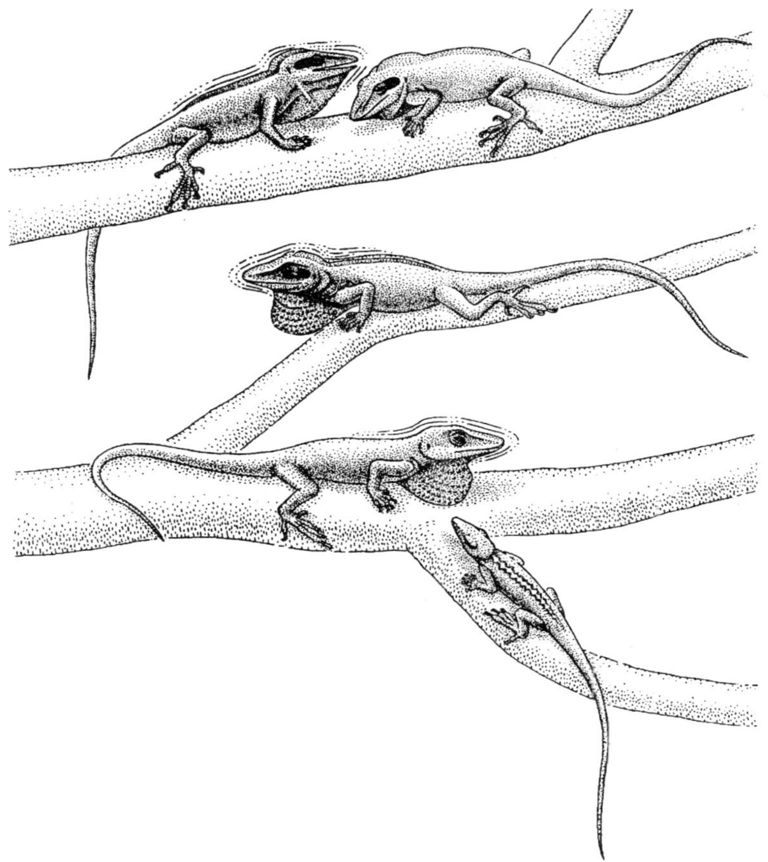

Fig. 13.13. Displays exhibited by male Anolis carolinensis.

this region prior to ejaculation. It has been suggested that the sexual segment is homologous to the seminal vesicles of male mammals.

Female reptiles

Reptiles have paired hollow ovaries with little stromal tissue. Oogonia are present in the mature ovary as described for anamniotes and give rise to primary oocytes throughout reproductive life. The developing oocyte becomes invested with granulosa cells, which are separated from the surrounding thecal cells by a connective tissue layer (membrana propria). The theca is differentiated into an inner, granular theca interna surrounded by a fibrous theca externa. The cells of the granulosa are considered the primary source of follicular estrogen during ovarian recrudescence, although some histochemical evidence implies that thecal cells also may be steroidogenic. Histochemical changes in cholesterol-positive lipid inclusions and 3β-HSD activity parallel estrogen-dependent oviductal growth and changes in other sex accessory structures as well as changes in the gonadotropes in the adenohypophysis. As the oocyte enlarges it begins to project into the ovarian cavity.

The squamate granulosa contains a unique flask-shaped cell, the *pyriform cell*, that is in direct contact with the developing oocyte. These cells apparently are involved with early steps in oocyte development as they either degenerate or transform into typical granulosa cells at the onset of vitellogenesis.

As ovulation approaches the granulosa cells begin to accumulate cholesterol-positive lipids, and, following ovulation, they proliferate and luteinize to form corpora lutea. These corpora lutea are well vascularized, exhibit 3β-HSD activity and synthesize progesterone. They persist throughout egg laying in oviparous species or throughout gestation in viviparous forms. Corpora lutea of viviparous species synthesize greater amounts of progesterone than do those of oviparous species. Plasma progesterone levels are greatest following ovulation and are maintained at elevated levels throughout gestation in viviparous lizards and snakes, whereas preovulatory peaks of progesterone are found in oviparous turtles.

Generally speaking only a few follicles reach maturity at a given time; the majority undergo atresia. Follicular atresia is a common occurrence in reptilian ovaries as in other vertebrates, and formation of corpora atretica is also common. The importance of the latter structures is unknown, but the absence of 3β-HSD activity in corpora atretica makes it unlikely they would have an endocrine function. However, steroidogenic cells from atretic follicles may give rise to an "interstitial gland" that is believed to be a major source of ovarian estrogens.

Ovaries of reptiles show different patterns of follicular maturation and ovulation. Some produce several eggs simultaneously from each ovary (most reptiles). Others may alternate production of a single egg from each ovary (anoline lizards). Still others (turtles) may produce most eggs in one ovary during one season and most from the other ovary the next season. Differences in follicular atresia rather than in the number of oocytes beginning development may be responsible for these patterns.

Oviductal development apparently is under the influence of ovarian estrogens, and progesterone is without effect. In oviparous species, estrogens probably influence the secretion around the egg of albumin and shell from the anterior end of each oviduct. Estrogens also stimulate synthesis of vitellogenic proteins by the liver and cause increases in serum calcium of snakes, lizards and turtles. Crocodiles produce yolk proteins that are biochemically similar to those of birds. Measurement of calcium may be used as an indirect quantitative measurement of plasma vitellogenin. Undoubtedly the details of this process are very similar to those described for amphibians and for birds, although vitellogenesis has not been investigated as extensively in reptiles.

Oviposition or birth of live young is controlled by AVT in turtles, lizards and snakes. In the American chameleon, *Anolis carolinensis*, the sensitivity of the uterus to AVT is determined by presence or absence of a corpus luteum in the adjacent ovary.

Environment, behaviour and reproduction of reptiles

The role that physical and biological components of the environment play in sexual behaviour and reproduction has been extensively studied in reptiles. Species living in temperate climates exhibit distinct seasonal patterns. Reproduction in tropical reptilian species varies from cyclic patterns to continuous breeding. There is a strong tendency for an observed increase in the incidence of viviparity among species inhabiting colder climates (altitude or latitude), but it is not clear which is cause or consequence.

Among temperate lizards, temperature is the dominant environmental factor influencing reproduction. Photoperiod, humidity, and nutritional status play decisive roles in some species. Other groups of reptiles have not been studied as extensively as lizards.

Class Aves

Avian reproductive organs reflect a general anatomical adaptation to flight. In females of most species only the left ovary and its attendant oviduct develop, whereas the right-hand components remain in a rudimentary state. This asymmetry noted in the female is reflected in the male where the left testis usually is larger than the right although both are functional. Should the left ovary be removed surgically or destroyed by disease, the right rudiment may develop, but it will usually form an ovotestis or a testis.

Avian gonads develop from a pair of undifferentiated primordia associated with the embryonic nephrotome. These primordia are invaded by primordial germ cells that migrate through the blood from the splanchnopleure and develop into the germinal epithelium. The embryonic gonad goes through a bipotential state in which both cortical and medullary components are present. Differentiation of cortical tissue is necessary for ovarian development and the medullary portion is suppressed. The reverse condition prevails in male birds. In contrast to mammals it is the male bird that is the homogametic sex (similar sex chromosomes), and it is the female that has unlike sex chromosomes. Developing a female phenotype requires estrogens. Castration of a young female causes development of male plumage.

Development or suppression of the mullerian and wolffian ducts eventually depends on the direction of gonadal development as it does in other vertebrates. In females in which only the left half of the reproductive system usually develops, the left ovary receives the larger proportion of germ cells that migrate to the gonads. The mechanism behind this disproportionate distribution of germ cells is not known.

All birds are oviparous but display significantly more parental behaviour than any other nonmammalian group of vertebrates. Birds are endothermiclike mammals and use their body heat to support development of the embryo within the egg much as the mammal does in utero. Consequently birds can breed successfully under conditions that are too cold for their reptilian relatives. The adaptation of long-distance flight allows utilization of polar and subpolar regions where winter conditions are too severe for survival but where during the summer months there is adequate warmth and food to breed and rear young birds to a size sufficient to migrate to warmer latitudes for winter.

Reproduction is decisively cyclic in adult birds and is closely attuned to environmental factors. Migratory and nonmigratory polar and temperate species typically exhibit seasonal cycles with breeding occurring in the spring and possibly continuing through much of the summer. On the other hand, species occupying arid regions may show irregular cycles cued to the availability of water, which may not occur with any seasonal regularity. One tropical species, *Zonotrichia capensis*, has been reported to breed every 6 months regardless of rainfall as long as food is available.

Both ovaries and testes remain small in nonbreeding birds but may undergo tremendous hypertrophy in a very short time. This is especially advantageous for migratory species or species relying on particular stimuli for breeding where gonadal recrudescence can await arrival on the breeding grounds or appearance of suitable conditions such as abundant food.

Male birds

The testes are permanently located in the body cavity, and each testis consists of a mass of convoluted seminiferous tubules lined with a germinal epithelium and surrounded by connective tissue. Both developing germ cells and sustentacular cells can be seen in the germinal epithelium as well as numerous fibroblasts. As in mammals and other vertebrates the cytoplasm of the sustentacular cells completely envelops the germ cells and cytologically and histochemically appear to be steroidogenic. Typical steroidogenic interstitial cells occur between the seminiferous tubules.

In nonbreeding birds, testes are very small, and histologically these quiescent testes appear to be composed largely of interstitial cells. However, this is only an artifact produced by a marked postbreeding regression of spermatogenetic tissue. The onset of spermatogenesis (recrudescence) results in rapid and marked increase in testicular size. Such rapid and extreme growth (to as much as 500 times the resting gonad weight) results in considerable strain and damage to the tunica albuginea surrounding the testis, and it must be replaced each year during the postnuptial phase of the testicular cycle. Replacement is accomplished through differentiation of fibroblasts and formation of a new tunica directly beneath the damaged one. It is often possible to distinguish between juvenile birds and postnuptial birds by the presence of two connective tissue capsules around the testis in the latter.

Testicular recrudescence may involve a single synchronous spermatogenetic event or separate spermatogenetic waves, depending whether a given species produces successive clutches during a particular breeding season. In either event, following spermiation spermatozoa migrate to expanded distal ends of the vasa deferentia known as *seminal sacs*. Spermatozoa will be ejaculated from the seminal sacs during mating.

The annual testicular cycle of temperate birds has three more or less distinct phases: (1) the regeneration or *preparatory phase*, (2) the acceleration or *progressive phase*, and (3) the *culmination phase*. Similar phases can be identified in all birds regardless of the seasonal nature of their reproductive cycles or what environmental factors control testicular events. The preparatory phase begins immediately after the reproductive period and is characterized by marked collapse of the testis. The most common environmental factor influencing the avian testis is photoperiod. Placing quiescent, temperate birds on long photoperiods will typically stimulate testicular recrudescence, whereas maintenance of birds on short photoperiods into the normal breeding season represses anticipated testicular events. Animals in the preparatory phase are insensitive to effects of long photoperiod and are termed *refractory*. The end of the preparatory phase is heralded by restoration of photosensitivity. The endocrinological basis for the photorefractory period in birds is not clear and more than one mechanism may be involved in different species. Some studies suggest that feedback of testosterone is responsible for induction of the photorefractory period and for low levels of LH during the photorefractory period. Other investigations point to changes in hypothalamic sensitivity and/or steroid metabolism.

During the progressive phase there is an increase in gonadotropin secretion brought about by actions of lengthening photoperiod on the hypothalamo-hypophysial axis. Increased circulating gonadotropins stimulate both spermatogenesis and androgen secretion by the interstitial cells. An increasingly intensive period of sexual activity and song occurs, and males of some species may begin exhibiting territorial behaviour and mate selection. This effect of long photoperiod can be blocked by low temperatures.

The culmination phase coincides with the time of ovulation in females and includes the time of insemination. The male typically is ready for breeding before the female, and his testes will be bulging with spermatozoa. Successful breeding involves a complex, hormonally dependent series of events involving precise male-female interactions.

Interstitial cells

A characteristic lipid cycle occurs in avian interstitial cells similar to that described for other vertebrates. There is accumulation of lipid in young birds followed by rapid depletion coincident with onset of the first breeding season and spermatogenesis. The interstitial cells of adult birds are small and sparsely lipoidal in winter although they occupy a large proportion of the testis because of the regressed nature of the seminiferous tubules. There is gradual accumulation of lipids, including cholesterol, throughout the progressive phase as well as an increase in 3β-HSD activity. At the time of maximum sexual display there is rapid depletion of interstitial cell lipid. Cholesterol disappears completely, but 3β-HSD activity remains strong, indicating lipid depletion is a consequence of rapid synthesis and secretion of androgens. The activity of 17α-hydroxylase is also high at this time. A massive disintegration of interstitial cells occurs during the preparatory phase, and new interstitial cells differentiate from fibroblasts.

Sustentacular cells

Cyclical changes in lipid content are characteristic of avian sustentacular cells that ultrastructurally resemble steroidogenic cells. Both 3β-HSD and 17β-HSD activities have been reported for these cells. They become densely lipoidal following the breeding season, and no detectable 3β-HSD activity remains. The lipid is depleted with the onset of the next period of spermatogenesis.

Endocrine control of testicular function

The hypothalamus contains two gonadotropic centers that separately control release of LH and FSH from the adenohypophysis. Hyperplasia of interstitial cells is caused by LH, and they become lipoidal and exhibit increased 3β-HSD activity. Avian testes are much more sensitive to avian LH than to mammalian LH. Androgens secreted by these cells stimulate sex accessory structures and secondary sexual characters. Purified mammalian FSH is less effective than avian FSH in stimulating spermatogenesis. Local effects of androgens from sustentacular cells are responsible for stimulating meiosis. Androgens are known to maintain spermatogenesis even in hypophysectomized birds.

Prolactin is present in the male pituitary and has been reported to inhibit FSH release and block spermatogenesis in some species. The formation of incubation patches on males of certain species is induced in part by PRL working cooperatively with testicular steroids.

Sex accessory structures in male birds

Wolffian ducts give rise to paired vasa deferentia, vasa efferentia and the epididymides, all of which exhibit hypertrophy with the onset of sexual activity. These events are all prevented by castration. A testis is connected to the vasa efferentia by small rete tubules in the tunica albuginea that become enlarged during the breeding season. The vasa efferentia show increased secretory activity during the breeding season and coalesce to form a long, coiled tube and epididymis. Hypertrophy of the epididymis is accompanied by secretion of seminal fluid. Spermatozoa are not stored in the epididymis but enter the enlarged vas deferens. The distal end of each vas deferens (seminal sac) fills with spermatozoa. The posterior walls of the seminal sacs protrude into the cloaca as erectile papillae that facilitate transfer of spermatozoa. During copulation the cloaca is everted and these papillae are brought into contact with the vagina of the female. In some species the cloaca is modified into a penis-like intromittent organ.

Avian ovary

Few studies of avian reproduction have employed wild species, and much of our knowledge of ovarian reproductive events has been gleaned from domestic species, in particular the hen. However, these studies have provided relatively complete assessment of ovarian function in domestic species and provide a basis for comparison to wild species. In many respects, ovarian function in birds is like that of their oviparous ancestors, the reptiles.

The domestic hen differs most importantly from wild birds by being a continuous breeder. Prior to hatching there is a proliferation of oogonia to produce thousands of primary oocytes that will serve the hen throughout her long and busy reproductive life. No new oocytes will be formed after hatching, unlike the situation for many anamniotes. Most of these oocytes will undergo atresia during early maturational stages. The primary follicle consists of an oocyte surrounded by a layer of granulosa cells. As the follicles grow, thecal layers are added, and the follicles become highly vascularized. Both granulosa and thecal cells are steroidogenic and possess 3β-HSD activity. Estrogens secreted from the follicle cells cause the liver to produce large quantities of plasma phosphoprotein (vitellogenin) that are sequestered from the blood by growing oocytes. Vitellogenin binds free calcium ions in the blood, and the complex is incorporated into growing oocytes. The additional demand for calcium to construct the shell stimulates appropriate endocrine mechanisms to meet this need. Vitellogenin is enzymatically hydrolyzed to produce phosvitin and lipovitellin. The liver also synthesizes large quantities of triglycerides that are transported in the blood as β-lipoproteins. Micropinocytosis by the oocytes of these phosphoproteins and triglycerides is stimulated by FSH. The growing avian oocyte completely fills the follicle, and antrum formation does not occur.

Developing follicles bulge conspicuously from the surface of the avian ovary, giving it the appearance of a bunch of grapes. The largest follicles become suspended from the ovary only by a narrow band of tissue. At this stage the follicle is highly vascularized except for a rough, avascular spot, the *stigma*, where the follicle will rupture at ovulation.

Atresia of developing follicles may occur at any time during follicular development. These atretic follicles can be easily recognized by an influx of fibroblasts that phagocytize the yolk materials. Granulosa and thecal cells are lipoidal and contain cholesterol. There are many corpora atretica at all times in the ovary, but their importance, if any, is not recognized. As they disintegrate, some of the cells of the corpora atretica may become stromal interstitial cells and secrete estrogens.

Birds are characterized by the absence of persistent corpora lutea following ovulation. Collapsed follicles consist largely of granulosa cells containing progesterone, abundant smooth endoplasmic reticulum and considerable 3β-HSD activity. The only evidence for a functional role, however, is the observation that surgical removal of these ruptured follicles increases the time that the ovulated egg is retained in the oviduct.

Endocrine control of ovarian function in birds

There is a close correlation between pituitary gonadotropin content and ovarian function in both domestic and wild birds. Hypophysectomy causes ovarian regression and extensive follicular atresia, which can be prevented by gonadotropin replacement therapy. Follicular development is stimulated by FSH, and FSH will maintain oviducts in hypophysectomized birds. Estrogen secretion is controlled by both LH and FSH. Mammalian gonadotropins, however, are not always as effective in birds as are avian pituitary or gonadotropin preparations. Furthermore, avian FSH is very effective at stimulating follicle development in lizards, emphasizing the close similarity between reptilian and avian pituitary hormones and the cautions necessary when interpreting the effects of mammalian hormones in birds.

As is the case for certain reptiles, growth of follicles and ovulation is a continual process throughout the breeding season. Ovarian function is regulated so that typically only one egg is discharged at a time. This condition is reminiscent of the human and the lizard *Anolis carolinensis*, in which only one ovum is discharged, and the ovaries alternate in providing the ovum. A hierarchy of graded follicle size is maintained. The endocrinological basis for establishment of such a hierarchy, however, is not known for either reptiles or birds, and it represents one of the major unanswered questions in reproductive biology.

The synthesis of vitellogenin by the liver is induced by estrogens. Total serum calcium concomitantly increases, which is related to the binding of calcium by vitellogenin. In addition to incorporation of vitellin proteins into the oocyte, circulating calcium is sequestered by the shell glands of the "uterus" (expanded region of the oviduct) for construction of the egg shell. Thus estrogens provide for transfer of calcium from storage sites (bone) to the egg shell as well as for stimulation of vitellogenin production.

Pituitary LH is responsible for triggering ovulation of the fully mature follicle. Plasma LH peaks about 6 to 8 hours before ovulation in domestic hens as well as in Japanese quail, but the magnitude of the avian LH surge is considerably smaller than that observed in mammals. This lower surge of LH might be an adaptation to ensure only sufficient LH for ovulating the largest follicle.

Calcium availability may be a potent factor regulating reproduction in female birds. Production of shelled eggs in domestic species directs as much as 10% of the body calcium stores per day into eggs. If large amounts of calcium are not available in the diet, the reproductive axis is shut down before damage to the skeleton occurs. When sufficient calcium becomes available, the birds resume laying. It is possible that calcium depletion in wild birds contributes to cessation of breeding and induction of the refractory period.

Another pituitary hormone, PRL, plays an essential role in development of a specialized, defeathered region in some species known as an incubation patch, which aids in incubating eggs. The action of PRL on secretion of crop milk by the pigeon crop sac for use in feeding young birds has resulted in development of a most useful biological assay for PRL activity in all tetrapod pituitaries. Prolactin does not affect steroidogenesis in cultured chick granulosa cells and probably has no effect on progesterone synthesis.

Oviduct

Estrogens are secreted by the growing follicle as a result of the actions of FSH on thecal cells and probably on granulosa cells as well. Their primary influences are on liver (vitellogenesis) and on the oviduct. Estrogens stimulate hypertrophy of the oviduct and differentiation of secretory regions. Five differentiated regions can be identified in the mature oviduct: *infundibulum*, *magnum*, *isthmus*, *shell gland* and *vagina*. After ovulation the ovum enters the infundibulum and is fertilized in the upper end of the oviduct before albumen is added. The middle portion of the oviduct or magnum becomes highly glandular under the influence of estrogens, forming tubular glands and goblet cells. Estrogens stimulate synthesis of *ovalbumen* protein by the tubular glands, whereas progesterone stimulates the goblet cells to produce the other major egg white protein, *avidin*. After accumulation of several coatings of albumen the egg passes to the muscular isthmus where two shell membranes are applied. These membranes are composed of protein fibers cemented together with albumen. The shell consists largely of calcium salts supported by a fibrous protein matrix deposited on the outermost shell membrane by the shell gland or "uterus." After the shell has been applied, contraction of a powerful sphincter muscle causes the egg to rotate in the vagina and enter the cloaca pointed-end first. Movement of the egg into the cloaca as well as its extrusion into the nest (oviposition) is controlled by a neurohypophysial hormone. An increase in plasma arginine vasotocin together with a concomitant decrease in neurohypophysial arginine vasotocin coincides with oviposition, and treatment with either arginine vasopressin or oxytocin can cause premature oviposition.

Incubation patches

In many avian species a ventral region (apterium) becomes defeathered, highly vascularized and edematous just prior to or during egg laying. In addition, the epidermis of this region may become hyperplastic. This specialized region is termed an *incubation* or *brood patch*, and when in contact with the eggs provides an efficient transfer of warmth from the parent bird to the eggs. Incubation patches may form in females, males or both, depending upon the species and which sex is responsible for incubating eggs. However, mere possession of an incubation patch is not proof of incubating behaviour. Male house sparrows (*Passer domesticus*) have no incubation patch yet exhibit incubating behaviour, whereas male flycatchers (genus *Empidonax*) develop an incubation patch but do not show incubating behaviour.

Formation of incubation patches involves cooperative actions of both estrogens and PRL. Estrogens seem to stimulate vascularization of the patch region, and PRL stimulates defeathering and epidermal hyperplasia. Both hormones are necessary for normal patch development. Furthermore, the response of epidermis in forming a patch is both site specific and tissue specific. After transplant to the dorsal surface ventral skin will still respond to PRL but not to estrogen. Vascularization of the ventral skin occurs only when it is in its normal location. On the other hand, dorsal skin transplanted to the normal patch site will not respond to either estrogens or PRL.

Androgen-dependent secondary sex characters in males and females

Androgens play important roles in both male and female birds. In a number of species a change in bill colour is associated with breeding. Such changes are induced in both sexes by androgens but not by estrogens or progesterone. However, there is at least one case in which bill colour change occurs only in the female, and the colour change is induced by estrogens.

Plumage colour changes may also be controlled by androgens. This is the case for phalaropes in which the females possess the more colourful plumage. One cannot presume this to be the case unless specific studies have been performed, because androgens are not always responsible for nuptial plumage. For example, development of nuptial plumage in castrated male weaver finches, *Euplectes orix*, has long been known as the classical bioassay for LH. Estrogens apparently inhibit formation of nuptial dress, and castrated females will develop male plumage. Assumption of nuptial dress in males can also be blocked with estrogens.

In some strains of chicken both sexes have female type plumage and castration causes development of male plumage. Treatment of castrated males with testosterone causes a return to female type plumage, but growth of the comb and wattle are stimulated (a normal male trait). If castrated males are treated with DHT, the growth of the comb and wattle are stimulated but there is no reversion to female plumage. In these strains of chicken, the skin aromatizes testosterone to estrogens and stimulates female plumage. In the comb and wattle, 5α-reductase converts testosterone to DHT. Since DHT can not be aromatized, the plumage of DHT-treated castrated males does not revert to the female type.

Reproductive behaviour in birds

Each avian species exhibits a precise sequence of endocrine-dependent behaviours such as migration, acquisition of territory, advertisement by song, attraction of mate, pairing, nest building, egg laying, incubating and rearing young birds. The actual sequence of events and their endocrinological bases are species specific and cannot easily be generalized. Successful breeding involves a complex interaction of male and female birds in precise sequences (that is, if male does A, then female does B, which stimulates male to do C, etc.) as well as presence of suitable environmental cues such as proper nesting material, availability of water, etc. Little experimental work has been done with wild birds since it is difficult to get them to perform under laboratory conditions although several descriptive studies on hormone levels and behaviour are available.

Androgens appear to be responsible for territorial display and aggression in wild birds. Aggressive behaviour can be stimulated also by FSH but not LH in males. Courtship appears to involve negative feedback of testosterone on FSH levels, which results in reduction in circulating androgens and allows for subsequent, less aggressive behaviours. Androgen levels are low during mating. Androgens antagonize incubation patch development, and a reduction in circulating testosterone may be necessary for patch development in males of certain species. In domestic chickens and ring doves, testosterone does stimulate courtship and copulation.

Bowing behaviour in feral pigeons coincides with maximum androgen synthesis but decreases prior to egg laying, coincident with an increase in progesterone levels. Progesterone is a well-known stimulus for incubation behaviour in laying pigeons. Removal of the postovulatory follicle from chickens blocks nesting behaviour. Prolactin can induce incubation behaviour, but this action is presumed to be due to increased release of progesterone.

14

DEVELOPING VERTEBRATE GONAD

A study of the endocrine functions of the embryonic gonad of the vertebrates constitutes a considerable chapter in the division of endocrinology concerned with the sex glands and the sex hormones. The problem was opened by the investigations of Lillie (1917) and of Keller and Tandler (1916) on the spontaneous free-martin. Since that time, the experimental investigations have been multiplied in all of the vertebrate groups. It is impossible to give a substantial coverage in a brief review; consequently, I shall limit myself to certain investigations in which I have participated and which concern the bird. My expose shall center about the following question: What is the actual role that the gonads play in normal development?

DEMONSTRATION OF GONADAL SECRETIONS IN THE BIRD

The following describes two series of experiments which demonstrate the existence of gonadal hormonal secretions in the developing chick.

Coelomic Grafts of Gonads

Female gonads, taken from embryos 6 to 10 days of age, were grafted in the coelom of embryos 2 to 3 days of age. They feminized the gonads of the embryonic males so that they evolved into ovotestes.

Reciprocally, male gonads, grafted into the coelom of embryonic females 2 to 3 days of age, effected the complete regression of the mullerian duct.

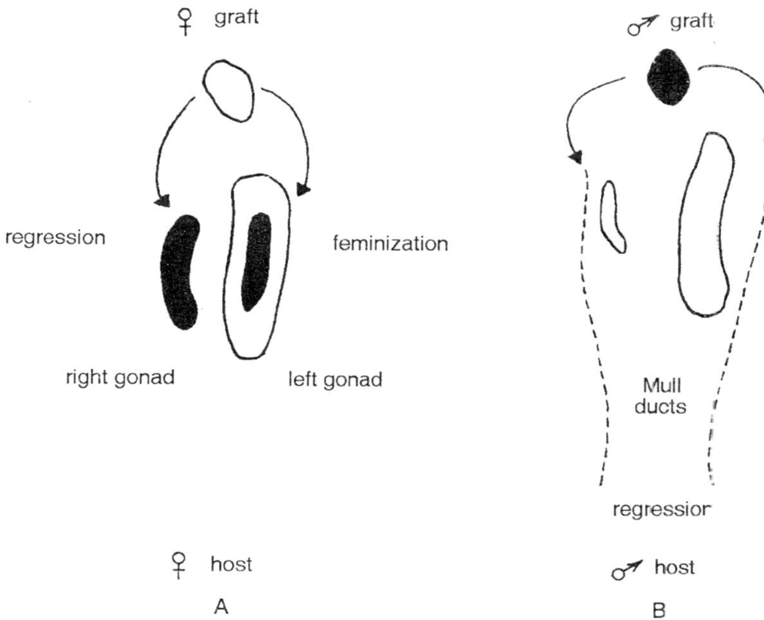

Fig. 14.1. A—Effect of a coelomic graft of ovary on a male embryo: feminization of the left gonad. B—Effect of a testicular coelomic graft on a female embryo: regression of the mullerian ducts.

The effect of these male and female gonadal secretions is similar to that exercised by crystalline sex hormones when the latter are injected into the embryo.

Parabiosis of Gonads Cultivated *in Vitro*

Parabiotic experiments in the Amphibia have demonstrated that hormonal substances elaborated by the gonads of one of the partners act on the gonads of the other partner.

Experiments having the same significance have been performed in the duck, by means of the method of *in vitro* culture, that allows one to parabiose directly two gonads of each sex at 8 to 9 days of age. Under these conditions, the male gonad is feminized by the secretions of the female gonad. These results recently have been verified and extended by Weniger in studies with heterosexual associations of very young gonads taken from chicks on the fifth to the sixth day of incubation. The results demonstrate that the gonads are capable of elaborating hormones at the very onset of the period of sexual differentiation.

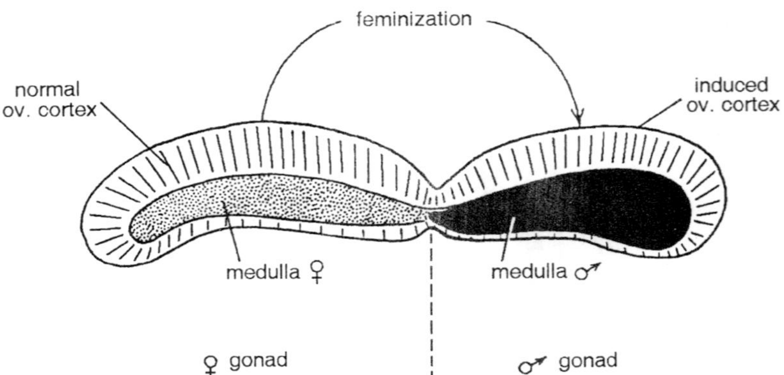

Fig. 14.2. Parabiosis between potential male and female gonads, cultivated in vitro: feminization of the male gonad.

Role of Gonadal Hormones in the Differentiation of the First Somatic Sexual Characters of the Embryo

Several precocious somatic sexual characters are the duct of Muller in the embryo of the chicken and of the duck, and the syrinx and the genital tubercle in the embryo of the duck.

These organs effect their sexual differentiation immediately after the gonads have initiated their differentiation in either the male or the female direction, that is, between the ninth and the twelfth day of

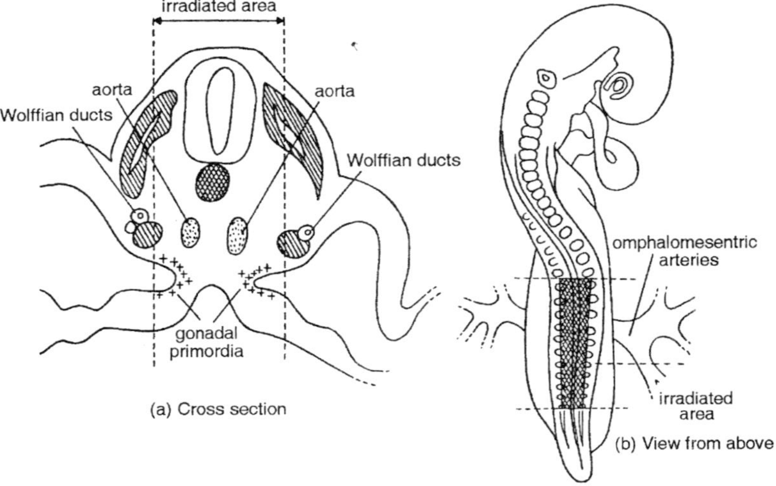

Fig. 14.3. Localization of the area irradiated with X-rays in a $2^1/_2$-day-old chick embryo, in order to obtain the embryonic castration.

incubation in the duck. The relationship of these sexual differentiations to the gonads has been demonstrated by experiments employing castration and experiments employing *in vitro* culture.

Experiments Employing Castration

We have effected castration of embryos of birds by destroying the germinal ridges of young embryos 2 to 4 days of age by means of X-rays.

Action on the duct of Muller

The two mullerian ducts persist in the castrated chick embryo in the male as well as in the female. Both conserve the same length in the castrate, whereas the distal part of the right duct regresses in the female. With the exception of this difference, the duct of Muller in the castrate is more characteristic of the female type than of the male type.

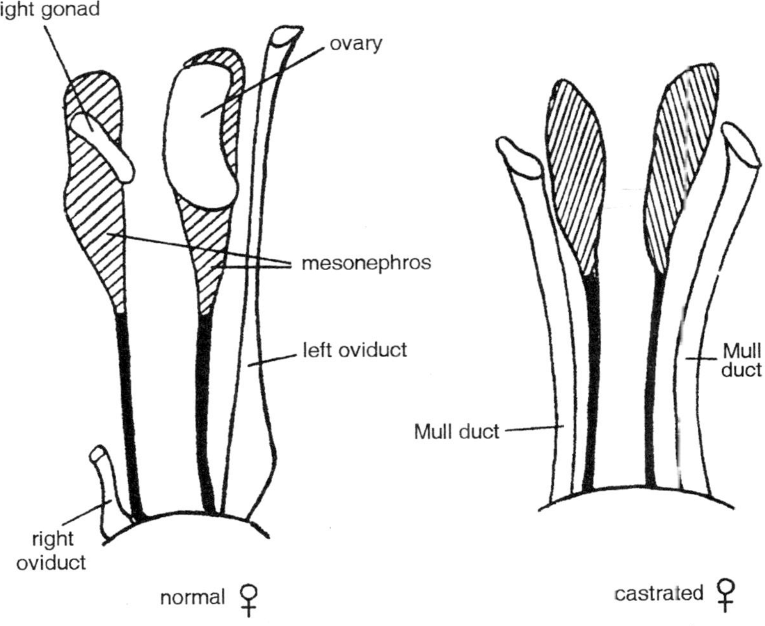

Fig. 14.4. Result of a total castration, in male as well as in female embryo: maintenance of both mullerian ducts, which conserve the same length.

Action on the syrinx and on the genital tubercle

These two characters have the same behaviour in the castrates. The syrinx normally evolves into a voluminous and very asymmetric organ in the male; it is reduced in size and has a symmetric form in the female. The syrinx of completely castrated embryos is identical to that of the male; that of incompletely castrated female embryos is intermediate between the male syrinx and the female syrinx. In fact, if only small rudiments of the ovary persist, the syrinx evolves into an intermediate form. These results demonstrate that the male type of development is a basic "asexual" type, capable of self-differentiation in the absence of hormones; this is the nonhormonal type of development. It is the hormone secreted by the female gonad that is responsible for the difference between the two types inasmuch as it inhibits the male, or neutral, form of differentiation. The small quantity of hormone secreted by the ovarian rudiments is sufficient to partially inhibit the neutral form.

The secretion of the female gonad plays the same role in the embryonic differentiation of the genital tubercle. In the complete absence of male or female gonads, the tubercle evolves according to the male type that represents the neutral form; in the presence of rudiments of female gonads, it shows a form of development intermediate between the male and female types. This demonstrates that, as in the case of the syrinx, the secretion of the ovary is responsible for sexual differentiation: it inhibits the male or neutral form.

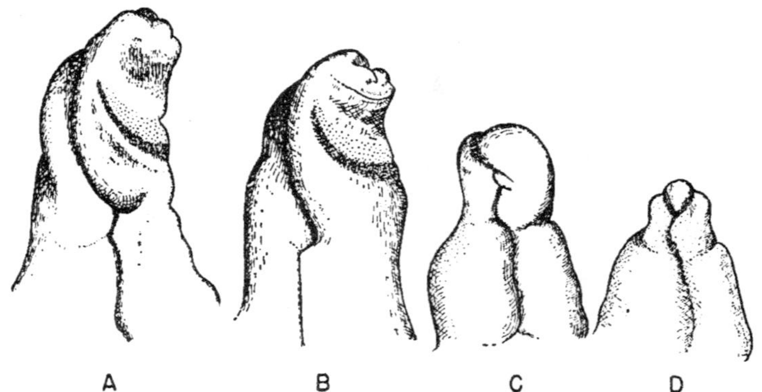

Fig. 14.5. Genital tubercles of normal and castrated embryos 16 days of age. A—Genital tubercle of a normal male embryo; B—Genital tubercle of completely castrated embryo; C—Genital tubercle of an incompletely castrated female; D—Genital tubercle of a female embryo.

Experiments Employing *in Vitro* Culture

The experiments with *in vitro* culture confirm clearly the results obtained with castration.

The method of organ culture used in my laboratory makes it possible to cultivate a number of embryonic organs of the bird. The syrinx, the genital tubercle of the duck, the duct of Muller of the chick are removed before the time of sexual differentiation of the gonads: before the ninth day of incubation for the syrinx and the genital tubercle: before the eighth day of incubation for the mullerian ducts.

Cultivated on a nonhormonal medium, the organs are maintained and develop. The syrinx and the genital tubercle evolve according to the *male type*, irrespective of the sex of the embryo from which they are taken. The duct of Muller subsists and continues its differentiation, irrespective of the sex of the embryo.

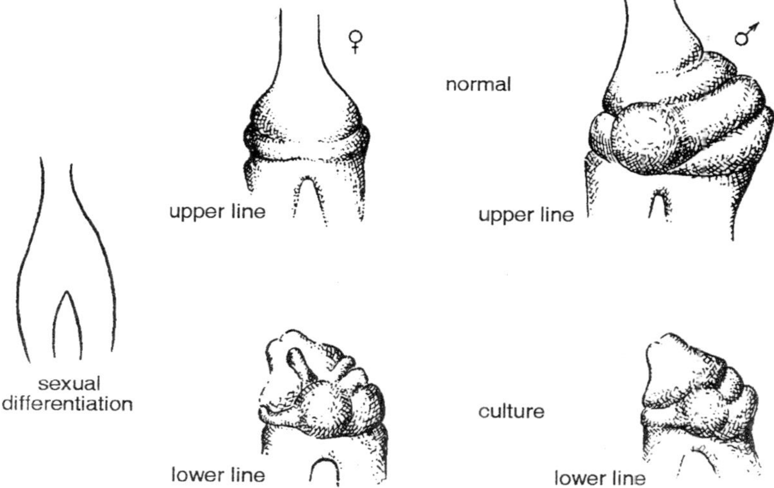

Fig. 14.6. In vitro development of the syrinx, taken before the time of sexual differentiation.

On the other hand, if these organs are taken from female embryos after gonadal differentiation, they exhibit a female evolution in culture. The syrinx remains symmetric and poorly developed, the genital tubercle atrophies, and the duct of Muller continues its differentiation. These results furnish the proof that the gonadal secretions condition the sexual differentiation of these organs.

The undifferentiated syrinx, cultivated in association with female gonads or on a medium enriched with estrogenic hormone, differentiates according to the female type. The duct of Muller, associated with male gonads or cultivated in the presence of an androgenic hormone, becomes necrotic and regresses.

All of these results lead to one conclusion: the most precocious somatic sexual characters are under the influence of hormones secreted either by the male gonads (for example, the mullerian ducts), or by the female gonads (for example, the syrinx and the genital tubercle).

The accompanying scheme summarizes the results of castration and *in vitro* cultures. It allows one to define the neutral type on the basis of its relationship to male and female types. It can be seen that, for the duct of Muller, the neutral type approaches the female type; for the development of the syrinx and the genital tubercle, it is identical to the male type.

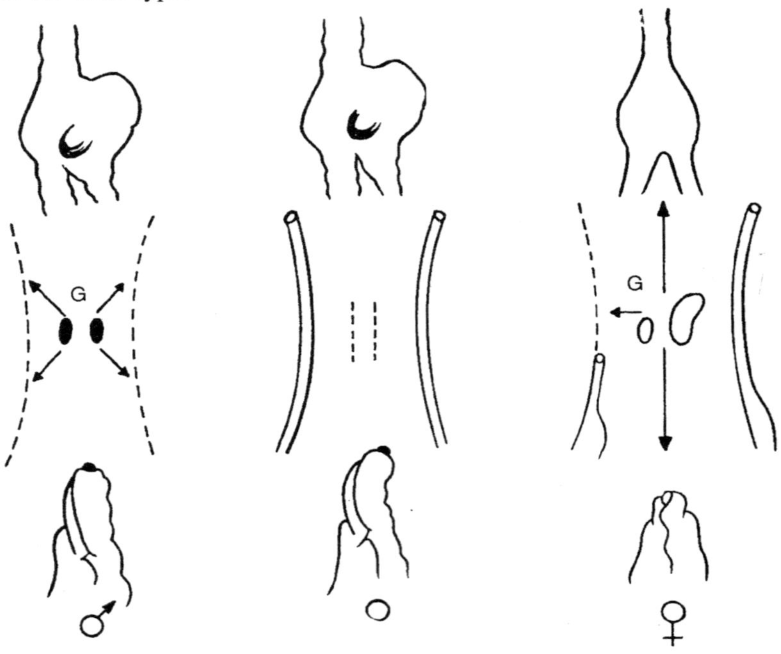

Fig. 14.7. Scheme summarizing the development of male, female, and neutral characters in the duck embryo. Arrows indicate the role of both male and female secretions of the gonads, G, during sexual differentiation.

If one compares these results to those obtained in the mammal, one notes that, in this group, the somatic sexual characters are conditioned, as in the bird, by sex hormones. The maintenance of the duct of Muller, as in the bird, is a neutral character; its regression is under the influence of secretions from the male gonad. The maintenance of the duct of Wolff is also under the influence of male secretions. As

for the genital tubercle, it evolves according to the male type only in the presence of the testicular secretion. The neutral type of the mammal is therefore much closer to the female type than is that of the bird.

MODE OF ACTION OF GONADAL SECRETIONS—THE PROBLEM OF REGRESSION

In the bird, the sexual differentiation of precocious somatic characters is due principally to an inhibitory action exercised by the gonads. The case of the duct of Muller is particularly interesting. It is subject not only to inhibition but also to regression, that is, it disappears completely.

We have studied the conditions of this regression by different methods.

Wolff, Ostertag, and Scheib (1949) have established that, in a physiological solution, the mullerian ducts of the male taken after the differentiation of the gonads, autolyze completely and disappear from the medium. They likewise disappear if cultivated on a nutritive medium. On the other hand, the undifferentiated mullerian ducts, male or female, survive and differentiate according to the female type.

It has been established that the duct of Muller of the normal embryonic male begins necrosis at the time its morphologic regression becomes evident. The same phenomenon is exhibited in culture if the duct of Muller is taken from a male embryo after the sexual differentiation of the gonads. One can induce a similar necrosis in the undifferentiated duct of the female cultivated *in vitro* by adding a crystalline male hormone to the culture medium, whereas the same treatment has no effect on other organs, for instance on the wolffian ducts.

However, necrosis alone cannot explain the disappearance of the mullerian ducts. Accompanying the necrosis is a rapid autolysis due to the production of proteolytic enzymes as Scheib (1955) has successfully demonstrated. An homogenate of male mullerian ducts at 9 days of age, incubated for several hours, liberates an enzyme capable of digesting the cellular proteins, whereas an homogenate of female ducts of the same age is inactive.

The autolytic enzyme is liberated under the influence of the male hormone. In fact, an homogenate of female duets of 9 days, incubated for several hours in the presence of an androgenic hormone, dehydroandrosterone, becomes active and digests the cellular proteins.

Thus, an androgenic hormone is capable of liberating an enzyme, or of activating an enzyme system in an extract of mullerian ducts under *in vitro* conditions. This establishes that the reaction can be initiated independently of all vital processes.

One can summarize the process of regression with respect to the mullerian ducts in the following manner. The secretion of the male gonad (or an androgenic hormone) has a necrotic action on the tissues of the mullerian ducts. That destructive action is exercised specifically on the duct of Muller and on other organ (in particular, it does not

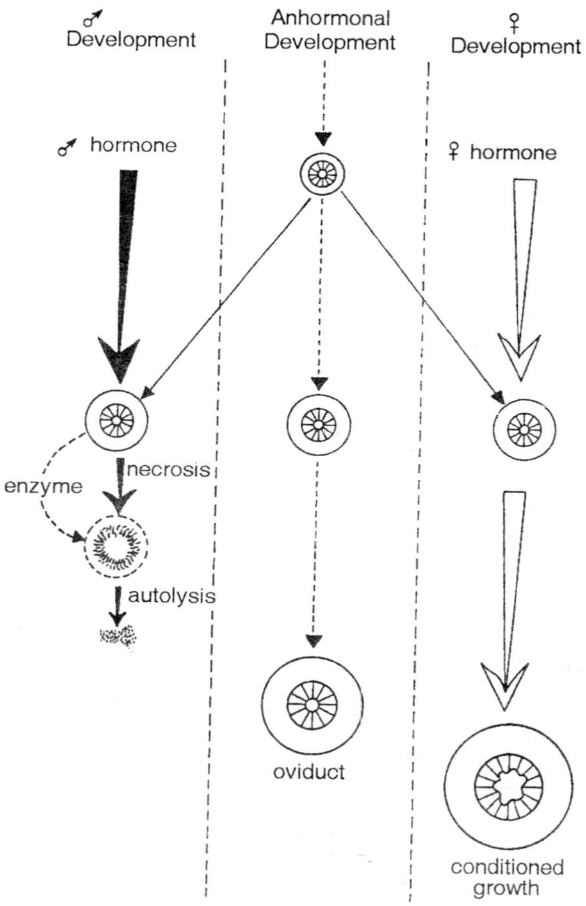

Fig. 14.8. Explanation of the regression of the mullerian ducts in the male embryo, of their evolution in castrated and female embryos.

act on the duct of Wolff cultivated *in vitro* on the same medium as the duct of Muller).

At the same time, the male hormone determines the formation of a very active proteolytic enzyme in the course of the necrosis of the mullerian ducts, and the enzyme perpetrates a very rapid autolysis of the organ. The two processes, necrosis and hydrolysis, effect a rapid disappearance of the duct of Muller and offer an explanation for the sudden and accelerated regression of this organ.

Similar mechanisms may be operating in other phenomena of atrophy that are encountered in embryonic development.

INDEX

D

E

T